# THE MILITARY LEADERSHIP HANDBOOK

# THE MILITARY LEADERSHIP HANDBOOK

Edited by
Colonel Bernd Horn and Dr. Robert W. Walker

Foreword by Major-General J.P.Y.D. Gosselin

CANADIAN DEFENCE ACADEMY PRESS

DUNDURN PRESS
TORONTO

Copyright © Her Majesty the Queen in Right of Canada, 2008

Catalogue No. D2-229/2008E

Published by Dundurn Press and Canadian Defence Academy Press in cooperation with the Department of National Defence, and Public Works and Government Services Canada.

All rights reserved. No part of this information (publication or product) may be reproduced or transmitted in any form or by any means, electronic, mechanical, photocopying, recording, or otherwise, or stored in a retrieval system, without the prior written permission of the Minister of Public Works and Government Services Canada, Ottawa, Ontario, K1A 0S5, or *copyright.droitauteur@pwsgc.ca*. Permission to photocopy should be requested from Access Copyright.

Editor: Michael Carroll
Copy-editor: Shannon Whibbs
Designer: Erin Mallory
Printer: Transcontinental

Library and Archives Canada Cataloguing in Publication

Horn, Bernd, 1959-
 The military leadership handbook / Bernd Horn, Robert W. Walker.

Includes index.
ISBN 978-1-55002-766-2

 1. Command of troops. 2. Military art and science. 3. Leadership.
I. Walker, Robert William, 1943- II. Title.

UB210.H67 2008            355.3'3041          C2007-904979-6

1   2   3   4   5       12   11   10   09   08

 Conseil des Arts   Canada Council
   du Canada        for the Arts

We acknowledge the support of the **Canada Council for the Arts** and the **Ontario Arts Council** for our publishing program. We also acknowledge the financial support of the **Government of Canada** through the **Book Publishing Industry Development Program** and **The Association for the Export of Canadian Books**, and the **Government of Ontario** through the **Ontario Book Publishers Tax Credit** program and the **Ontario Media Development Corporation**.

Care has been taken to trace the ownership of copyright material used in this book. The author and the publisher welcome any information enabling them to rectify any references or credits in subsequent editions.

*J. Kirk Howard, President*

Printed and bound in Canada
www.dundurn.com

Canadian Defence Academy Press
PO Box 17000 Station Forces
Kingston, Ontario, Canada
K7K 7B4

Dundurn Press
3 Church Street, Suite 500
Toronto, Ontario, Canada
M5E 1M2

Gazelle Book Services Limited
White Cross Mills
High Town, Lancaster, England
LA1 4XS

Dundurn Press
2250 Military Road
Tonawanda, NY U.S.A.
14150

# TABLE OF CONTENTS

FOREWORD *by Major-General J.P.Y.D. Gosselin*     9

INTRODUCTION *by Bernd Horn and Robert W. Walker*     11

1 — ATTITUDES     15
*Allister MacIntyre*

2 — CHANGE     31
*Jeffrey Stouffer*

3 — CHARACTER     48
*Robert W. Walker*

4 — COHESION     57
*Allister MacIntyre*

5 — COMBAT MOTIVATION     73
*Bernd Horn*

6 — COMBAT STRESS REACTIONS     91
*George Shorey*

7 — COMMAND     104
*Bernd Horn*

| 8 — Communication | 114 |
|---|---|
| *Allister MacIntyre and Danielle Charbonneau* | |
| 9 — Conflict Management | 129 |
| *Bradley Coates* | |
| 10 — Counselling | 140 |
| *Paul Pellerin* | |
| 11 — Courage | 158 |
| *Bernd Horn* | |
| 12 — Creativity | 168 |
| *Robert W. Walker* | |
| 13 — Cultural Intelligence | 187 |
| *Emily Spencer* | |
| 14 — Culture | 200 |
| *Karen D. Davis* | |
| 15 — Decision-Making | 215 |
| *Bill Bentley* | |
| 16 — Discipline | 228 |
| *Robert Edwards* | |
| 17 — Diversity | 244 |
| *Justin Wright* | |
| 18 — Ethics | 261 |
| *Daniel Lagacé-Roy* | |
| 19 — Fatigue | 276 |
| *Bernd Horn* | |

| | | |
|---|---|---|
| 20 — Fear | | 285 |
| *Bernd Horn* | | |
| 21 — Followership | | 302 |
| *Brent Beardsley* | | |
| 22 — Gender | | 309 |
| *Karen D. Davis* | | |
| 23 — Grief | | 320 |
| *Rhonda Gibson and Robert D. Sipes* | | |
| 24 — Influence | | 339 |
| *Bill Bentley and Robert W. Walker* | | |
| 25 — Interpersonal Feedback | | 355 |
| *Danielle Charbonneau and Allister MacIntyre* | | |
| 26 — Judgment | | 366 |
| *Robert W. Walker* | | |
| 27 — Mentoring | | 378 |
| *Daniel Lagacé-Roy* | | |
| 28 — The Military Professional | | 390 |
| *Bill Bentley* | | |
| 29 — Morale | | 402 |
| *Bernd Horn and Daniel Lagacé-Roy* | | |
| 30 — Motivation | | 414 |
| *Phyllis P. Browne and Robert W. Walker* | | |
| 31 — Physical Fitness | | 429 |
| *Bernd Horn* | | |

32 — The Professional Development Framework  436
   *Robert W. Walker*

33 — Professional Ideology  451
   *Bill Bentley*

34 — Self-Development  461
   *Brent Beardsley*

35 — Stereotypes  471
   *Allister MacIntyre*

36 — Stress and Coping  481
   *Allister MacIntyre and Colin Bridges*

37 — Teams  494
   *Robert W. Walker, Allister MacIntyre, and Bill Bentley*

38 — Theories  511
   *Emily Spencer*

39 — Trust  525
   *Jeffrey Stouffer, Barbara Adams, Jessica Sartori, and Megan Thompson*

Summary  543

Contributors  545

Index  551

# FOREWORD

I am delighted to introduce this seminal book, *The Military Leadership Handbook*. This volume represents an important addition to the body of literature both on leadership and the profession of arms. It provides an applied and readable compendium of ideas, concepts, and practices related specifically to military leadership. Although focused, to a large degree, on the Canadian context and experience, it is applicable to any military organization and, quite frankly, to anyone interested in the study of leadership.

Moreover, *The Military Leadership Handbook* adds to the innovative Strategic Leadership Writing Project created by the Canadian Forces Leadership Institute and the Canadian Defence Academy Press. This book continues to demonstrate our commitment to capturing key themes and operational topics of importance for military personnel serving in the complex security environment of today. This important work will be of great value to those leaders charged with the heavy responsibility of leading their nation's sons and daughters into harm's way, as well as those leaders with equally important responsibilities to develop and improve their institution's capabilities to enable mission success.

Accordingly, I wish to reiterate the importance of this book, as well as all others in the Strategic Leadership Writing Project series. These publications provide military professionals and the public at large with theoretical knowledge, as well as practical experience, to ensure a better understanding of leadership. They will assist both military leaders to better prepare to lead and command in the demanding environments in which they find themselves, and non-military professionals to comprehend the enormous challenges their nations' armed forces face in the current complex and lethal global security environment. At the Canadian Defence Academy, it is our firm belief that our efforts to provide well-researched, relevant, and authoritative publications on key operational topics both enlighten and empower those who serve, and those who interact with those who serve, in the profession of arms in Canada.

Major-General J.P.Y.D. Gosselin
Commander
Canadian Defence Academy

# INTRODUCTION

Leadership, in many ways, is an enigmatic concept. Although innumerable books and articles have been written on the subject, it still has many faces. Entire books are often written on the subject without even defining the term "leadership" itself. In addition, leadership is often confused with and/or used interchangeably with other concepts such as "command" and "management." Equally, there seems to be a sentiment that leadership may be hard to define, but "I certainly can recognize it when I see it."

This varied approach is not entirely hard to understand. After all, leadership touches everything we do across the entire spectrum of society. It is studied by scholars, businesses, industries, sports teams, and militaries, just to name a few interested actors. In the end, it is of interest and practised by all of us on a daily basis, whether we recognize it or not.

Nonetheless, this compendium, *The Military Leadership Handbook*, is strictly focused on military leadership. Furthermore, doctrinally the Canadian Forces (CF) has defined effective military leadership as "directing, motivating and enabling others to accomplish the mission professionally and ethically, while developing or improving capabilities that contribute to mission success."[1] As such, this handbook is not intended to provide a theoretical treatise on leadership, but rather to provide a concise and complete manual that identifies, describes, and explains all those concepts, components, ideas, and subjects that deal with, or directly relate to, military leadership.

As such, many of the topics range from the abstract and theoretical to the very applied. This is entirely deliberate as the intent was to make this compendium as comprehensive as possible. In the end, this handbook will be of invaluable assistance to military members, as well as others, who wish to have a better understanding of leadership, and military leadership specifically.

In addition, it should be noted that the manual is in consonance with the Canadian Forces (CF) Professional Development Framework (PDF). The PDF is described within this compendium, however, it is important to provide some context. The global security environment, modern technological innovation, and the changing nature of society that the world has undergone in the last fifteen years has necessitated an aggressive CF approach to understanding leadership, to ensuring leader effectiveness through a congruence of institutional demands and leader capabilities, and to providing continuous Professional Development (PD) of CF leaders. As a result, a model defining CF Effectiveness (Figure 1) was developed, which fuelled a requirement to determine leader characteristics, competencies, skills, and knowledge, which in turn generated a framework of relevant leader capacities, including those needed for the CF.

The result was a five-metacompetency or five-element cluster of requisite CF leader capacities: Expertise, Cognitive Capacities, Social Capacities, Change Capacities, and Professional Ideology. This cluster of five requisite

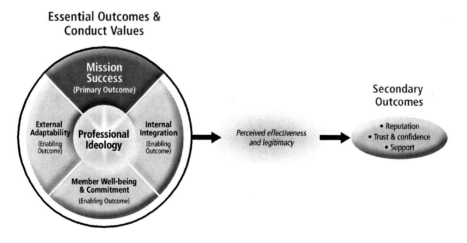

Figure 1: The Canadian Forces Effectiveness Framework

metacompetencies was integrated with a continuum of four leader levels (junior, intermediate, advanced, senior) to generate a PDF (See Figure 2). This framework constituted a template for both defining the necessary leader knowledge and expertise, the effective cognitive, social, and change capacities, and a military professionalism across the continuum of levels of leadership, as well as determining the most relevant subject matter and most effective learning strategies to develop these leader elements. The consequence is congruence between the CF Effectiveness model — representing the institution — and the CF Professional Development Framework — representing the CF leadership. Such congruence ensures effective CF real-world military applications with successful outcomes.

In sum, this handbook, which is in consonance with the PDF, can provide a comprehensive compendium of those concepts, components, ideas, and subjects that relate directly to military leadership. Only through a comprehensive understanding of all aspects of effective leadership, reinforced through experience and further self-development, can individuals maximize their own abilities to lead effectively. This handbook is one important source to assist military leaders, as well as all other interested people of any discipline or profession, with the challenges of effective leadership.

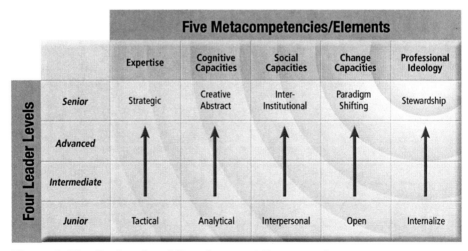

Figure 2: The Professional Development Framework

**NOTES**

1. Canada, *Leadership in the Canadian Forces: Conceptual Foundations* (Kingston: DND, 2005), 30.

# 1
# ATTITUDES
by Allister MacIntyre

Attitudes are an inescapable consequence of our humanity. We cannot function without attitudes because they infuse every element of our existence with a reality that is unique to each and every person on the planet. They drive our belief systems, fuel our emotions, and compel us to behave in ways that are consistent with the attitudes we hold. But why do leaders have to be aware of attitudes? This understanding is crucial if leaders expect to understand how attitudes influence their own behaviours, and the actions of their followers. Our understanding of the roles played by attitudes can help us to either curb undesirable behaviours, or shape attitudes in a manner that will motivate acceptable and desirable behaviours.

Our attitudes are nothing more than the subjective experience of our likes and dislikes, our passions and disgust, our obsessions and loathing, our attractions and aversions. We all possess an unlimited number of attitudes about virtually every person, place, and thing we encounter during our lives. Because attitudes exist inside people, they are not easy to see, but we can gain some insight into the attitudes of others if we pay attention to how they behave. Effective leaders need to recognize that attitudes will influence how followers will react, feel, and behave in response to an attitude object.[1]

It is well understood that our behaviour toward other people will be swayed in part by the impressions and expectations we hold about them.[2] The way in which we form these attitudes and stereotypes[3] has also been researched to a great extent. Leaders must be aware that their own behaviours, as well as the behaviours of their followers, will be influenced by the attitudes they hold. People are inclined to respond in a favourable or unfavourable manner due to attitudes.[4] As such, the attitudes are viewed as being either positive or negative, and estimates of their intensity (from extremely positive to extremely negative) are often estimated. Although we cannot physically view an attitude, it has been argued that our attitudes are reflected in our beliefs, feelings, or expressions of our intended behaviour. It is also worth noting that even though attitudes are important, they are not the only reason for how we behave. Although we may wish to act a certain way, our actual behaviour will be determined by additional things such as how others in our social group act, how society expects us to behave, the rules and regulations that are in place, our moral development, and our fear of consequences if we do something that is unacceptable.

## **FUNCTIONS OF ATTITUDES**

It is easy to appreciate the universal nature of attitudes, but do they serve any sort of practical function? The simple answer is yes. They enhance our ability to make sense of the world, allow us to express our values, ensure that we maintain effective relationships with those who share similar attitudes, and help to guide our behaviour. These purposes can be classified as functions. It has long been recognized that attitudes are not functionally alike, and they have been studied on the basis of the functions they serve.[5] The five major functions performed by attitudes are: first, a knowledge function (indicating a person's need to structure the world in a meaningful and consistent fashion); second, an instrumental function (reflecting an attempt to maximize rewards and minimize punishments); third, an ego-defensive function (to protect us from basic truths about ourselves); fourth, a value-expressive function (meaning that we express attitudes that are important to our self-concept or personal values); and fifth, a social adjustment function (we hold attitudes that are similar to others in our group to help us to fit in).

Daniel Katz, a well-known attitude theorist and researcher, states that "the functional approach is the attempt to understand the reasons people hold the attitudes they do. The reasons, however, are at the level of psychological motivations and not of the accidents of external events and circumstances."[6] Other theorists have expanded this idea by arguing that the ability to change an attitude largely depends upon the function that the attitude serves.[7] This is one of the most critical concepts that leaders must understand about attitudes. Effective leaders will not only understand their followers' attitudes, they will be able to influence these attitudes to make sure that they are in line with group ideals. By understanding the functions served by attitudes, leaders can use the most effective approaches to change attitudes in the right direction.[8]

In other words, we could attempt to change an instrumental attitude by altering the rewards and punishments associated with the attitude. For example, many serving members of the Canadian Forces (CF) express positive attitudes toward second language training and actively seek opportunities in this regard. After all, bilingualism is an important consideration for promotions, and selection for many postings and positions. If the CF were to discontinue placing an importance on bilingualism, we might expect that those with instrumental attitudes would experience a shift in how they perceive and endorse second-language training. However, for those whose favourable attitudes are not based on rewards, a shift would not be expected. Similarly, correcting misinformation about an attitude object can adjust knowledge attitudes relatively easily. However, leaders must understand that ego-defensive and value-expressive attitudes are the most difficult types of attitudes to change. These types of attitudes are highly resistant to change because they require fundamental changes to one's self-concept and an adjustment in basic values and/or beliefs. Nevertheless, CF leaders can have a dramatic impact on value-based attitudes by embodying the military ethos and ideology as encouraged by the doctrinal manual *Duty with Honour: The Profession of Arms in Canada*[9] and by reinforcing the core values of duty, courage, loyalty, and integrity in all that they do.

When attitudes are based upon ego-defensive needs, a threatened individual will either avoid an unpleasant situation, or exhibit hostility. This is because an ego-defensive attitude functions as a defence mechanism (such as denial, projection, or repression) to protect one's self-concept from either internal or external threats. For example, it has been suggested that prejudice may stem, for some people, from subconscious feelings of inferiority. Daniel Katz states that the "usual procedures for changing attitudes and behavior have little positive effect upon attitudes geared into our ego defenses. In fact they may have a boomerang effect of making the individual cling more tenaciously to his emotionally held beliefs … punishment is threatening to the ego-defensive person and the increase of threat is the very condition which will feed ego-defensive behavior."[10] The implication here is that a traditionally transactional style of leadership, with its basis in rewards and punishments, will not be an effective approach for altering ego-defensive attitudes.

Daniel Katz does offer three factors that may help to change ego-defensive attitudes. First, there must be a removal of threat; second, there must be an opportunity to ventilate feelings; third, the person must acquire insight into the reasons for the defensive attitude. Although Katz offered this advice to therapists, and it was not expressed in leadership terms, it is easy to appreciate that someone with a transformational leadership style would most effectively use the approach.

While ego-defensive attitudes serve to prevent people from revealing their true nature to either themselves or others, value-expressive attitudes provide positive expression to one's central values and self-concept. Katz offered two conditions as being relevant to changing value-expressive attitudes: first, people need to be dissatisfied with their self-concept or its associated values; and second, old attitudes must be viewed as being inappropriate.[11] These conditions sound similar, but there are conceptual differences. In the first case, the change originates in a shift in one's value system; in the second condition, a change does not occur in the value system, rather the attitudes held are assessed as being inconsistent with the existing value system.

Katz illustrated one method of actively transforming value systems by describing the brainwashing techniques of Chinese Communists in Korea.[12] A crack in an individual's belief system must be found, and then it is exploited by using appropriately directed influences. Although one might question the reality of brainwashing effects, there is no questioning the fact that hundreds of Allied soldiers cooperated with their captors and twenty-one American prisoners of war chose to remain even though they were given permission to return home.[13]

## **ATTITUDE FORMATION**

If leaders expect to successfully change attitudes, they must grasp how attitudes form in the first place. Many theorists feel that attitudes form either through direct experience with the attitude object or, alternatively, without the benefit of such direct experience.[14] It has been found that if attitudes stem from direct experience they will be more clearly, confidently, and stably maintained than those that result from more indirect means. However, there is greater attitude-behaviour consistency when people have direct experience with the attitude object. Finally, attitudes formed without personal experience appear to be fundamentally different from those formed as a result of direct experience. Psychologists Russell Fazio and Mark Zanna offer two ways of explaining why attitudes formed through direct behavioural experience are held more confidently.[15] First, more information is available for those with direct experience and, second, there may be an information-processing difference between direct and indirect experience.

## **ATTITUDE IMPORTANCE**

Once formed, some attitudes are more resistant to change than others. One critical aspect that needs to be appreciated by all leaders is the degree of importance associated with a given attitude. Many studies have examined attitude importance and resistance to change. Important attitudes are more resistant to change because: first, important attitudes are associated with other attitudes, beliefs, and values, and this becomes a stabilizing force; second, important attitudes are accompanied by stores of relevant knowledge which can be used to counter-argue competing information; third, people tend to be attracted to others with similar attitudes, hence these attitudes are reinforced by social norms; last, people are more likely to commit themselves in public to attitudes they consider important, which increases resistance to change.[16] Psychologist Jon Krosnick points out that the stability of important attitudes has interesting implications with regard to how people resolve attitude inconsistencies. He states that because "important attitudes are unlikely to change, inconsistency between an important attitude and an unimportant one is likely to be resolved by bringing the latter in line with the former."[17] In other words, it can be argued that attitude change follows the path of least resistance.

## **ATTITUDE CHANGE THROUGH CONTACT WITH OTHERS**

If we accept that attitudes formed through direct experience will be stronger, and appreciate that the importance of an attitude is influenced by things like the amount of knowledge we have, then perhaps having direct contact with an attitude object will change attitudes. Furthermore, the changes could be in either a positive or negative direction, given the nature of the interaction. This approach to attitude change was a concept developed by Israeli theorist Yehuda Amir.[18] Amir felt that increased contact with members of an out-group would improve understanding of the group and produce greater tolerance. The weaknesses in this argument can be demonstrated by the mixed results of racial-integration experiments, as well as countless research findings. For example, it has been pointed out that "the desegregated classroom has not produced many of the positive results initially expected by social scientists some 25 years ago."[19] Amir and his colleague Rachel Ben-Ari added that contact, as a tool to improve inter-group relations, must be strongly qualified by individual and situational factors.[20] Furthermore, it has been argued that an individual's improved attitude toward contact-group members will not necessarily generalize to the entire group.[21] In fact, out-group members who are eventually accepted are perceived as exceptions to the group from which they come. Finally, Yehuda Amir has cautioned that attitude change following inter-group contact may not be in the anticipated direction.[22] Favourable conditions may lead to improved attitudes, but unfavourable conditions might actually increase negative attitudes. Additionally, any changes produced by contact might be a change in intensity rather than direction.

Yehuda Amir presents some of the favourable conditions that tend to reduce prejudice (a specific type of attitude) as being: first, equal status between the members of the contact groups; second, contact is with higher status members of the minority group; third, an "authority" and/or the social climate are in favour of the inter-group contact; fourth, the contact is intimate rather than casual; fifth, the inter-group contact is pleasant or rewarding; and sixth, the members of both groups in the contact situation interact in functionally important activities or develop common goals that are that are higher in importance than the individual goals.[23] Amir also presents the unfavourable conditions that tend to strengthen negative attitudes. These are: first, the contact situation produces competition; second the contact is unpleasant, involuntary, or tension-laden; third, the prestige or status of one group is lowered because of the contact; fourth, the members of a group, or the group as a whole, are in a state of frustration; fifth, the groups in contact have moral standards which are objectionable to one another; and sixth, when the contact is between a majority and minority group and the members of the minority group are of a lower status.

These are extremely important considerations for leaders because they have the greatest amount of control over situational factors and conditions that can improve prejudicial attitudes. Furthermore, the conditions are especially important for military leaders interested in ensuring the cohesiveness of their followers. CF leaders are constantly called upon to command serving members from diverse backgrounds and must work closely with leaders from other elements, other government departments, and other nations. Leaders can use their influence to ensure that minority group members are viewed as having at least equal, if not higher, status. As an authority figure, they must convey that they are in favour of and promote the inter-group contact. They can ensure that the working relationships extend beyond a superficial level and contain some elements of reward. Finally, they can influence the nature of the activities being performed so that they are viewed as being functionally important and instilled with common goals that are that are higher in ranking in importance than the individual goals. Naturally, an effective CF leader will also make an effort to reduce or eliminate competition between the groups; make sure that the prestige or status of the group members is not lowered as a result of the contact situation; and take necessary steps to reduce frustrations and tensions.

## **GROUP CONFLICT AND COOPERATION**

The preceding discussion of contact between groups leads naturally to the consideration of whether different groups cooperate or experience conflict. Countless studies have been conducted in the areas of group conflict and cooperation.[24] Social psychologists usually differentiate between "in-groups," a social unit that we either belong to or identify with, and "out-groups," a social group that we either do not belong to or do not view as being relevant to our identity.[25] Competition and conflict between groups is normal due to the attitudes and perceptions held toward members in other groups. There tends to be a natural progression from perceived competition among groups to perceived hostility. As a result of the bonds within a group, there is a tendency for: in-group members to view themselves as being virtuous and superior; cooperation among in-group members; penalties for those who go against normal group behaviours; and, a willingness to fight for the in-group. Conversely, in-group members tend to: view out-groups as being contemptible and inferior, and perhaps immoral; maintain a social distance from out-groups; approve hatred of out-groups; and, distrust and fear out-groups.

Conflict between groups is normal but, with the introduction of common goals, cooperation is promoted.[26] This notion ties in nicely with one of the favourable conditions required to foster a reduction in prejudice in contact groups;[27] specifically, the development of common goals that rank higher than individual goals. Cooperation between groups often leads to a reduction in hostility and conflict between groups. However, in some circumstances, the inter-group relationships will increase negative attitudes.

## EDUCATION AND ATTITUDE CHANGE

In military settings, the mindset tends to focus on education and training whenever there is a need to exercise influence. With respect to intergroup relations there seems to be an assumption that education will eliminate prejudice. In terms of racial bigotry, psychologist James Vander Zanden points out that, despite over several decades of research into factual instruction, the only conclusion is that it tends to lessen some of the more extreme expressions of prejudice.[28] There are weaknesses in the arguments in favour of education to improve attitudes. First, in a study of attitude change through the use of audiovisual programs, differences between groups were found immediately following the viewing of videotapes, but a follow-up assessment five weeks later showed no significant differences.[29] This suggests that improvements in attitudes as a result of education may only be short-term. Additionally, it has been demonstrated that those who hold strong opinions on complex social issues are more likely to be biased when examining evidence. Confirming evidence will be accepted at face value, while disconfirming information will be subjected to critical evaluation

## COGNITIVE DISSONANCE AND ATTITUDE CHANGE

If education and training is not an effective means to change attitudes, then another approach might be warranted. One such alternative approach involves a theory known as "cognitive dissonance," first developed by psychologist Leon Festinger almost half a century ago.[30] This theory argues that one of the most powerful motives for humans is the drive for cognitive consistency. When we have two conflicting thoughts, ideas, or attitudes, then we will experience cognitive dissonance.[31] When we have this sort of discomfort we will attempt to resolve the dissonance by altering one of the cognitions to fit the other. An example of this kind of behaviour can be found in the habitual use of artificial sweeteners. When faced with evidence that the sweeteners may cause cancer, the reaction might be to rationalize that not enough of the product is consumed for this to be a real threat, or one might argue that they would rather die of cancer than obesity. In either case, the information that is inconsistent would be resisted.

Dissonance also occurs when there is an inconsistency between our attitudes and the way we behave.[32] Therefore, one could attempt to change attitudes by concentrating on behaviour change. Using the cognitive dissonance approach, it might be argued that there is no need to change attitudes toward out-groups, only a need to change behaviour toward out-groups. After all, if the behaviour is inconsistent with our attitudes, then result will be dissonance and perhaps a change in attitudes. Most military leaders have position power that allows them to direct and control their followers' behaviour. As a consequence, they are in a favourable position to implement type of approach for changing behaviour and influencing attitudes.

However, there are fundamental weaknesses in this argument as well. Dissonance will only be experienced if behaviours cannot be explained by reference to external demands. As stated by organizational psychologists Clay Hamner and Dennis Organ, when a behaviour "is fully justified by external circumstances (for example, the avoidance of pain or the acquisition of rewards), little or no dissonance is aroused by inconsistency between that behavior and the person's attitude."[33] Joel Cooper and Russell Fazio refer to this as the assumption of personal responsibility, a necessary link leading to cognitive dissonance. They add that people can avoid dissonance by denying responsibility for the unwanted behaviour or viewing the dissonance as an unforeseeable consequence of their behaviour.[34]

## **SOCIAL NORMS AND BEHAVIOUR**

There is no doubt that attitudes will have some bearing on behaviour; however, as indicated at the beginning of this chapter, our actions are only partly determined by attitudes, values, and other such attributes. Behaviour is also influenced by social norms and other factors in the psychological environment. Generally speaking, social norms typically involve some form of social sanction for members of a group who deviate, and rewards for those who comply.[35] One of the more famous illustrations of this line of reasoning is offered by the theory of reasoned action as developed by social psychologists Martin Fishbein and Icek Ajzen[36] and later refined by Ajzen as the theory of planned behaviour.[37] In both of these approaches, attitudes are viewed as just one of the possible predictors of a person's "intention" to behave a certain way. The second contender for influencing our behaviour is our perception of the typical behaviour that would be expected by others within our reference group. In his model of planned behaviour, Icek Ajzen incorporated our assessment of how much real control we have in a given situation as another predictor of behavioural intention.

However, conflicting norms are possible, especially when one is participating in more than one reference group. This is particularly evident within an organizational setting where the norms defined by the chain of command may not be in agreement with norms set by others on our own level. It has been argued that "a member is more likely to conform, the more valuable to him are the rewards he receives from other conforming members, relative to those he receives from alternative actions."[38] For example, people who conform to a management norm, despite the fact that they do not believe in the norm, are described as "skeptical conformers."

## **SUMMARY**

Because of the conditions required for success, a change in attitudes based upon the functional approach is not likely. Contact with out-groups may just as easily lead to increased negative attitudes, and the pursuit of common goals may not make a difference. The education argument is weak because most studies use college students and, in any case, the positive effects may only be short-term. Finally, conforming to a social norm need not be accompanied by an associated change in attitude. Change occurs only under very specific circumstances, and even then not without considerable resistance. Despite the apparent distortions they cause, attitudes play a valuable role in our day-to-day lives. They permit categorical thinking, which in turn allows us to cope with the enormous number of perceptions we experience. We simply do not have the capacity or time to examine every experience as a unique, special occurrence.

The message here for CF leaders is that they must first accept that attitudes will exist no matter what they do to try to change this certainty. Their own attitudes will influence their behavioural intentions, and their followers will be similarly swayed by these mostly subconscious drives. Leaders will be able to influence attitudes, but their success in this regard will be obstructed by things like the strength of the attitude, the functions served by the attitudes, and the knowledge and beliefs associated with the attitude object. In short, it is impossible for people to not hold a variety of attitudes, and therefore, it is equally impossible for people to not be prejudiced to some degree.

Nevertheless, leaders also need to realize that attitudes are only one possible determinant of behaviour. Social norms can have a strong influence on actual behaviour. Thus, even if followers experience uncomplimentary feelings about members of a particular out-group (e.g., another country, race, or culture; gender; or military element) the attitude should not translate into behaviour changes in the presence of strong leadership. The CF leaders' responsibility in this regard is to ensure that the behavioural norms for their subordinates are consistent with the military ethos and values.

## NOTES

1. An attitude object is any person, place, or thing that generates an attitudinal response.
2. R.H. Fazio and M.P. Zanna, "Direct Experience and Attitude-Behavior Consistency," *Advances in Experimental Social Psychology* 14 (1981):161–202.
3. The tripartite view of attitudes holds that an attitude is comprised of three components. The cognitive component contains everything we

know, or believe to be true, about the attitude object (stereotypes). The affective component refers to our emotional response to the attitude object — how the person, place, or thing makes us feel. Finally, the behavioural component makes us want to act a certain way in the presence of the attitude object.

4. See J.E. Alcock, D.D. Carment and S.W. Sadava, *A Textbook of Social Psychology* (Scarborough, ON.: Prentice Hall, 1988); and L.A. Penner, *Social Psychology: Concepts and Applications* (New York: West Publishing Company, 1986).
5. For additional information on this aspect of attitudes, see G.M. Herek, "Can Functions be Measured? A New Perspective on the Functional Approach to Attitudes," *Social Psychology Quarterly* 4 (1987): 285–303; D. Katz, "The Functional Approach to the Study of Attitudes," *Public Opinion Quarterly* 24 (1960): 163–204; and K.G. Shaver, *Principles of Social Psychology* (Hillsdale, NJ: Lawrence Erlbaum, 1987).
6. Katz, "The Functional Approach to the Study of Attitudes," 170.
7. See, Shaver, *Principles of Social Psychology*.
8. The function that an attitude serves can differ from person to person, even though the attitude itself may be identical in all practical aspects.
9. Canada, *Duty with Honour: The Profession of Arms in Canada* (Kingston: DND), 2003.
10. Katz, "The Functional Approach to the Study of Attitudes," 182.
11. *Ibid.*
12. *Ibid.*
13. D.G. Meyers, *Social Psychology*, 3rd ed. (New York: McGraw-Hill, 1990).
14. An expansion of this concept can be found at: R.H. Fazio and M.P. Zanna, "On the Predictive Validity of Attitudes: The Role of Direct Experience and Confidence," *Journal of Personality* 46 (1978): 228–243; R.H. Fazio, M.P. Zanna, and J. Cooper, "Direct Experience and Attitude-Behavior Consistency: An Information Processing Analysis," *Personality and Social Psychology Bulletin* 4 (1978): 48–51; D.T. Regan and R. Fazio, On the Consistency Between Attitudes

and Behavior: Look to the Method of Attitude Formation," *Journal of Experimental Social Psychology* 13 (1977): 28–45; and C. Wu and D.R. Shaffer, "Susceptibility to Persuasive Appeals as a Function of Source Credibility and Prior Experience with the Attitude Object," *Journal of Personality and Social Psychology* 52 (1987): 677–688.

15. R.H. Fazio and M.P. Zanna, "On the Predictive Validity of Attitudes: The Role of Direct Experience and Confidence," 228–243.
16. J.A. Krosnick, "Attitude Importance and Attitude Change," *Journal of Experimental Social Psychology* 24 (1988): 240–255.
17. Ibid., 252.
18. See Y. Amir, "Contact Hypothesis in Ethnic Relations," in *The Handbook of Interethnic Coexistence*, ed. E. Weiner (New York: Continuum Publishing, 1998), 565–585; and Y. Amir and R. Ben-Ari, "International Tourism, Ethnic Contact, and Attitude Change," *Journal of Social Issues* 41 (1985): 105–115.
19. E. Aronson and D. Bridgeman, "Jigsaw Groups and the Desegregated Classroom: In Pursuit of Common Goals," *Personality and Social Psychology Bulletin* 5 (1979): 438–446.
20. Y. Amir and Ben-Ari, "International Tourism, Ethnic Contact, and Attitude Change," 105–115.
21. S.W. Cook, "Interpersonal and Attitudinal Outcomes in Cooperating Interracial Groups," *Journal of Research and Development in Education* 12 (1978): 97–113.
22. Y. Amir, "Contact Hypothesis in Ethnic Relations."
23. Ibid.
24. W.C. Hamner and D.W. Organ, *Organizational Behavior: An Applied Psychological Approach* (Dallas: Business Publications, Inc., 1978); and M. Sherif, O. Harvey, B.J. White, W.R. Hood, and C.W. Sherif, *Intergroup Conflict and Cooperation: The Robbers Cave Experiment* (Norman, OK: Institute of Group Relations, University of Oklahoma, 1961).
25. E. Aronson, *The Social Animal*, 9th ed. (New York: Worth Publishers, 2004).
26. Sherif et al., *Intergroup Conflict and Cooperation: The Robbers Cave Experiment*.

27. Y. Amir, "Contact Hypothesis in Ethnic Relations."
28. J.W. Vander Zanden, *Social Psychology* (New York: Random House, 1977).
29. R. Goldberg, "Attitude Change Among College Students Toward Homosexuality," *Journal of American College Health* 30 (1982): 260–268.
30. L. Festinger, *A Theory of Cognitive Dissonance* (Evanston, IL: Row-Peterson, 1957).
31. J.F. Calhoun and J.R. Acocella, *Psychology of Adjustment and Human Relationships* (New York: Random House, 1978).
32. Hamner and Organ, *Organizational Behavior: An Applied Psychological Approach*.
33. *Ibid.*, 115.
34. J. Cooper, and R.H. Fazio, "A New Look at Dissonance Theory," *Advances in Experimental Social Psychology* 17 (1984): 229–266.
35. See J.M. Levine, "Reaction to Deviance in Small Groups," in *Psychology of Group Influence*, ed. P. B. Paulus (Hillsdale, NJ: Lawrence Erlbaum Associates, 1980).
36. M. Fishbein and I. Ajzen, *Belief, Attitude, Intention and Behavior: An Introduction to Theory and Research* (Reading, MA: Addison-Wesley, 1975).
37. I. Ajzen, "The Theory of Planned Behavior," *Organizational Behavior and Human Decision Processes* 50 (1991): 179–211.
38. G.C. Homans, *Social Behavior: Its Elementary Forms* (New York: Harcourt Brace Jovanovich, 1974), 103.

## SELECTED READINGS

Amir, Y. "Contact Hypothesis in Ethnic Relations," in *The Handbook of Interethnic Coexistence*. Edited by E. Weiner. New York: Continuum Publishing, 1998.

Eagly, Alice A. and Shelley Chaiken. "Attitude Structure and Function," in *The Handbook of Social Psychology*, 4th edition, Volume 1. Edited by Daniel T. Gilbert, Susan T. Fiske and Gardner Lindzey. New York: McGraw-Hill, 1998.

Eagly, Alice A. and Shelley Chaiken. *The Psychology of Attitudes.* New York: Harcourt Brace College Publishers, 1993.

Petty, Richard E. and Jon A. Krosnick. *Attitude Strength: Antecedents and Consequences.* Hillsdale, NJ: Lawrence Erlbaum Associates, Publishers, 1995.

Petty, Richard E. and Duane T. Wegener. "Attitude Change: Multiple Roles for Persuasion Variables," in *The Handbook of Social Psychology,* 4th edition, Volume 1. Edited by Daniel T. Gilbert, Susan T. Fiske, and Gardner Lindzey. New York: McGraw-Hill, 1998.

Pratkanis, Anthony R., Steven J. Breckler, and Anthony G. Greenwald. *Attitude Structure and Function.* Hillsdale, NJ: Lawrence Erlbaum Associates, Publishers, 1989.

Shaver, K.G., *Principles of Social Psychology.* Hillsdale, NJ: Lawrence Erlbaum, 1987.

# 2
# CHANGE
by Jeffrey Stouffer

> ... *to respond to the challenges of external chaos, the management of change has become the prime occupation of those who inhabit the executive suites of the world's leading enterprises.*[1]
>
> — James O'Toole

Most people would agree that change represents a truism of the human experience. In fact, the Greek philosopher Heraclitus (540–475 BC) was the first to acknowledge the universality of change and he claimed that the nature of everything is change itself. We tend, however, to think of change as a 20th-century phenomenon likely because of the complexity and the unprecedented tempo of change resulting from social and technological innovations. As Rosabeth Kanter, former editor of the *Harvard Business Review* and business consultant, notes, "organizational change has become a way of life."[2] Indeed, leading change has become a critical organizational reality, requiring considerable attention and thoughtful nurturing to effectively manage. The military is no exception. Canadian Forces (CF) transformation, as an example, is to some extent predicated on the need to adjust to a changing world that is rife with uncertainty and previously unimagined threats.

This uncertainty, combined with organizational dynamics, highlights the need for militaries to become increasingly capable of adapting to the changing nature of conflict, societal shifts, technology, and the myriad other issues that face defence institutions. As an example, in discussing today's complex and ambiguous environment, U.S. General David Fastabend explains that if we were to choose one advantage over our adversaries, it would be "to be superior in the art of learning and adaptation."[3] Fastabend remarks that our adversaries are adaptive and that "our choice is quite clear, adapt or die."[4]

These comments stress the need for leaders to care about change. Leaders are the principal agents of change and their actions largely determine the success or failure of any change effort. As John P. Kotter, a professor of leadership at the Harvard Business School and, arguably, the leading authority on change reports, "… successful transformation is 70 to 90 percent leadership and only 10 to 30 percent management."[5]

Change, however, is difficult to accomplish. It is one of the most difficult leadership challenges. Research has shown that two-thirds of Total Quality Management efforts normally are abandoned; re-engineering efforts have garnished a 70 percent failure rate; and 50 percent of transformation efforts fail during their initial phases.[6] When change does fail, the cause is frequently attributed to the inappropriate behaviours, attitudes, and beliefs of leaders, and leaders who lack the requisite change competencies to recognize and understand the implications and subsequent impact that can result prior, during, and following change initiatives. Indeed, as John Kotter explains, "The combination of cultures that resist change and managers who have not been taught how to create change is lethal."[7] It is therefore incumbent on leaders at all levels to have expert knowledge of the change process and to reinforce the understanding that, although change evokes pain, it represents a necessary component of both individual and organization success. Thus, the intent of this chapter is to present what is known about change, specifically, the leader's role in achieving successful change.

## **CHANGE DEFINED**

Common workplace phrases such as "the only constant is change" or "change for change's sake" remind us of just how familiar people are with this concept. These phrases hint of some degree of dissatisfaction, frustration, or change fatigue. The *Canadian Oxford Dictionary* provides a simple but useful definition of change: "the act or an instance of making or becoming different."[8] Within the context of organizations, the act of making or becoming different is generally in response to changes in technology, clients, competitor strategies, and the social, economic, and political environment. How the organization reacts and what it does in response to these pressures is what successful change is all about.

## THE HUMAN RESPONSE TO CHANGE

Without a doubt, change is disruptive. Change can or will shift or affect the organizational power structures, the relations between individuals, allies and adversaries, the opportunities and payoffs, and personal pride. In doing so, it will evoke myriad human responses. Some people will embrace change while others will engage in acts of omission, display counterproductive behaviours, and/or respond in emotional ways (e.g., being cynical, depressed, angry, frustrated, frightened). Impending change will cause people to think in terms of gains and losses, security or threats, opportunity or death. They will contemplate whether the change is relevant, if it can be ignored, and how they can best cope with and regulate the change. With certainty, however, "Whenever human communities are forced to adjust to shifting conditions, pain is ever present."[9] As Lee Iacocca, Chrysler's most notable CEO, revealed, "From Wall Street to Washington, from boardrooms to union halls, what anybody with power is most scared of is change. Any kind of change. Especially change that is forced on them."[10]

To achieve successful change demands an understanding of the potential human response to change. This awareness enables leaders to develop strategies to minimize the impact or pain associated with change and can serve to inoculate leaders against the surprise or shock when they encounter employee denial, anger, depression, decreased performance, and/or an unanticipated time period to adapt to the change.

## A STRATEGY FOR SUCCESSFUL CHANGE

Unfortunately, no off-the-shelf formula or recipe exists that can be dusted off and applied generically to change efforts. Evaluations and lessons learned from organizational case studies do highlight, however, several key steps that help to generate successful change. Arguably, John Kotter presented the best-known strategy or process for creating successful change. Kotter outlined an eight-stage process that includes:

* establishing a sense of urgency;
* creating a guiding coalition;
* developing a vision and strategy;
* communicating the change vision;
* empowering broad-based action (removing obstacles, encouraging risk-taking, and changing systems that may undermine the change effort);
* generating short-term wins;
* consolidating gains and producing more change; and
* anchoring new approaches in the culture.[11]

While the above steps hold merit and appeal to common sense, some may argue that junior leaders play a limited role in several of the key steps proposed by Kotter. Certainly, the extent to which and the challenges that senior vice junior officers experience in leading change vary somewhat. For example, most would agree that creating vision is largely a senior leader responsibility. The vision is meaningless, however, without subordinate leaders who ensure that the vision is put into action. Thus, leaders at all levels play a key role in ensuring successful change.

So what then is the role of the leader in effecting successful change? Adopting the steps provided by Kotter, blended with the framework for successful change reported in *Leadership in the Canadian Forces: Leading the Institution*,[12] the following section provides a template or guide that, if applied, should help to generate successful change. It is important to note that this process is not necessarily intended as sequential and that leader influence or activities may be exerted across the various process steps at the same time. For example, while building their coalition teams, leaders can also develop their plan and establish short-term change targets.

*Creating Vision and Establishing a Sense of Urgency*

Without a clear vision, organizations will drift, lack purpose, and become increasingly unable to respond efficiently to new challenges. As detailed in *Leadership in the Canadian Forces: Leading the Institution*, creating vision is a critical institutional leadership activity that provides the ultimate sense of purpose, direction, and motivation for an organization. A carefully crafted vision provides the expression, verbal or written, of where an organization is headed. It is a representation of what the organization should become and it serves to create inspiration and a sense of identity with the organization, and to generate a focus of effort toward reaching the desired end state. Through the vision, leaders strive to create energy and, as well, to attract commitment to the change effort.

That creating vision is largely an institutional leadership responsibility in no way marginalizes the contributions of subordinate leaders. Since subordinate leaders are often closest to those most affected by change, they must strive to fully understand the intent of the vision and be capable of communicating it to their subordinates with clarity. This requires that subordinate leaders collect all available information associated with the vision through asking questions, conducting research, confirming leadership intent, participating in change focus groups, attending town halls and/or other forums, and anticipating subordinate questions to mitigate potential concerns. Anticipating and preparing for subordinate questions is a critical leader responsibility at this stage.

Beyond vision, leaders at all levels can achieve or hasten commitment to change by establishing a sense of urgency to catch people's attention and signal that the change is important. This will begin the process of shifting people from their comfort zones. Indeed, research confirms that the biggest mistake people make when trying to change organizations is to "plunge ahead without establishing a high enough sense of urgency in fellow managers and employees."[13] Leaders can generate urgency through communications that explain the need for and advantages of change, but also, through careful observation and the identification of readily apparent crises or organizational pain that demonstrates that "business as usual" is no longer an acceptable response to future challenges. As Craig M. McAllaster, a professor of management, notes: "… seize the moment when pain exists because when pain goes away, so does the motivation and energy for change associated with it."[14]

*Building a Coalition Team*

Once the vision has been established, leaders must build a change or coalition team whose influence will help to set the conditions for change, shape the behaviour of others, and generate additional commitment for the change effort. In choosing team members, leaders must ensure that those selected:

- \*   have the requisite expertise relevant to the change effort;
- \*   are open and tolerant of new ideas;
- \*   are committed to the vision; and
- \*   have the power and credibility to drive change in their spheres of influence.

Involving others will help to create an atmosphere of trust and transparency that will promote the free exchange of information. Eliciting the support of others in the change process will also help to generate additional ideas, develop a more comprehensive change plan, increase buy-in, help identify unforeseen concerns, and reduce the potential for resistance to change. As business consultant and author Suzanne Slattery asserts, "When people respond to change, they do so on the basis of whether the change was done by them or to them."[15] Change is much easier to accept when it is not perceived as forced compliance. Leaders, at every opportunity, must attend to expanding the breadth of support for the change effort.

Without exception, selected team members must be fully committed to the vision, both in terms of words and actions. Because leading change occurs in a very public forum, leaders must carefully monitor their actions. As John Kotter remarks, "Nothing undermines the communication of a change vision more than the behavior on the part of key players that seems inconsistent with the vision."[16] Any indication of leader ambivalence can lead to subordinate perceptions that the change is not critical or urgent, or that it will not result in the desired outcome. Leader ambivalence can be created through adherence to the old value system while trying to convince others of the need to change, observed contradictions between the rhetoric of change and daily work practices, and/or bragging about past performances and successes while trying to promote change.[17] Without argument, commitment to the change effort will not be fully achieved in the presence of leader ambivalence.

*Creating a Plan to Achieve the Vision — Stakeholder and Impact Analysis*

With the assistance of the coalition team, leaders must formulate realistic plans to execute the vision or change. This typically begins with conducting a stakeholder and impact analysis to gather as much information as possible that is relevant to the change effort. This takes considerable time and effort, however, the collected information will be invaluable and will translate into a more effective plan, as well as help mitigate future risk. The analysis should reveal:

- what needs to change;
- what can remain intact;
- potential obstacles to change;
- personal and team strengths and weaknesses;
- the capacity of the team/organization to change;
- the resources required to make the vision possible; and
- potential allies (enablers) as well as those who may undermine the initiative.

Leaders should take the time to hear and embrace the views of dissenting personnel as these views may offer unique perspectives, help to identify additional concerns or potential obstacles, and offer alternative pathways to follow. Listening to the concerns of dissenters may also help get them on track. Be aware, however, that skeptics with power and influence over others can cause considerable damage. Under some circumstances, personnel unwilling to commit to the change process may have to be transferred.

In the absence of an impact analysis, a gut feel or inappropriate course of action may be adopted that could damage the credibility and/or reputation of the leader, as well as slow down or halt the change process. For a leader who has collected the necessary information, the best or preferred course of action should be evident. This course of action should define the way ahead in terms of identifying objectives, potential paths to follow, how communications will proceed, how progress will be monitored, the type of feedback to be provided, ways to reduce or eliminate barriers or obstacles to change, and the resources required to reach the end state. Once the plan is forged, it must be communicated.

*Communicating the Vision/Change*

Successful change requires effective communications — communications that occur early and frequently. As John Kotter explains, "The real power of a vision is unleashed only when most of those involved in an enterprise or activity have a common understanding of its goals and direction."[18] If the reasons for and mechanisms of change are not understood, it will be difficult to effectively execute the desired change. Thus, the objective of the communication plan is to ensure that the message reaches those it needs to reach. The more the communication plan resonates with followers, the greater the likelihood it will be endorsed by subordinates. This requires that leaders know the target audience and what motivates its members. The communication plan must be aggressive, simple, accurate, clear, and delivered in a manner that evokes enthusiasm and motivation.

To aid with the transmission of information, leaders should take full advantage of the numerous available communication platforms (i.e., e-mail, letters, internet/intranet, in-house publications, presentations, posters, etc.). To the maximum extent possible, however, face-to-face communications are recommended, as this form of interaction should alleviate the impact of resistance to change more than that of other communication methods.

*Implementing Change and Maintaining Momentum*

With change being difficult, it is essential that leaders aggressively monitor progress and strive to achieve and sustain momentum. Failure to do so may result in complacency and the tendency to revert to previous practices. Leaders can draw on a number of techniques to help maintain the necessary momentum. These include:

* leader persistence in communications;
* capitalizing on intermediate goals;
* creating a culture of continuous learning and adaptation; and
* engaging in professional development opportunities.

Arguably, the key ingredient to maintaining momentum is the leader's commitment to persistent and timely communications throughout the change process. People need to know what is happening and if their efforts are valued and consistent with the objectives of change. Frequent and timely communications will reinforce commitment to the change effort, ensure that it is not forgotten, continue to remind people that it is still important, serve to build trust in both the leader and the change plan, mark progress, and identify future objectives. Conversely, failure to communicate will generate ambivalence, create the impression that the effort is no longer viewed as important or valued, raise suspicions and, ultimately, result in considerable resistance to change.

Short-term and intermediate goals should be used to maintain momentum during the change process. Intermediate goals that are clearly observable and achievable help to dilute the change process into more realistic and manageable steps. As such, these intermediate goals can be used to maintain momentum in several ways:

* To reward progress early and often. Achieving intermediate goals also provides evidence that the change effort is progressing as planned and these successes will help to evoke additional commitment as well as encourage those less committed to alter their opinions.
* To provide the opportunity to consolidate gains and adjust plans as required without compromising the integrity of the primary or ultimate objective, as well as the credibility of the leader. The impact of not meeting or having to adjust intermediate goals will be far less damaging than that of not meeting or changing the desired end state.
* To provide the means to measure and report on progress and success, as well as provide necessary feedback.

Another way for leaders to sustain momentum is to actively support and encourage continuous learning and innovation. Effective transformation, as U.S. Navy Commander Steven W. Knott asserts, can only come from an organizational culture that "promotes free thinking and a critical exchange of ideas."[19] This can be accomplished by creating an atmosphere that is open to new ideas, encourages risk-taking and creative thought, tolerates failure, promotes trust and confidence in others, and encourages lifelong learning. Subordinates must come to understand that they can shape the future as well as take advantage of emerging opportunities.

Regardless of the best efforts of leaders, some degree of chaos and uncertainty will undoubtedly occur. This uncertainty can result in conflict that will require immediate intervention. Failure to act quickly can foster workplace divisions, reduce leader credibility, and create confusion as to what is the preferred course of action to achieve the desired end state. Momentum gained can be lost quickly if leaders fail to act. Through professional development, notably, the development of conflict resolution and counselling skills, leaders will become more adept in dealing with potential conflict. In doing so, leaders will increase their ability to maintain change momentum.

*Sustaining Change*

Even after the desired end state has been achieved, leaders must continue to exercise vigilance to ensure that the change becomes firmly embedded into the emerging culture. As people have the tendency to shift back to old ways of doing business or to develop redundant systems, continued monitoring will enable the leader to gauge the extent to which the change has been fully implemented. At this point, the pressing leadership challenge may still involve encouraging others to let go of the past. Indeed, the willingness or ability to let go of the old culture may still seem counterintuitive to many, especially for those who benefited under the old system. As Dr. Peter Drucker, author of 31 management and business books, indicates, "… you have to build organized abandonment into your system."[20]

Leaders are in the obvious position to promote a culture that is more conducive to change. This can be achieved through rewards, establishing observable linkages between change efforts and mission success, and instilling values that support a culture of continuous learning, professional development, and innovation. In referring to organizational success, former General Electric CEO Jack Welch asserted that: "Our behavior is driven by a fundamental core belief: The desire and the ability of an organization to continuously learn from any source — and to rapidly convert this learning into action — is its ultimate competitive advantage."[21] Creating a culture of learning and innovation, although frequently espoused by senior leaders, also becomes a critical junior leader responsibility. Once established, however, this culture will support future change efforts.

*Summary: The Change Process*

Leaders at all levels play a critical role throughout the various stages of the change process. As such, they require the requisite competencies and knowledge to effectively drive change. They must, however, also be aware of the many obstacles to effective change, albeit, most of which can be predicted and effectively managed through awareness and effective leadership practices.

## **OBSTACLES TO CHANGE: WHAT LEADERS NEED TO KNOW**

Niccolo Machiavelli, in his seminal 1513 work, *The Prince*, identified that "… there is nothing more difficult to take in hand, more powerless to conduct or more uncertain in its success than to take the lead in the introduction of a new order of things because the innovators have for enemies all those who would have done well under the old conditions and lukewarm defenders in those who may do well."

Not much has changed. Leaders must expect and prepare for some resistance to change but guard against the tendency to think that subordinates will always fight change, and that resistance is purely a negative response or a personal shortcoming of an individual or team. Indeed, resistance may simply represent devotion or loyalty to the old culture or value system that has been consistently rewarded and reinforced in the past. Resistance is a healthy human instinct and, as such, leaders should assist others to understand that "change is an opportunity and not a threat."[22] Of note, it has been argued that the label "resistance to change" is sometimes used "to dismiss potentially valid employee concerns about proposed changes."[23]

Research has shown that people resist change for a variety of complex and diverse reasons. For example, resistance to change has been associated with:

* perceived threats to one's security, loss of status, and fear of the unknown;[24]
* threats to employee core identity;[25]
* loss of control, routines and traditions;[26] and
* lack of adequate resources to implement change.[27]

It is also generally accepted that people resist change because it involves effort and uncertainty, and it moves them away from their comfort zones. Unfortunately, "the basic tension that underlies many discussions of organizational change is that it would not be necessary if people had done their jobs right in the first place."[28] Thus, organizational change is often predicated within the context of failure. Not a great place to start.

Leaders should also acknowledge that most subordinates have been exposed to considerable workplace change and "quick-fix" remedies that frequently were abandoned prematurely or that resulted in limited success. This is an important consideration, as research has demonstrated that when people have been exposed to significant change, they become less capable to deal with additional change.[29] This incapacity to deal with change can be further exacerbated if the change is perceived as self-serving or pointless. This perception can lead to resistance, as well as serve to undermine trust in the leader.

The message for leaders is that resistance can be somewhat mitigated if leaders focus on the people, as well as the desired end state. In the end, change is all about people.

## SUMMARY

Change is inevitable and necessary for the survival of any organization. Leading change is a complex process that requires extensive planning, execution, and vigilant monitoring to sustain momentum, as well as mitigation against potential obstacles that can serve to undermine change efforts. As such, it represents a significant challenge for organizations, leaders, and subordinates. Attention to the human response to change, as well as the critical role played by the leader, will help to reduce potential resistance to change and increase the probability of success. Leaders must benefit from lessons learned and identify the necessary conditions to secure success during change endeavours. To achieve successful change, leaders must recognize the cost of change (to be mitigated when possible), prepare employees for the change process, and recognize that acceptance of change takes considerable time. In the end, those leaders considering change should always confirm why change should be considered, whether that change is necessary, and, whether people are ready for it.

## NOTES

1. James O'Toole, *Leading Change: The Argument for Values-Based Leadership* (New York: Ballantine Books, 1996), xvi.
2. Rosabeth M. Kanter, *The Enduring Skills of Change Leaders,* in *On Leading Change*, eds. Frances Hesselbein and Rob Johnston (San Francisco: Jossey-Bass, 2002), 48.
3. Brigadier-General David A. Fastabend and Robert H. Simpson, "Adapt or Die: The Imperative for a Culture of Innovation in the United States Army," *Army Magazine*, February 2004, 2.
4. L. Wong, S. Gerras, W. Kidd, R. Pricone, R. Swengros, *Strategic Leadership Competencies* (Carlisle Barracks, PA: Strategic Studies Institute, U.S. Army War College, 2003), http://www.carlisle.army.mil/research_and_pubs/research_and_publications.shtml.
5. John P. Kotter, *Leading Change* (Boston: Harvard Business School Press, 1996), 26.
6. Peter Senge, *The Dance of Change* (New York: Doubleday, 1999), 5–6.
7. Kotter, 29.
8. Katherine Barber, ed., *The Canadian Oxford Dictionary* (Don Mills, ON.: Oxford University Press, 1998), 237.
9. Kotter, 4.
10. Lee Iacocca, quoted in James O'Toole, *Leading Change: The*

*Argument for Values-Based Leadership* (New York: Ballantine Books, 1996), xvi.
11. Kotter, 21.
12. Canada. *Leadership in the Canadian Forces: Leading the Institution* (Kingston: DND, 2007).
13. Kotter, 4.
14. Craig M. McAllaster, "Leading Change by Effectively Utilizing Leverage Points within an Organization," *Organizational Dynamics*, Vol. 33 (2004): 319.
15. Suzanne M. Slattery, *Engaging Leadership: How to Empower People to Reshape the Pyramid* (Patron Publishers, 1992), 87.
16. Kotter, 27.
17. Gregory S. Larson and Phillip K. Tompkins, "Ambivalence and Resistance: A Study of Management in a Concertive Control System," *Communications Monographs* 72 (2005): 1–21.
18. Kotter, 85.
19. Steven W. Knott, "Knowledge Must Become Capability: Institutional Intellectualism as an Agent for Military Transformation," Chairman of the Joint Chiefs of Staff Strategy Essay Competition: Essays 2004 (Washington, DC: National Defense University Press, 2004): 52–54.
20. Peter F. Drucker and Peter M. Senge, "Strategies for Change Leaders," in *On Leading Change*, eds. Frances Hesselbein and Rob Johnston (San Francisco: Jossey-Bass, 2002), 9.
21. Jack Welch, *General Electric Annual Report*, 1996, letter to stakeholders.
22. Statement by Peter F. Drucker, in *On Leading Change*, eds. Frances Hesselbein and Rob Johnston (San Francisco: Jossey-Bass, 2002), 11.
23. Sandy K. Piderit, "Rethinking Resistance and Recognizing Ambivalence: A Multidimensional View of Attitudes Toward an Organizational Change," *Academy of Management Review*, Vol. 25 (2000): 784.
24. Christopher P. Neck, "Thought Self-Leadership: A Self-Regulatory Approach Towards Overcoming Resistance to Organizational Change," *The International Journal of Organizational Analysis*, Vol. 4 (1996): 202–212.

25. Edgar H. Schein, *Organizational Culture and Leadership*, 2nd ed. (San Francisco: Jossey-Bass, 1992).
26. Lynn Isabella, "Evolving Interpretations as a Change Unfolds: How Managers Construe Key Organizational Events," *Academy of Management Journal*, Vol. 33 (1990): 7–41.
27. Tony Proctor and Ioanna Doukakis, "Change Management: The Role of Internal Communication and Employee Development," *Corporate Communications: An International Journal*, Vol. 8 (2003): 268–277.
28. Karl E. Weick and Robert E. Quinn, "Organizational Change and Development," *Annual Review of Psychology* 50 (1999): 362.
29. Patricia Sikora, David Beaty, and John Forward, "Updating Theory on Organizational Stress: The Asynchronous Multiple Overlapping Change (AMOC) Model of Workplace Stress," *Human Resource Development Review*, Vol. 3 (2002): 3–35.

## SELECTED READINGS

Canada. *Leadership in the Canadian Forces: Leading the Institution.* Kingston: DND, 2007.

Canada. *Leadership in the Canadian Forces: Leading People.* Kingston: DND, 2007.

Elrod II, P.D. and D.D. Tippett. "The 'Death Valley' of Change." *Journal of Organizational Change Management* 15 (2002): 273–291.

Hesselbein, Frances and Rob Johnston, eds. *A Leader to Leader Guide on Leading Change.* San Francisco: Jossey-Bass, 2002.

Kotter, John. *Leading Change.* Boston: Harvard Business School Press, 1996.

Lee, William and Karl Krayer. *Organizing Change.* San Francisco: John Wiley & Sons, , 2003.

O'Toole, James. *Leading Change: The Argument for Values-Based Leadership*. New York: Random House, 1995.

Senge, Peter. *The Dance of Change*. New York: Doubleday, 1999.

Weick, Karl E. and Robert E. Quinn. "Organizational Change and Development." *Annual Review of Psychology* 50 (1999): 361–386.

# 3
# CHARACTER
## by Robert W. Walker

From the earliest of times, numerous human qualities have been represented as the most virtuous, admired, and respected human characteristics required of leaders in order for them to be effective. This handbook elaborates on a number of these qualities relevant to effective leaders, e.g., judgment, trust, intelligence, versatility, expertise, ethics, motivation, and professionalism, among many others. The military, much like other professional institutions such as medical organizations, justice systems, legal firms, and church clergy, all have distinct structures with their hierarchies of authoritative relationships, administrative/bureaucratic internal networks, and varying emphases on achieving organizational outcomes. Moreover, these unique institutions share a common attribute. They can only advance their missions through the strengths and values of the highly qualified, formally recognized, relatively self-regulated professionals embedded in their organizations.

A nation's military organization has such a profession, the profession of arms, embedded within it and, therefore, the military's leaders need to balance and successfully address the integrated demands that evolve from its organizational attributes and its professional responsibilities. They need character! The issue, then, becomes the identification of the specifics of that character needed for effective military leadership in an "organizational + professional" institution, specifically, characteristics such as emotional stability, conscientiousness, agreeableness, tolerance, flexibility/adaptation, and openness to ideas. So, leaders must be responsible as stewards of the profession of arms and for fundamental organizational roles.

With respect to this issue of character, military leaders are accountable in a bifurcated environment with one part being a formal accountability to the government and the people for the effective leadership of a large, federal bureaucratic ministry,[1] and for the stewardship of the profession of arms spread across the nation and internationally via a unique expertise, internalized ethos, identity, values, and professional ideology. This circumstance of responsibility by military leaders for both organizational and professional expectations creates a challenge to identify the premier requisite human qualities that, collectively, would constitute character. This chapter addresses character as a collection or synthesis of attributes — human behaviours, capacities, knowledge, dispositions, and values that, when clustered together, provide a concept for character.

## **MILITARY RESEARCH ON CHARACTER**

Character has been described at different times by various academics, military researchers, and other subject matter experts as a pattern of behaviour or a moral constitution, and also as various collections of individual qualities or traits or personality attributes. As significant examples, two books on leadership by authors/researchers/academics James Kouzes and Barry Posner reflect extensive research on the credibility of leaders and the character that people seek and admire in their leaders before committing to a common cause.[2] They identified four qualities of character that they ranked above all others — honest, forward-looking, inspiring, and competent. Kouzes and Posner maintain that a leader, to possess these four qualities, requires an inner exploration of his/her principles or beliefs ("credos"), competence, and self-confidence, the sum of which, the authors emphasize, is character. They see these qualities as necessary for leaders to have dependency as leaders in the eyes of others. Kouzes and Posner see the quest for character as a noble and worthwhile one and not a simple one, but a quest that also is baffling, frustrating, and difficult.

Character as a leader requisite has been touted in the military doctrine of numerous countries, e.g., United States, New Zealand, Australia, and Canada. It is explored as a critical collection of strengths, virtues, and values required for strong leadership. A number of armed forces projects include:

*United States*

One American military study by Michael Mathews and colleagues, actually focusing on the greatest strengths evident among military samples in an international comparison of character, identified those strengths as honesty, hope, bravery, industry, and teamwork.[3] U.S. Army War College researcher Leonard Wong and associates identified "moral character … the profession's ethic … requiring moral courage …"[4] Donald Snider, professor of leadership at the U.S. Military Academy at West Point, defined character as, "those moral qualities that constitute the nature of a leader, and shape … decisions and actions."[5] He identified truth, right (professional military ethics, institutional values), and actions (walk the talk, exemplify the congruence of truth and right) as requisite for a leader with character. Finally, an extraordinary and most relevant book published in 2005 by the Center for Creative Leadership (one of the co-authors having served ten years as a full professor of leadership at the United States Air Force Academy), expressed credibility's two broad dimensions as expertise and character, with character having the attributes of integrity, trust, truthfulness, fairness, responsibility, plus respect and concern for others.[6]

*Australia and New Zealand*

Commemorating the 90th anniversary of the landing by members of the Australian and New Zealand Corps (ANZAC) on the Gallipoli peninsula, General Peter Cosgrove, chief of the Australian Defence Force, described the military character as having steadfastness, courage, "mateship," and mutual support, with a commitment to sacrifice.[7] Related to this, the Australian Army has described character as possessing those inner qualities of a member evident in positive and constructive behaviour applied to self-development, relationships, and community. It also integrated the importance of personal character and competence, of values and attitudes, into their individual-team-organization paradigm, and then integrated that with the U.S. military concept of Be, Know, Do that addresses the crucial components of service, character, and leadership.[8] Finally, a current New Zealand Defence Force initiative is aimed at integrating the 2007 Canadian Forces Professional Development Framework of metacompetencies and attributes, including the professional attributes of an internalized ethos (values, beliefs), moral reasoning, and credibility (including character), into its force-wide leadership development.[9]

*Canada*

A Canadian Forces (CF) leadership doctrine manual included the qualities of integrity, fairness, and dependability in their definition of character.[10] Further, it identified trust in leadership as evolving and resulting predominantly from the presence of these listed qualities of character in the leader, along with a demonstrated leader competence, and a leader's care and consideration of others.[11]

**DEFINING CHARACTER**

The probability of effective leadership is increased by:

* a leader's positive personal attributes, most of which contribute to character;

* the successful application of these personal qualities, (i.e., character) to leader interactions and relationships with others in the process of leading people, i.e., qualities or attributes such as team-building skills and communication competence; and,

- for those leaders at senior levels, the successful application of these personal qualities (i.e., character) to leader interactions and relationships with others for leading-the-institution responsibilities. These responsibilities include stewardship of the military profession, shaping organizational change, and partnering external to the military institution.[12]

Essentially, then, character, when present, is anchored and evident in those personal qualities demonstrated by the leader and experienced by others in interpersonal and group/team interactions. The research on the military institution and the profession of arms supports an understanding of character as a cluster of human qualities, capacities, and effects.

The evidence, once synthesized, supports the position that character brings together the components of Integrity, Competence, Respect, and Inspiration, each with descriptors (samples provided) that define them:

- <u>Integrity</u>: Honest, fair, ethical, just, moral, authentic, principled, dutiful, dedicated, trusted;

- <u>Competence</u>: Expertise, knowledge, self-aware, professional, dependable, proficient, wise;

- <u>Respect</u>: Loyal, responsible, consistent, egalitarian, approachable, sincere, understanding;

- <u>Inspiration</u>: Motivating, imaginative, creative, enabling, influential, courageous, evolving.

## <u>SUMMARY</u>

The purpose of this chapter was to describe character, which was challenging since the boundaries surrounding the qualities of leadership are elastic and porous. Moreover, arguments exist as to which qualities fit within other more expansive qualities, or do not, or in themselves should be the overarching concepts; and human characteristics are not mutually exclusive but are overlapping, interconnected, and interdependent, like puzzle pieces. Clarity may exist only after sufficient and correctly connected puzzle pieces are engaged, and empirical cause-and-effect experimental research on character is completed.

Possessing character, and being perceived by others as possessing character sufficient to be an effective leader, requires an abiding integrity and honesty that generates a pervasive trust among others, a confidence among others of the leader's fundamental competence to generate impact, an expectation by others of respectful exchanges with a leader in interactions reflecting loyalty, approachability, predictability, and responsibility, and a responsiveness to the needs of others to be inspired, motivated, and enabled by the leader to serve a common cause. These are the interdependent and requisite components of character.

## NOTES

1. See the Government of Canada's *The Leadership Network* with its four "Key Leadership Competencies" for bureaucracies: Values & Ethics (integrity and respect); Strategic Thinking (analysis and ideas); Engagement (people, organizations, partners); and Management Excellence (action management, people management, financial management). This is distributed across a framework of six levels: deputy ministers, assistant deputy ministers, directors general, directors, managers, and supervisors. This 4 x 6 framework supports government leaders in their ministries and departments, and is not significantly different from the Professional Development Framework's (PDF) four metacompetencies of Expertise, Cognitive Capacities, Social Capacities, and Change Capacities. However, it varies substantially with respect to the PDF's Professional Ideology with attributes such as an internalized ethos, moral reasoning, and,

credibility and impact necessary for circumstances of unlimited viability.
2. James M. Kouzes and, Barry Z. Posner, *Credibility* (San Francisco: Jossey-Bass Publishers, 1993); James M. Kouzes and Barry Z. Posner, *The Leadership Challenge*, 3rd ed. (San Francisco: Jossey-Bass Publishers, 2002), 13.
3. Michael. D. Matthews, Jarle Eid, Dennis Kelly, Jennifer K.S. Bailey, and Christopher Peterson, "Character Strengths and Virtues of Developing Military Leaders: An International Comparison" in *Military Psychology* Vol. 18 (Supplement), 2006, S57–S58.
4. L. Wong, S. Gerras, W. Kidd, R. Pricone, R. Swengros, Strategic Leadership Competencies (Carlisle Barracks, PA: Strategic Studies Institute, U.S. Army War College, 2003), http://www.carlisle.army.mil/research_and_pubs/research_and_publications.shtml.
5. Donald M. Snider, "Leaders of Character, Officership, and the Army Profession," A presentation at the Canadian Conference of Ethical Leadership, Royal Military College of Canada, 28 November 2006.
6. Richard L. Hughes and Katherine C. Beatty, *Becoming a Strategic Leader* (San Francisco: Jossey-Bass with the Center for Creative Leadership, 2005).
7. General Peter Cosgrove, "At the Going Down of the Sun ..." *Defence Magazine*, April 2005. Available at http://www.defence.gov.au/defencemagazine/editions/20050401/columns/cosgrove.htm.
8. For a U.S.–Canada comparison for the Be-Know-Do concept, see Robert W. Walker, *The Professional Development Framework: Generating Effectiveness in Canadian Forces Leadership*, CFLI Technical Report 2006-01. (Kingston: Canadian Forces Leadership Institute, 2006), 29–30. Available at http://www.cda-acd.forces.gc.ca/cfli.
9. Personal communications with NZDF members, September–October 2007. See also Robert W. Walker, *The Professional Development Framework: Generating Effectiveness in Canadian Forces Leadership* CFLI Technical Report 2006-01.
10. Canada, *Leadership in the Canadian Forces: Doctrine* (Kingston: DND, 2005), 5. Available at http://www.cda-acd.forces.gc.ca/cfli.

11. *Ibid.* Other Canadian sources wrestling with definitions and attributes appropriate for inclusion in the character cluster are: the CF manual *Duty with Honour: The Profession of Arms in Canada* (Kingston: DND, 2003). Available at http://www.cda-acd.forces.gc.ca/cfli. It identifies the four attributes of the profession of arms — responsibility, expertise, identity, military ethos — and the military values expressed within that Canadian military ethos — duty, loyalty, integrity, and courage. And also "The CF Statement of Defence Ethics," described in Canada, *Leadership in the Canadian Forces: Leading People* (Kingston: DND, 2007), 16, (available at http://www.cda-acd.forces.gc.ca), lists the six principled obligations of military members (all of which would appear to have high relevance to character) as integrity, loyalty, courage, honesty, fairness, and responsibility.
12. Robert W. Walker, *The Professional Development Framework*, 29–30.

**SELECTED READINGS**

Canada. *Duty with Honour: The Profession of Arms in Canada*. Kingston: DND, 2003. http://www.cda-acd.forces.gc.ca/cfli.

Canada. *Leadership in the Canadian Forces: Conceptual Foundations*. Kingston: DND, 2005. http://www.cda-acd.forces.gc.ca/cfli.

Hughes, R.L. and K.C. Beatty. *Becoming a Strategic Leader*. San Francisco: Jossey-Bass and the Center for Creative Leadership, 2005.

Kouzes, J.M. and B.Z. Posner. *Credibility*. San Francisco: Jossey-Bass Publishers, 1993.

Kouzes, J.M., and B.Z. Posner. *The Leadership Challenge*, 3rd ed. San Francisco: Jossey-Bass Publishers, 2002.

Matthew, L.J., ed. *The Future of the Army Profession*. New York: McGraw-Hill, 2002.

Walker, R.W. *The Professional Development Framework: Generating Effectiveness in Canadian Forces Leadership*, CFLI Technical Report 2006-01. Kingston: Canadian Forces Leadership Institute, 2006. Also available http://www.cda-acd.forces.gc.ca/cfli.

Wenek, K.W.J. "Defining Effective Leadership in the Canadian Forces." Unpublished Paper. Kingston: Canadian Forces Leadership Institute, 2003. http://www.cda-acd.forces.gc.ca/cfli/.

Wong, L., S. Gerras, W. Kidd, R. Pricone, R. Swengros. *Strategic Leadership Competencies*. Carlisle Barracks, PA: Strategic Studies Institute, U.S. Army War College, 2003. http://www.carlisle.army.mil/research_and_pubs/research_and_publications.shtml.

Zaccaro, Stephen J. *The Nature of Executive Leadership: A Conceptual and Empirical Analysis of Success*. Washington, DC: American Psychological Association, 2001.

# 4
# COHESION
by Allister MacIntyre

It has long been recognized that cohesion plays an important role when it comes to ensuring military success in a variety of domains. After all, military units can only function effectively when all of the members carry out their responsibilities in a collective manner. This means that every person in uniform, regardless of occupation, rank, or operational status, has a role to play to ensure mission success. This also means that the most proficient militaries will be those with a cohesiveness that bonds them together in a unified sense of purpose and belongingness. Leaders have an important role to play in creating the conditions that lead to cohesion, leading by example, and doing whatever is necessary to sustain high levels of interpersonal bonding within their groups and to instill an orientation that is mission-focused.

**COHESION DEFINED**

Social scientists most often view cohesion as both a group characteristic and as a determinant of performance. Alexander Makalachki, a professor of organizational behaviour, has referred to cohesion as the "stick togetherness" of a group.[1] Though admittedly quaint, this view is not inconsistent with the more formal definitions that have been offered. Cohesion has been variously defined as: a basic bond or uniting force,[2] the force that acts on members to remain in a group,[3] and the tendency for groups to stick together in the pursuit of goals and objectives.[4] Despite some slight variations in these views of cohesion, it is evident that there is an underlying agreement that cohesion is a bonding force within groups. For example, in a recent *Armed Forces and Society* article, U.S. Army Research Institute military psychologist Guy L. Siebold states that the most widespread and meaningful approach to understanding cohesion comes from social psychology, "... with its focus on bonding among group members and with their organization and military service."[5]

Cohesion is often linked to morale, but it is important to understand that the terms are not synonymous. According to Israeli psychologist Rueven Gal, cohesion and morale are considered to be two aspects of a higher order factor, which might be labeled "unit climate."[6] Morale is also based in such factors as pride in group achievements and aims, trust in group leadership, and a sense of devotion and loyalty to other group members.[7] As expressed by organizational psychologist Craig Pinder, "... the term morale is often used to describe the overall attitudes of a work group, rather than of a single individual."[8] For a more complete synopsis of how morale functions within groups, see Chapter 29 on morale in this handbook.

Within the military context, cohesion has been used interchangeably, and often incorrectly, with many other terms including camaraderie, bonding, brotherhood, morale, esprit, and will-to-fight.[9] Although cohesion tends to be viewed in a favourable light, it does not always lead to beneficial outcomes. If the goals of a cohesive group are not in line with organizational objectives, then these organizational objectives may not be fulfilled.[10] With this understanding in mind, military cohesion has been defined as "... the bonding together of members of an organization or unit in such a way as to sustain their will and commitment to each other, their unit, and the mission."[11] This definition implies that the group members will not only be bonded to each other, they will recognize the need to remain committed to the mission.

Few people would disagree with the perception of cohesion as a general bonding force amongst group members. However, it would be short-sighted to assume that the bonding takes place exclusively amongst one's peers, colleagues, and buddies within a group. The cohesion that takes place in a military unit, in particular, is not limited to relationships between peers (horizontal cohesion), but also includes relationships developed between subordinates and superiors (vertical cohesion).[12] As described by Lieutenant-Colonel Christian Cowdrey of the U.S. Marine Corps, horizontal cohesion refers to the caring that takes place among soldiers, and the mutual support. Vertical cohesion implies a bonding up and down the chain of command and can be described as leader-led support.[13] Naturally, both types of cohesion will play a critical role with respect to military effectiveness. Some theorists add organizational cohesion as a third level of bonding (taking pride in, and sharing the values of, an organization)[14] while others even add institutional cohesion (a bonding between personnel and their military branch) as a fourth form.[15] Group cohesion approaches have also differentiated between task cohesion (reflecting the feelings the group members have about the group objectives and goals) and social cohesion (a focus that is directed more toward feelings of personal involvement in social interactions with other group members).[16]

## **WHY COHESION IS IMPORTANT**

Although cohesion is not an easy concept to define or measure,[17] extensive research, carried out over a number of years, has yielded substantial support for a stable and positive relationship between group cohesion and performance.[18] Cohesion has been shown to not only have a positive impact on group morale,[19] it can help to improve group functioning and performance by buffering the effects generated by high-stress situations.[20] In their extensive review of 39 studies of cohesion in military units, researchers at the U.S. Army Research Unit concluded that unit cohesiveness contributed to a variety of desirable outcomes.[21] A slightly contrasting, yet not totally conflicting, perspective can be found in the writings of military sociologist Anthony King. He suggests that the real relationship may be that repetitive drills and military efficiency may lead to a strengthening of cohesion in groups rather than cohesion leading to performance.[22] King argues that the regimented procedures developed by military units serve to "… sustain social cohesion, even though the actions they demand of individuals often run against the instincts of self-preservation."[23] Conversely, the bonds of friendship will become strained and undermined when a group member exhibits military or professional incompetence.

Nevertheless, the importance of cohesion to combat-effectiveness has been extensively documented. Examples include: an examination into command style, group cohesiveness, and performance in Israeli tank crews;[24] an assessment of the demoralization of American troops in Vietnam;[25] an exploration into primary group relationships in the Wehrmacht during the Second World War;[26] and a case study in military cohesion during the South Atlantic Conflict of 1982 (Falklands).[27]

Canadian defence analyst Anthony Kellett also provides historical accounts of military conflicts where cohesion, or its absence, played important roles[28] These incidents include: the defence of Calais by the 30th Brigade in 1940, where "regimental pride, more than any other factor, sustained the men of the Thirtieth Brigade in the face of overwhelming firepower;"[29] the 1944 Burma operations of Merrill's Marauders, in which high morale and esprit de corps gradually dwindled as a result of exhaustion, false promises, and increasing distrust of commanders; the 1951 prolonged resistance of the vastly outnumbered Gloucestershire Regiment against the Chinese during the Imjin battle in Korea; and the defensive and offensive actions on the Golan Heights in 1973 by the Israeli Seventh Armored Brigade, where effective leadership proved to be a key factor in maintaining high morale despite a surprise attack by the Arabs, shortages of ammunition, mental and physical exhaustion, and entrapment by the enemy.

Of particular relevance for combat operations, Canadian military historian and infantry officer, Colonel Bernd Horn, asserts that a "... powerful tool for controlling fear is strong group cohesion or primary group relationships ... the greatest fear felt by most combat soldiers is the fear of letting down their comrades."[30] Similarly, Leonard Wong, a professor at the U.S. Army War College, contends that it is the sense of loyalty generated by cohesion that compels soldiers to face additional hardships and danger to recover a fallen comrade.[31] More recent positive examples include an in-depth analysis of the potential impact on unit cohesion caused by reassignments of personnel to other units while being deployed during the

Gulf War.[32] In her review of the literature, and quantitative analysis of the Falklands War, military sociologist Nora Stewart noted five consequences of military cohesion. Specifically, with higher levels of military cohesion, there will be:

* less non-battlefield casualties in combat;
* more soldiers who will fire their weapons in combat;
* less desertion in time of war;
* more soldiers who will fight valiantly; and
* less AWOL, drug addiction, alcoholism, and sick calls in peacetime.[33]

## **COHESION: THE DARK SIDE**

From the foregoing, there seems to be a clear linkage between cohesion and performance, although the traditional view of the relationship tends to be a simplistic one. Although a high level of cohesion is important, its positive impact almost demands a correspondingly high degree of group morale. If morale is not positive, then cohesiveness can even lead to adverse consequences.[34] For example, if a group is highly cohesive, but morale is low because of its work or living conditions (as was the case for many soldiers in Vietnam), then this cohesive group may engage in collective actions that are in conflict with the higher order goals established for that group.[35]

When the evidence is examined with a critical eye, it becomes clear that the presence of cohesion in a unit can enhance performance and the pursuit of organizational goals; but cohesion can just as easily jeopardize organizational aims as well. This is primarily because, although popular opinion focuses on the positive aspects, cohesion is not a simple one-dimensional construct. The traditional emphasis on the positive aspects of cohesion is due in large part to the mystique explored by military sociologists Edward Shils and Morris Janowitz in their discussion of the primary group,[36] and more recent dissertations such as Stewart's discussion of the Falklands War, where the Argentine forces are described as lacking cohesion, and the British as being very cohesive.[37] Many military scholars have been able to illustrate that cohesion is no guarantee for combat-effectiveness.[38]

Under certain circumstances, cohesive groups may pursue goals that are different from, or even antithetical to, those of their superiors, the unit, or the larger organization.[39] For example, it has been estimated that, during the Vietnam conflict, less than 20 percent of the fraggings of officers were carried out by individuals.[40] The more common type of fragging in Vietnam was group-engendered, involving covert knowledge and collusion on the part of other members of the group.[41] The Vietnam experience was also characterized by polarization between the races, and the "near mutinous actions of black servicemen in the early 1970s."[42] In point of fact, it has been generally stated that "… mutiny is a more common occurrence than one might suppose from the number of mutinies officially reported."[43] In arguing that primary groups can be highly cohesive, yet impede the goals of a military organization, Janowitz and Little point to the defensive norms (which were incompatible with the requirements of the military) developed by segregated Negro units during the Second World War.[44]

Finally, military psychologist Rueven Gal admits that even the highly cohesive Israeli Defence Force is not without problems of this nature. As an example, he describes a letter of protest sent to Prime Minister Menachem Begin by 350 officers and enlisted men. The letter warned of increasing difficulties in supporting the Israeli government's policy with respect to the occupied territories.[45] Most of these soldiers were employed with combat units.

## **VERTICAL AND HORIZONTAL COHESION**

To truly understand these negative consequences of cohesion, the distinction between vertical and horizontal cohesion becomes of critical importance. Vertical cohesion, defined as the positive bond, or favourable sentiments, that soldiers have for their leaders, is the mechanism by which group objectives are articulated with the goals of the larger organization. Vertical cohesion is fostered by exemplary leadership, which is characterized by a sense of fairness in superiors, and the willingness of competent superiors to lead their soldiers into combat and share equitably in the risk of death. This is referred to by Gal as the "follow-me" leadership characteristic that is imprinted in all junior Israeli commanders from their earliest training.[46] As such, vertical cohesion positively affects fighter spirit, the willingness to seek success in combat regardless of the risk of injury or death.[47]

In the absence of vertical cohesion, primary-group goals can be thrown out of alignment with those of command and can result in combat refusals, mutinies, and even fraggings. Canadian military psychologist Peter Bradley states that "Military life naturally creates strong horizontal cohesion ... and this lateral loyalty can become so powerful that it leads to leadership failures .... Vertical cohesion is the glue that ensures that the values and norms of lower-level units are consistent with unit, service, and national interests."[48] For example, when American involvement in Vietnam was being de-escalated, the perceived lack of legitimacy of army authority peaked. As a result, the levels of vertical cohesion declined rapidly, peer influence worked against fighting in combat, and deviant behaviour such as fragging and mutiny was at it highest level of occurrence.[49]

In his analysis of fighter spirit (using data collected from 1155 American soldiers following basic training in 1952), author Robert B. Smith found that vertical cohesion had a much stronger relationship with fighter spirit than horizontal cohesion.[50] Furthermore, exemplary leadership was shown to have both a strong effect on vertical cohesion and a moderately strong direct effect on fighter spirit. This is not to say that horizontal cohesion is unimportant; leadership and vertical cohesion are simply the key elements in combat motivation and effectiveness.

One of the primary functions of horizontal cohesion is to provide social support, which itself has instrumental value in times of extreme stress, and the positive stress-buffering effects of social support networks and groups have been extensively documented.[51] The Israeli Defence Forces also have investigated the buffering effects of horizontal cohesion on combat stress. Gal reports that studies conducted following the 1973 war found that 40 percent of the men who were diagnosed with battle shock reported minimal group cohesion and identification with their units, as compared with only 10 percent of a control group (men not suffering battle shock) reporting the same.[52] Obviously, cohesion is not only important for the achievement of group and organizational goals, but it also plays a crucial role in easing the negative consequences of extreme stress and, consequently, in the overall well-being and operational effectiveness of the unit.

It has been noted that vertical cohesion will be more important for mission success than horizontal cohesion. But, what do leaders need to do to ensure vertical cohesion? Faris R. Kirkland, a military academic and veteran of both Korea and Vietnam, succinctly sums up this requirement as: "Leaders who behave competently, tell the truth, keep their word, and take care of their troops earn trust and build vertical cohesion. This is not easy. Sometimes it may appear to be easier and more appropriate to withhold information from subordinates or even lie to them. Leaders who yield to this temptation lose their believability and compromise vertical cohesion in their units."[53]

In short, competent leaders who display genuine integrity and caring will both earn respect and build vertical cohesion.

## **DEVELOPING COHESIVE GROUPS**

Several of the discussions in the preceding sections explore the negative picture of cohesion, but this is only one side of it. The positive link between cohesion and performance is not only robust; it is undeniable. In this regard, the link is even strong when cohesion leads to negative consequences. The negative aspects only come into play because the objectives of the cohesive group are not consistent with organizational goals. The challenge for leaders rests with understanding that the right kinds of cohesion are necessary to ensure mission success. For example, Israeli military psychologist Shabtai Noy emphasizes that leadership and cohesion are the two critical factors that will influence the delicate balance determining whether the unit will suffer combat stress reactions or exhibit valour.[54] He states that, "A soldier keeps fighting for his comrades rather than against the enemy. He is afraid to lose his comrades if he lets them down. If he does, he may remain without their support against the prevailing anxiety as well as feel ashamed and guilty."[55] Thus, it is easy to see that cohesion will remain an important concept for militaries, and leaders will have to take the appropriate steps to develop cohesion within groups and units.

With this need in mind, it is fortunate that the cohesion literature is not confined to investigations into the impact of cohesion on group functioning. There is also a considerable body of work that identifies the factors that are known to contribute to group cohesion. In other words, it is not only important for leaders to understand how cohesion functions; leaders must appreciate the conditions that contribute to the right kind of cohesion. Military leaders also need to be aware that levels of unit cohesion will fluctuate over time in response to changes in conditions and group composition. For example, U.S. Army researchers Paul Bartone and Amy Adler were able to demonstrate that, during an operational deployment, unit cohesiveness develops in an inverted-U pattern by "... starting out low, reaching a high point around mid-deployment, and then decreasing again toward the end of the 6-month mission."[56]

There are many different ways for leaders to exert their influence to enhance unit cohesiveness. Specifically, it has been established that the amount of time group members spend together reinforces cohesion;[57] cohesion is stronger when group members like one another;[58] cohesion is stronger when group participation is rewarding;[59] cohesion can be strengthened by external threats;[60] and leaders can enhance levels of cohesion through encouraging feelings of warmth and acceptance;[61] or by simply providing a positive role model.[62]

Leaders also need to monitor and competently manage the conflicts that take place within their units. If conflicts do emerge, and these conflicts lead to negative outcomes, then the bonds that exist amongst the group members will be threatened. Any weakening of the interpersonal bonds will result in detrimental consequences to group cohesiveness. Yet, in military units in particular, leaders must also be aware of the cohesion building aspects of shared hardships. Establishing high standards for entry into a group, and the presence of collective unique challenges, struggles, and danger will lead to mutual respect, affiliation, and strong group cohesion.[63]

## SUMMARY

Political scientist and soldier, Colonel William Darryl Henderson, offers several guidelines that characterize leadership in cohesive units.[64] As per his thorough summary, leaders in cohesive units are:

* not managerial in approach;
* granted sufficient authority to meet their responsibilities;
* focused on personal, empathic, and direct contact with soldiers;
* charismatic, rather than glib, in nature;
* viewed as being protective of group members;
* capable of not only building on success, but neutralizing the effects of failure;
* able to use indoctrination to maximize leadership effectiveness;
* capable of demonstrating that all members, especially the leaders, share in the hardships and dangers;
* seen as being consistently competent;
* able to create group norms that are consistent with organizational objectives;
* capable of bringing about internalized values and discipline that will enable soldiers to overcome fears; and
* able to effectively put into use both personal and legitimate power.[65]

Cohesion comes in many shapes and sizes and can have both positive and negative outcomes. More than anything else, cohesion is a leadership issue. Effective leaders will understand the consequences of cohesion, will appreciate how to instill the right type of cohesion, and will actively put this knowledge to use by creating a unit that embodies both vertical and horizontal cohesion. They will lead by example, share in hardships, maintain a high level of competence, personify organizational values, and act with integrity and honour.

**NOTES**

1. A. Mikalachki, *Group Cohesion Reconsidered: A Study of Blue Collar Work Groups* (London, ON: University of Western Ontario School of Business Administration, 1969).
2. W.E. Piper, M. Marrache, R. Lacroix, A.M. Richardson, and B.D. Jones, "Cohesion as a Basic Bond in Groups," *Human Relations* 36 (1983): 93–108.
3. S.J. Zaccaro, "Nonequivalent Associations Between Forms of Cohesiveness and Group-Related Outcomes: Evidence for Multidimensionality," *The Journal of Social Psychology* 131 (1990): 387–399.
4. A.A. Cota, R.S. Longman, C.R. Evans, K.L. Dion, and L. Kilik, "Using and Misusing Factor Analysis to Explore Group Cohesion," *Journal of Clinical Psychology* 51 (1995): 308–316.
5. Guy L. Siebold, "The Essence of Military Group Cohesion," *Armed Forces & Society* 33, No. 2 (2007): 287.
6. R. Gal, *A Portrait of the Israeli Soldier* (Westport, CT: Greenwood Press, 1986).
7. A. Kellett, *Combat Motivation: The Behavior of Soldiers in Battle* (Boston: Kluwer Nijhoff Publishing, 1982).
8. C.C. Pinder, *Work Motivation: Theory, Issues, and Applications* (Glenview, IL: Scott, Foresman and Company, 1984).
9. See Gal, *A Portrait of the Israeli Soldier* and N.K. Stewart, *South Atlantic Conflict of 1982: A Case Study in Military Cohesion*, Research Report 1469. (Alexandria, VA: U.S. Army Research

Institute for the Behavioral and Social Sciences, 1988).
10. This aspect of cohesion will be expanded upon later in this chapter.
11. J.H. Johns et al., *Cohesion in the US Military* (Washington, DC: National Defense University Press, 1984), 4.
12. Stewart, *South Atlantic Conflict of 1982: A Case Study in Military Cohesion, Research Report 1469*. (Alexandria, VA: U.S. Army Research Institute for the Behavioral and Social Sciences, 1988).
13. C.B. Cowdrey, *Vertical and Horizontal Cohesion: Combat Effectiveness and the Problem of Manpower Turbulence*, A Monograph (Fort Leavenworth, KS: School of Advanced Military Studies, 1995).
14. A. Ahronson and C. Eberman, *Understanding Leadership and Teams in the Military Context* (Kingston, ON: Canadian Forces Leadership Institute, 2002). Contracted Paper at http://www.cda-acd.forces.gc.ca/cfli.
15. Siebold, 287.
16. Ahronson and Eberman.
17. G.L. Siebold, "The Evolution of the Measurement of Cohesion," *Military Psychology*, Special Issue 11, No.1 (1999): 5–26.
18. See C.R. Evans, and K.L. Dion, "Group Cohesion and Performance: A Meta-analysis," *Small Group Research* 22 (1991): 175–186; and B. Mullen and C. Copper, "The Relation between Group Cohesiveness and Performance: An Integration," *Psychological Bulletin* 115 (1994): 210–227.
19. R. Gal, "Unit Morale: From a Theoretical Puzzle to an Empirical Illustration — An Israeli Example," *Journal of Applied Social Psychology* 16 (1986): 549–564.
20. Z. Soloman, M. Mikulincer, and S.E. Hobfoll, "Effects of Social Support and Battle Intensity on Loneliness and Breakdown During Combat," *Journal of Personality and Social Psychology* 51 (1986): 1269–1276.
21. L.W. Oliver, J. Harman, E. Hoover, S.M. Hayes, and N.A. Pandhi, "A Quantitative Integration of the Military Cohesion Literature," *Military Psychology*, Special Issue 11, No. 1 (1999): 57–83.

22. A. King, "The Word of Command: Communication and Cohesion in the Military," *Armed Forces and Society*, Vol. 32, No. 4 (2006).
23. *Ibid.*, 505.
24. A. Tziner, and Y. Vardi, "Effects of Command Style and Group Cohesiveness on the Performance Effectiveness of Self-Selected Tank Crews," *Journal of Applied Psychology* 67 (1982): 769–775.
25. C.C. Moskos, "The American Combat Soldier in Vietnam," *Journal of Social Issues* 31 (1975): 25–37.
26. E.A. Shils and M. Janowitz, "Cohesion and Disintegration in the Wehrmacht in World War II," The *Public Opinion Quarterly* 12 (1948): 280–294, 314–315.
27. Stewart, *South Atlantic Conflict of 1982: A Case Study in Military Cohesion, Research Report 1469*.
28. Kellett, *Combat Motivation: The Behavior of Soldiers in Battle*.
29. *Ibid.*, 23.
30. B. Horn, "'But … It's Not My Fault!' — Disobedience as a Function of Fear," in *The Unwilling and the Reluctant: Theoretical Perspectives on Disobedience in the Military*, ed. C. Mantle (Toronto and Kingston: Dundurn Press and Canadian Defence Academy Press, 2006), 184.
31. L. Wong, "Leave No Man Behind: Recovering America's Fallen Warriors," *Armed Forces and Society*, Vol. 31, No. 4 (2005): 606.
32. J. Griffith and J. Greenlees, "Group Cohesion and Unit Versus Individual Deployment of U.S. Army Reservists in Operation Desert Storm," *Psychological Reports* 73 (1993): 272–274.
33. Stewart, *South Atlantic Conflict of 1982: A Case Study in Military Cohesion*, Research Report 1469.
34. Kellett, *Combat Motivation: The Behavior of Soldiers in Battle*.
35. *Ibid.*
36. Shils and Janowitz, "Cohesion and Disintegration in the Wehrmacht in World War II."
37. Stewart, *South Atlantic Conflict of 1982: A Case Study in Military Cohesion, Research Report 1469*.
38. See Kellett, *Combat Motivation: The Behavior of Soldiers in Battle*; Moskos, "The American Combat Soldier in Vietnam"; U. Ben-Shalom, Z. Lehrer, and E. Ben-Ari, "Cohesion During Military Operations:

A Field Study on Combat Units in the Al-Aqsa Intifada," *Armed Forces and Society* 32 (2005): 63–79; K. Lang, "American Military Performance in Vietnam: Background and Analysis," *Journal of Political and Military Sociology* 8 (1980): 269–286; and K. Lang, "The Dissolution of Armies in the Vietnam Perspective," in *The Military, Militarism, and the Polity: Essays in Honor of Morris Janowitz*, eds. M. L. Martin and E. S. McCrate (New York: The Free Press, 1984).

39. J. Griffith and M. Vaitkus, "Relating Cohesion to Stress, Strain, Disintegration, and Performance: An Organizing Framework," *Military Psychology,* Special Issue 11, No. 1 (1999): 27–55.
40. Kellett, *Combat Motivation: The Behavior of Soldiers in Battle*; Moskos, "The American Combat Soldier in Vietnam."
41. Moskos, "The American Combat Soldier in Vietnam."
42. *Ibid.*, 33.
43. J. Lovell, "Military Service, Nationalism, and the Global Community," in *The Military, Militarism, and the Polity: Essays in Honor of Morris Janowitz,* eds. M. L. Martin and E. S. McCrate (New York: The Free Press, 1984), 66; see also C. Mantle, ed., *The Unwilling and the Reluctant: Theoretical Perspectives on Disobedience in the Military* (Toronto and Kingston: Dundurn Press and Canadian Defence Academy Press, 2006).
44. M. Janowitz and R.W. Little, *Sociology and the Military Establishment* (Beverly Hills: Sage Publications, 1974).
45. Gal, *A Portrait of the Israeli Soldier.*
46. *Ibid.*
47. R.B. Smith, "Why Soldiers Fight. Part I: Leadership, Cohesion, and Fighter Spirit," *Quality and Quantity* 18 (1983): 1–32.
48. P. Bradley, "Obedience to Military Authority: A Psychological Perspective," in *The Unwilling and the Reluctant: Theoretical Perspectives on Disobedience in the Military,* ed. C. Mantle (Toronto and Kingston: Dundurn Press and Canadian Defence Academy Press, 2006), 35.
49. R.B. Smith, "Why Soldiers Fight. Part II: Alternative Theories," *Quality and Quantity* 18 (1983): 33–58.
50. Smith, "Why Soldiers Fight. Part I: Leadership, Cohesion, and Fighter Spirit."

51. See S. Cohen and T.A. Wills, "Stress, Social Support, and the Buffering Hypothesis," *Psychological Bulletin* 98 (1985): 310–357; and D.R. Williams and J.S. House, "Social Support and Stress Reduction," in *Job Stress and Blue Collar Work,* eds. C. L. Cooper and M. J. Smith (New York: Wiley and Sons, 1985).
52. Gal, *A Portrait of the Israeli Soldier.*
53. F.R. Kirkland, "Honor, Combat Ethics and Military Culture," in *Military Medical Ethics Volume 1,* ed. L.R. Sparacino (Washington, DC: The Borden Institute, 2003), 163.
54. S. Noy, "Combat Stress Reactions," in *Handbook of Military Psychology,* eds. R. Gal and A.D. Mangelsdorff (New York: John Wiley & Sons, 1991).
55. *Ibid.,* 513.
56. P.T. Bartone and A.B. Adler, "Cohesion over Time in a Peacekeeping Medical Task Force," *Military Psychology,* Special Issue 11, No. 1 (1999): 85–107.
57. Griffith and Greenlees, "Group Cohesion and Unit Versus Individual Deployment of U.S. Army Reservists in Operation Desert Storm."
58. A.J. Lott, and B.E. Lott, "Group Cohesiveness as Interpersonal Attraction: A Review of Relationships with Antecedent and Consequent Variables," *Psychological Bulletin* 64 (1965): 259–309.
59. J.P Stokes, "Components of Group Cohesion: Inter-member Attraction, Instrumental Value, and Risk Taking," *Small Group Behavior* 14 (1983): 163–173.
60. A.A. Harrison and M.M. Connors, "Groups in Exotic Environments," in *Advances in Experimental Social Psychology,* Vol. 8, ed. L. Berkowitz (Orlando, FL: Academic Press, 1984).
61. K.R. Westre and M.R. Weiss, "The Relationship Between Perceived Coaching Behaviors and Group Cohesion in High School Football Teams," *Sport Psychologist* 5 (1991): 41–54.
62. Piper et al., "Cohesion as a Basic Bond in Groups."
63. B. Horn, "A Law unto Themselves? — Elitism as a Catalyst for Disobedience," in *The Unwilling and the Reluctant: Theoretical Perspectives on Disobedience in the Military,* ed. C. Mantle (Toronto

and Kingston: Dundurn Press and Canadian Defence Academy Press, 2006).
64. William D Henderson, *Cohesion: The Human Element in Combat* (Washington, DC: National Defense University Press, 1985).
65. *Ibid.*, 107–115.

**SELECTED READINGS**

Gal, R. *A Portrait of the Israeli Soldier.* Westport, CT: Greenwood Press, 1986.

Henderson, William D. *Cohesion: The Human Element in Combat.* Washington, DC: National Defense University Press, 1985.

Janowitz, M. and R.W. Little. *Sociology and the Military Establishment.* Beverly Hills: Sage Publications, 1974.

Kellett, A. *Combat Motivation: The Behavior of Soldiers in Battle.* Boston: Kluwer Nijhoff Publishing, 1982.

Mantle, C., ed. *The Unwilling and the Reluctant: Theoretical Perspectives on Disobedience in the Military.* Toronto and Kingston: Dundurn Press and Canadian Defence Academy Press, 2006.

Military Cohesion. *Military Psychology*, Special Issue 11, No. 1 (1999).

Siebold, G.L. "The Essence of Military Group Cohesion." *Armed Forces & Society,* Vol. 33, No. 2 (2007): 286–295.

Smith, R.B. "Why Soldiers Fight — Part I: Leadership, Cohesion, and Fighter Spirit." *Quality and Quantity* 18 (1983): 1–32.

# 5
# COMBAT MOTIVATION
by Bernd Horn

Combat is arguably the most dramatic and dangerous of human experiences. Paradoxically, it brings out the best and worst in human beings. It is rife with danger and generates fear, if not terror. Yet, for those who survive, it can undoubtedly be one of the most exhilarating experiences they have ever undertaken. However, more often than not, it also carries a legacy of anguish, pain, and sorrow. Undeniably, for all who participate, combat can entail death, the infamous element of "unlimited liability" that all military members must accept. "The transition from life to death," observed one combat veteran, "was very swift."[1]

Faced with challenges and consequences of such magnitude, how then, does one convince and motivate individuals to fight? Research has shown that there is no one single "silver bullet" to explain combat motivation. Rather, there are a number of factors, some more important than others, which contribute to combat motivation and effectiveness. All play their part and each component affects individuals differently. Nonetheless, factors such as primary group relationships, cohesion, morale, discipline, the level of training received, self-confidence, clear communication, and the belief in the concept of duty with honour (military ethos), as well as good equipment, food, rest, and a desire to get the job done, are the key factors that motivate individuals to fight.

## COMBAT MOTIVATION DEFINED

In the simplest of terms, combat motivation refers to a soldier's willingness to fight. A Canadian military psychologist and former serving officer, Anthony Kellett, defines combat motivation as "the conscious or unconscious calculation by the combat soldier of the material and spiritual benefits and costs likely to be attached to various courses of action arising from his assigned combat tasks."[2] Kellett explains that such motivation is driven by "the influences by which soldiers make choices, the degree of commitment to, and persistence in effecting, a certain course of action."[3]

## FACTORS AFFECTING COMBAT MOTIVATION

As stated earlier, combat motivation is a function of myriad factors, many of which are interrelated and mutually supporting. In addition, the impact of any one, or combination of, these factors varies with each individual.[4] Nonetheless, factors impacting combat motivation are:

- leadership;
- primary group relationship;
- cohesion;
- morale;
- discipline;
- training;
- self-confidence;
- communications;
- belief in mission;
- duty/military ethos;
- get the job done;
- equipment; and
- food and rest.

Having identified the factors, it is prudent to examine each in detail. Again, it must be reiterated that many are overlapping and mutually reinforcing. Also, each individual in combat will react differently to any one, or combination of, these factors.

*Leadership*

In the simplest of terms, leadership is about influencing people to achieve some objective that is important to the leader, the group, and the organization. It is the human element — leading, motivating, and inspiring, particularly during times of crisis, chaos and complexity when directives, policy statements, and communiqués have little effect on cold, exhausted, and stressed subordinates. Strong leadership will encourage subordinates to go beyond the obligation to obey and commit to the mission in a way that maximizes their potential. It is the very individualistic, yet powerful, component that allows commanders and leaders at all levels to shape and/or alter the environment or system in which people function and thereby, influence attitudes, behaviour, and the actions of others. As such, it is a key factor in motivating individuals to fight.

Researcher John Dollard noted that 89 percent of combat veterans surveyed emphasized the importance of getting frequent instructions from leaders when in a tight spot.[5] Furthermore, evidence clearly indicates that leaderless groups normally become inactive.[6] Not surprisingly, Samuel Stouffer found that "cool and aggressive leadership was especially important" in pressing troops forward in dangerous and fearful situations such as storming across a beach raked by fire.[7]

The evidentiary importance of leadership is based on the fact that "role modelling" has an extremely important influence on a person's reaction to threatening situations. With regard to the evocation of courageous behaviour, American enlisted men in the Second World War told interviewers that leadership from in front was very important.[8] Most research has reinforced this intuitive deduction. It has been repeatedly proven that the presence of strong thoughtful leadership creates "a force which helps resist fear" and in essence provides a strong component of combat motivation.[9]

*Primary Group Relationship*

Sociologists have defined primary groups as "those small groups in which social behaviour is governed by intimate face-to-face relations (i.e. association and cooperation)."[10] Samuel Stouffer, in his seminal study on the American soldier in the Second World War, found that the primary group serves two principal functions in combat motivation. First, it sets and emphasizes group standards of behaviour. Second, it supports and sustains the individual in situations of stress that the individual would not otherwise have been able to withstand. Therefore, Stouffer concluded that a primary group generates its own norms, and joins or withholds its own informal sanction to enforce organizational commands and expectations.[11] As such, its impact on being able to motivate an individual to fight is immense.

Anecdotal evidence continues to reinforce these findings. Specialist Matthew Eversmann, during the battle in Mogadishu, Somalia, acknowledged that "seeing the men [in his squad] perform gave me the confidence and reassurance that I needed."[12] More to the point, troops in Iraq underscored the importance of primary-group relations to combat motivation. "Me and my loader were talking about it," conceded one soldier fighting in Iraq, "and in combat the only thing that we really worry about is you and your crew."[13] Another soldier revealed, "I know that as far as myself, sir, I take my squad mates' lives more important than my own."[14]

*Cohesion*

Cohesion is defined as "the bonding together of members of an organization/unit in such a way as to sustain their will and commitment to each other, their unit, and the mission."[15] Former U.S. Army chief of staff, General Edward Meyer, explained that cohesion is "the bonding together of soldiers in such a way as to sustain their will and commitment to each other, the unit, and mission accomplishment, despite combat or mission stress."[16] Intuitively, it becomes obvious that cohesion within a group or organization will act as an instrumental catalyst to motivate individuals to fight.

One researcher concluded, "The most significant persons for the combat soldier are the men who fight by his side and share with him the ordeal of trying to survive."[17] S.L.A. Marshal, in his seminal battlefield studies, observed, "I hold it to be one of the simplest truths of war that the thing which enables an infantry soldier to keep going with his weapons is the near presence or presumed presence of a comrade."[18]

The peer pressure involved within a cohesive group alone acts as a power prompt to fight. No one in the tight circle wishes to lose credibility, reputation, or to be the individual to let down their comrades. "I had the usual new boy's dreadful fear of failing and I was much more frightened of that," conceded one officer, "than any of the horror going on around me."[19] Private Joe Dimasi explained, "You've got to get this in your mind. It is death before dishonour — that's it. I went through the whole war that way."[20]

*Morale*

Scholars have defined morale "as the mental, emotional, and spiritual state of the individual. It is how he feels — happy, hopeful, confident, appreciated, worthless, sad, unrecognized, or depressed."[21] The link of morale to combat motivation is as intuitive as it is vital. "It has been the Israeli experience that company morale correlates significantly with personal morale and that the components of the latter include confidence in the soldier's own military skills, in his weapons, and in himself."[22] Napoleon Bonaparte insisted, "Morale makes up three-quarters of the game; the relative balance of manpower accounts only for the remaining quarter."[23] Similarly, Major Syd Thomson, speaking within the context of bitter fighting in Italy during the Second World War, insisted, "During a sticky battle morale is as important, if not more important than good tactics."[24] Quite simply, the state of mind is a powerful catalyst in determining willingness to fight.

*Discipline*

In the same manner, discipline and response to leadership are crucial variables shaping attitudes among combat veterans. The role of discipline is one of providing a psychological defence that helps the soldier to control fear and ignore danger through technical performance. "It is a function of discipline," extolled Field Marshall Bernard Montgomery, "to fortify the mind so that it becomes reconciled to unpleasant sights and accepts them as normal everyday occurrences.... Discipline strengthens the mind so that it becomes impervious to the corroding influence of fear.... It instils the habit of self control."[25] It also builds confidence in self, comrades, and the fighting organization — all of which compels individuals to fight.

The duty of obedience that discipline instils, as well as the imposition of discipline and the impossibility of escaping it, all add to the motivation of individuals to go into harm's way. Discipline itself also depends on moral pressure that prompts men to advance from sentiments of fear or pride, as well as the sense of unity of a group and doing one's part. Moreover, it is also a function of the surveillance of both peers and superiors who know each other well. In sum, discipline guides the uncertain and controls the unwilling.

*Training*

Training is defined as "a predictable response to a predictable situation."[26] Training instils confidence and hones technical skills to the point of instinctive reaction/response. Furthermore, it prepares individuals for what they will see, experience, and have to accomplish in varied environments and situations. In short, training helps soldiers to know what to expect and how to react, thus creating confidence in their ability to fight and survive. "Training," professed General William Slim, "was central to the discipline soldiers needed to control their fear, and that of their subordinates in battle; to allow them to think clearly and shoot straight in a crisis, and to inspire them to maximum physical and mental endeavour."[27] Kellett concluded, "One of the most valuable assets that training can confer on a soldier is confidence, not only in his own military skills and stamina, but also in his weapons and equipment."[28]

The value of training is also derived from its ability to create an element of developing instinctive responses. Drill, for instance, is utilized to teach the instinctive reaction of a body of troops to commands. "What is learnt in training," insisted commando commander Lord Lovat, "is done instinctively in action — almost without thinking down to the last man."[29] Overall, training is one of the most vital components in providing soldiers with the motivation to willingly undertake combat operations.

*Self-Confidence*

Confidence is perhaps the greatest source of emotional strength that a soldier can draw upon. "With it," argued behavioural expert Bernard Bass, "he willingly faces the enemy and withstands deprivations, minor setbacks, and extreme stresses, knowing that he and his unit are capable of succeeding."[30] Research has shown that "the general level of anxiety in combat would tend to be reduced insofar as the men derived from training a high degree of self-confidence about their ability to take care of themselves ... troops who expressed a high degree of self-confidence before combat were more likely to perform with relatively little fear during battle."[31] One special operations forces officer acknowledged, "We wouldn't be able to do the things we do if a guy knew he was going to be faced with a degree of danger and didn't have the confidence to confront that and carry out the task regardless."[32]

*Communications*

Clear communications is an extremely important component to combat motivation. Everyone wants to know anything and everything that impacts their current or future circumstances. Ignorance breeds anxiety, uncertainty, doubt, and rumours. All feed hesitation and fear. In contrast, accurate, up-to-date information that provides situational awareness and allows all ranks to understand the situation, mission, threats, operational plan, and any other important detail that can affect them, helps to curb hesitation and therefore breeds confidence. "If a soldier knows what is happening and what is expected of him," explained a combat experienced officer, "he is far less frightened than the soldier who is just walking towards unknown dangers."[33] Similarly, another combat veteran officer asserted, "Fear can be checked, whipped and driven from the field when men are kept informed."[34] In the end, accurate, clear and timely communications are an important factor in motivating individuals to fight.

*Belief in Mission*

The actual belief in the mission is another key factor in combat motivation. Tasks and missions must be perceived, by those executing them, to be worthwhile and achievable. Intuitively, it is understandable that no one wishes to risk his/her life for frivolous or inconsequential ventures or gains. Similarly, although research data has shown that patriotism and ideology have long been minimized as motivators to incite individuals to fight, a just cause is very important for both the combatant and the support network, namely, the public back home.[35] As such, the belief in the cause/mission, and the support that is generated, or conversely, the controversy that is unleashed, does play a significant role in motivating individuals to fight.

*Duty/Military Ethos*

The Canadian Forces doctrine states, "The military ethos embodies the spirit that binds the profession of arms. It clarifies how members view their responsibilities, apply their expertise, and express their unique military identity. The ethos incorporates fundamental institutional values that distinguish members of a professional military from ill-disciplined irregulars or mercenaries."[36] As such, the military ethos of an institution plays a pivotal role in motivating individuals to fight. In combination with all other factors, the military ethos weighs on the conscience, integrity, and character of an individual. It is the driving force behind a powerful internal voice that encourages individuals to do their duty; not let down their comrades, regiment or country; and stand by their sworn oath.

*Get the Job Done*

As noted earlier, the belief in the legitimacy or justification for the war or conflict, specifically that it is a just cause; as well as an understanding of the true nature and intent of the enemy, are important catalysts in motivating individuals to fight. These elements generate a realization and conviction that action must be taken and, as such, create an attitude of "getting the job done," which fuels motivation to fight.

*Equipment*

Good equipment is another vital factor in motivating individuals to fight. If military personnel believe that their equipment is as good as, or superior, to that of their enemy, they will have no hesitation to go into harm's way since they know that their training, skill, and equipment will allow them to prevail and survive contact. Anecdotally, the confidence Canadian soldiers have in their vehicles in Afghanistan, whether the Light Armoured Vehicle III (LAV III) or the RG 31 Nyala, allow them to unceasingly seek contact with insurgents since they know the probability of winning a firefight or surviving a direct hit (i.e. mine, improvised explosive devices/IED, or rocket-propelled grenades/RPG) are in their favour. Good equipment creates confidence, which in turn steels individual resolve to undertake combat.

*Food and Rest*

Such simple factors as good health, good food, adequate rest and sleep, clean dry clothes, protection from the elements, and adequate washing facilities all have impact on morale, as well as emotional, mental, and physical well-being, all of which impacts an individual's motivation to go into harm's way. Obviously, all soldiers need sleep, food, and drink regardless of their level of physical fitness. Practical experience in conflicts since the Second World War has demonstrated that the physical and psychological factors that lowered morale and sapped the courage of participants were fatigue, hunger, and thirst.[37] Paradoxically, there is a symbiotic relationship between fatigue and fear. The more fatigued a person is — the more susceptible to fear they become. And, the greater their fear, the greater is the drain on their energy. "Tired men fright more easily," observed Colonel S.L.A. Marshall in his decades of battlefield studies, and he concluded, "frightened men swiftly tire."[38] As a result, attention to the basic necessities such as rest, food, hygiene, and good health are key to providing the conditions under which individuals will be motivated to conduct combat operations.

## **LEADERSHIP AND COMBAT MOTIVATION**

Effective military leadership, defined as "directing, motivating, and enabling others to accomplish the mission professionally and ethically, while developing or improving capabilities that contribute to mission success," is critical to nurturing and supporting combat motivation.[39] In the end, the leadership is about influencing people to achieve some objective that is important to the leader, the group, and the organization. It is the human element — leading, motivating, and inspiring, particularly during times of crisis, chaos and complexity when directives, policy statements, and communiqués have little effect on cold, exhausted, and stressed subordinates. Strong leadership will encourage subordinates to go beyond the obligation to obey and commit to the mission in a way that maximizes their potential. It is the very individualistic, yet powerful component that allows commanders and leaders at all levels to shape and/or alter the environment or system in which people function and thereby, influence attitudes, behaviour, and the actions of others.

As such, Dollard observed that 89 percent of those combat veterans surveyed emphasized the importance of getting frequent instructions from leaders when in a tight spot.[40] Furthermore, evidence clearly indicates that leaderless groups normally become inactive.[41] As noted earlier, Stouffer found that "cool and aggressive leadership" was key to motivating frightened individuals in tough spots.[42] Much research has reinforced the intuitive deduction that "men like to follow an experienced man ... [he] knows how to accomplish objectives with a minimum of risk. He sets an example of coolness and efficiency which impels similar behaviour in others." In this regard, the presence of strong, thoughtful leadership creates "a force which helps resist fear," and thus enabling men to conduct combat operations.[43]

But, this effect is only present if there is trust in the leadership. Soldiers must believe that leaders mean what they say. Body language, tone, and eye contact can all betray insincerity. Actions must match words. In the end, it comes down to setting the example. A leader must never ask, or expect, troops to do that which leaders are unwilling to do themselves. Stouffer's study showed that what the officers did, rather than what they said, was important. "The commander can win the trust of his subordinates very quickly," explained Field Marshal Erwin Rommel, "if his orders are clear and prudent, if he takes constant care of them and if he is hard with himself, living under the same austere conditions as his men."[44] In short, those who impose risk must be seen to share it. It has been long said that a sense of equality of sacrifice is fundamental in cementing a fighting force. Leadership based on example and demonstrated competence reduces the need to rely on commands and sparks confidence and motivation to fight. As such, clearly, the role of the commander and leader is essential to combat motivation and combat effectiveness. They provide the encouragement, direction, and example for all to follow.

In summary, to maximize combat motivation of subordinates, leaders must:

* Demonstrate professional capability (i.e. technical/tactical competence) that demonstrates that they will not waste the lives of their subordinates through incompetence;

* Provide clear, accurate timely communication. The credibility of information passed is vital. Leaders must keep personnel informed as much as possible about virtually everything. It is not only the content of the message that is important but also the process itself. Regular communications ensures that everyone knows that they are not alone — that they are still part of a team. It is for this reason that communications should always be maintained at all cost;

* Know their men and women. Care for them and back them up. The leaders need not be popular but must have (i.e. earn) their respect;

* Be fair and provide firm discipline. The leader must hold individuals responsible and accountable, just as they themselves are responsible and accountable;

* Have a sense of responsibility and humanity. American General John A. Lejeune stated, "leaders must have a strong sense of responsibility of their office, the resources they expend in war are human lives";[45]

* Have a sense of humour. Humour is the most important form of self-discipline and acts to release tension;

* Demonstrate coolness and calm under pressure. "The commander strolled across the battlefield issuing orders in an icily calm voice," reported one

historian. "By setting such a courageous example [Lieutenant-Colonel A.A. "Bert"] Kennedy stiffened the determination of his regiment. The riflemen advanced again into the face of the enemy fire."[46] Another example provides a similar effect, "As always the Brigadier spoke as though from some profound and superhuman depth of knowledge. It was impossible to doubt the calm assurance of the Brigadier's voice, and now despite themselves both men felt some of their anxiety lifted." [47]

The observations of Field Marshal Viscount Montgomery provide an example conclusion. "In no case will good results be obtained," he asserted, "unless the leader is a man who can be looked up to, whose personal judgement is trusted, and who can inspire and warm the hearts of those he leads — gaining their trust and confidence, and explaining what is needed in a language that will be understood." Having achieved those ends, the effect of leadership on motivating individuals to fight is all encompassing. "I personally recall when in the advance in Germany one day, our Platoon was 'on point' and we suddenly came under small arms fire from our front and my men all took to the ditches," recalled one veteran. "I was peering about, under some cover to get a fix on the enemy when in a matter of minutes, I felt a poke in my back from a walking stick and it was the Brigadier with a smile. His comment was simply, 'not to hold up the entire Division,' so 'press-on' which is what we did. The point is, that you have no idea what confidence is carried to the troops when you have great leadership."[48]

## **SUMMARY**

Convincing individuals to go into harm's way and face potential death, wounding and/or dismemberment is a daunting challenge. It is one that cannot be taken lightly for the consequences are enormous. So, how then does one convince and motivate individuals to fight? Quite simply, there is no one single answer. Combat motivation is a function of a number of factors, some more important than others, which contribute to convincing or compelling individuals to fight. All play their part; many are overlapping and interrelated and each component affects individuals differently. Nonetheless, the key factors for combat motivation are primary group relationships, cohesion, morale, discipline, the level of training received, self-confidence, clear communication, and the belief in the concept of duty with honour (military ethos), as well as good equipment, food, rest, and a desire to get the job done.

## NOTES

1. Reginald Roy, ed., *The Journal of Private Fraser* (Victoria, BC: Sono Nis Press, 1985), 205.
2. Anthony Kellett, *Combat Motivation: The Behavior of Soldiers in Battle* (Boston: Kluwer Nijhoff Publishing, 1982), 6.
3. *Ibid.*, 6.
4. Readers are encouraged to read the specific chapters in this handbook related to these factors in order to gain deeper insights into the respective concepts.
5. John Dollard, *Fear in Battle* (Westport, CT: Greenwood Press Publishers, 1944), 44.
6. Elmar Dinter, *Hero or Coward* (London: Frank Cass, 1985), 92.
7. Samuel Stouffer, *Studies in Social Psychology in World War II — The American Soldier: Combat and Its Aftermath* (Princeton, NJ: Princeton University Press, 1949), 68.
8. Anthony Kellett, *Combat Motivation* — Operational Research and Analysis Establishment (ORAE) Report No. R77 (Ottawa: DND, 1980), 299.
9. Dollard, 44.
10. Alexander George, "Primary Groups, Organization, and Military Performance," in *Handbook of Military Institutions*, ed. R.W. Little (Thousand Oaks, CA: Sage, 1971), 297.

11. *Ibid.*, 309.
12. Russell W. Glenn, *Capital Preservation: Preparing for Urban Operations in the Twenty-First Century* (Santa Monica, CA: RAND Arroyo Center, 2001), 423.
13. Leonard Wong, Thomas A. Kolditz, Raymond A. Millen, Terrence M. Potter, "Why They Fight: Combat Motivation in the Iraq War" (unpublished paper, Carlisle, PA: Strategic Studies Institute, U.S. Army War College, May 2003), 3.
14. *Ibid.*, 10.
15. W.D. Henderson, *Cohesion: The Human Element in Combat* (Washington, DC: National Defense University, 1985), 4.
16. Frederick Manning, "Morale, Cohesion, and Esprit de Corps," in *Handbook of Military Psychology,* eds. David A. Mangelsdorff and Reuven Gal (London: John Wiley 1991), 457.
17. Henderson, 5.
18. S.L.A. Marshal, *Men Against Fire* (New York: William Morrow & Company, 1947), 42.
19. Lieutenant-Colonel Colin Mitchell, *Having Been a Soldier* (London: Mayflower Books, 1970), 50.
20. Captain T.M. Chacho, "Why Did They Fight? American Airborne Units in World War II," *Defence Studies,* Vol. 1, No. 3 (Autumn 2001): 81.
21. Manning, 454.
22. Anthony Kellett, "The Soldier in Battle: Motivational and Behavioral Aspects of the Combat Experience," in *Psychological Dimensions of War,* ed. Betty Glad (London: Sage Publications, 1990), 217.
23. Kellett, *Combat Motivation*, 3.
24. Daniel G. Dancocks, *The D-Day Dodgers* (Toronto: McClelland & Stewart, 1991), 180.
25. Field Marshall Bernard L. Montgomery, "Discipline from Morale in Battle: Analysis," in Canada, *The Officer: A Manual of Leadership for Officers in the Canadian Forces* (Ottawa: DND, 1978), 66.
26. Professor Ronald Haycock, "Clio and Mars in Canada: The Need for Military Education," presentation to the Canadian Club, Kingston, Ontario, 11 November 1999.

27. Robert Lyman, *Slim, Master of War* (London: Constable, 2004), 78.
28. Anthony Kellett, "The Soldier in Battle: Motivational and Behavioral Aspects of the Combat Experience," 217.
29. Will Fowler, *The Commandos at Dieppe: Rehearsal for D-Day* (London: HarperCollins, 2002), 55.
30. B.M. Bass, *Leadership and Performance Beyond Expectations* (New York: Free Press, 1985), 69.
31. S. J. Rachman, *Fear and Courage* (San Francisco: W.H. Freeman and Company, 1990) 63–64.
32. Major Reg Crawford, Australian SASR, Phil Mayne, "Professionals Accept High-Risk Employment," *Army*, No. 907, 27 June 1996, 3.
33. Lieutenant-Colonel Colin Mitchell, *Having Been a Soldier*, 41.
34. Canada, *CDS Guidance to Commanding Officers* (Ottawa: DND, 1999), 230.
35. For instance see Samuel Stouffer, *Studies in Social Psychology in World War II — The American Soldier: Combat and its Aftermath* (Princeton, NJ: Princeton University, 1949); John McManus, *The Deadly Brotherhood* (San Marino, CA: Presidio Press, 1998), 230; and Alexander George, "Primary Groups, Organization, and Military Performance," in *Handbook of Military Institutions*, ed. R.W. Little (Thousand Oaks, CA: Sage, 1971), 303.
36. See Canada, *Duty with Honour: The Profession of Arms in Canada* (Kingston: DND, 2003).
37. Major P.B. Deb, "The Anatomy of Courage," *Army Quarterly*, Vol. 127, No. 4 (October 1997): 405.
38. S.L.A. Marshall, *The Soldier's Load and the Mobility of a Nation* (Quantico, VA: The Marine Corps Association, 1950), 46.
39. Canada, *Leadership in the Canadian Forces: Conceptual Foundations* (Kingston: DND, 2005), 30.
40. Dollard, 44.
41. Dinter, 92.
42. Stouffer, 68.
43. Dollard, 44.

44. David Fraser, *Knight's Cross: A Life of Field Marshal Erwin Rommel* (London: HarperCollins, 1993), 39.
45. Quoted in Colonel B.P. McCoy, *The Passion of Command: The Moral Imperative of Leadership* (Quantico, VA: Marine Corps Association, 2006), 35.
46. Dancocks, 182.
47. Colin McDougall, *Execution* (Toronto: MacMillan of Canada, 1958), 9.
48. Letter, Sergeant Andy Anderson to Colonel Bernd Horn, 10 January 2003.

**SELECTED READINGS**

Canada. *Leadership in the Canadian Forces: Leading People*. Kingston: DND, 2007.

Capstick, Colonel Mike, Lieutenant-Colonel Kelly Farley, Lieutenant-Colonel Bill Wild (Ret'd), and Lieutenant-Commander Mike Parkes. *Canada's Soldiers: Military Ethos and Canadian Values in the 21st Century*. Ottawa: DND, 2005.

Grossman, Lieutenant-Colonel Dave. *On Killing: The Psychological Cost of Learning to Kill in War and Society*. Boston: Little, Brown and Company, 1995.

Henderson, William Darryl. *Cohesion: The Human Element in Combat, Leadership and Societal Influence in the Armies of the Soviet Union, the United States, North Vietnam, and Israel*. Washington, DC: National Defense University Press, 1985.

Holmes, Richard. *Acts of War: The Behaviour of Men in Battle*. New York: The Free Press, 1985.

Keegan, John. *Soldiers*. New York: Konecky & Konecky, 1985.

Kellett, Anthony. *Combat Motivation: The Behavior of Soldiers in Battle.* Boston: Kluwer Nijhoff Publishing, 1982.

Marshall, S.L.A. *Men Against Fire.* New York: William Morrow and Company, 1947.

McCoy, Colonel B.P. *The Passion of Command: The Moral Imperative of Leadership.* Quantico, VA: Marine Corps Association, 2006.

Shalit, Ben. *The Psychology of Conflict and Combat.* New York: Praeger Publishers, 1988.

Wong, Leonard, Thomas A. Kolditz, Raymond A. Millen, and Terrance M. Potter. "Why They Fight: Combat Motivation in the Iraq War." Carlisle Barracks, PA: U.S. Army War College, Strategic Studies Institute, 2003.

# 6
# COMBAT STRESS REACTIONS
## by George Shorey

This chapter is focused on combat-related stress reactions and is primarily directed at leaders at the tactical level. The causes, symptoms, and key actions immediate leaders can take to lessen the destructive potential of combat stress reactions (CSR) will be addressed. Information on the associated issues of combat guilt and responses to loss are also included. As this discussion is limited to select aspects of combat stress, suggested supplementary readings are offered at the closing of the chapter.

**DEFINITION AND BRIEF HISTORY**

> The view that only the "weak" collapse under the strain of war or that psychiatric collapse in battle is rare can only be maintained if one is prepared to ignore a great deal of evidence. Closer to the truth is the view that psychiatric breakdown in war is all too common; that given enough stress and exposure to battle, almost all soldiers will suffer some degree of psychiatric debilitation ... it is an inevitable result of the nature of war.[1]

As highlighted in the above quote, significant stress reactions resulting from the extremes of combat are an inevitable consequence of warfare and all individuals are susceptible. The term "Combat Stress Reactions" covers a range of *reversible* effects or symptoms caused by the stressors of combat, and refers to the *temporary* psychological distress that results in an inability to function normally.

The understanding of how to most effectively respond to and treat CSR has developed and matured considerably through the 20th century. Early attempts to explain battle-related dysfunctional reactions and severe anxiety states has ranged from nerve damage to the brain caused by bombing shock waves ("shell shock") to theories suggesting deficiencies in personality or character, such as "lack of moral fibre" (LMF) as coined by some military psychiatrists during the First World War.

During the Second World War, the term "combat exhaustion" surfaced and acknowledged the significant psychological impact of intense, prolonged warfare on soldiers who became stress casualties as a result of their cumulative "battle fatigue." The subsequent CSR term received considerable recognition during the latter part of the 20th century, when the critical support roles of immediate leaders and peers were emphasized (i.e., sub-unit social support network). Furthermore, stress reactions stemming from the experience of combat were regarded as "normal reactions to extreme and abnormal demands" as opposed to a sign of mental illness or personal shortcoming.

## **CSR AS DISTINGUISHED FROM PTSD**

Too often, CSR, a temporary, short-term reaction to the stress of combat, is confused with the long-term, clinically diagnosed condition known as Post Traumatic Stress Disorder (PTSD). While in the context of combat, both are related to the same extreme stress events but the likelihood of developing the significantly debilitating disorder PTSD can be lessened for many, provided pre-combat preventive measures and proper CSR in-theatre casualty management take place.

CSR will manifest itself during or generally shortly after the event and, properly managed, will allow the soldier to rotate back into the line and resume active duty in a short period of time. Mismanaged or ignored, the symptoms associated with CSR can worsen and have the potential to develop into the extremely debilitating and life-altering PTSD.

## PRIMARY AND SECONDARY CAUSES OF CSR

> The patrol picked its way through jungle so thick that by noon it was dark. A dead, midnight kind of darkness. Fifty men threaded their way. The first ten began to cross a river. The soldier walking point touched something with his boot. It was not a twig, not a root, not a rock. It was a trip wire to oblivion. In an instant the wire triggered a huge, fifty-pound Chinese mine. There was an enormous roar, like the afterburner of a jet, as it exploded, instantly ripping the point man apart. Shrapnel flew for yards.[2]

Combat stress reactions are linked to both primary and secondary causes. The lead cause of CSR is a perception of imminent external threat to survival. It is principally the fear of death, but is also associated with such fears as mutilation or the loss of a buddy. Situations considered "highly stressful" are those that have a "perceived significant potential outcome" or major consequence from the individual's perspective. Stress results from an imbalance between the perceived demand (or threat), and the person's perceived ability to meet that demand. Troop percentages experiencing CSR are dependent on both the intensity and duration of combat, and on the troops' evaluations of their ability to survive it.

In combat, there is little difference between the objective reality of the threat and its perception. However, the subjective perception of one's ability to survive combat will be influenced by such factors as the degree of confidence and trust held toward self, peers, and leaders. A soldier's coping ability is also compounded when the threat is novel or unpredictable. This interrelationship between stress and perception is fundamental and holds implications for leaders and unit personnel in addressing CSR.

The effects of the primary stressor can be further intensified by the extreme hardship and deprivation often associated with combat. These are termed "secondary causes" and include such stressors as physical exhaustion; an inability to distinguish friend from foe; separation or isolation from normal supports; intense noise and fire levels; reduced visibility; having no clearly defined front; sleep deprivation and/or disrupted sleep cycles; dehydration; cold injury; and, deficiencies in nutrition. The combined effect of these sample secondary stressors can significantly erode a soldier's resources for coping with, and enduring, the principal threat to survival.[3]

Combat stress reactions are most prevalent under conditions of high intensity combat and when taking physical casualties (killed and wounded). These conditions intensify the anxiety, fear, and collective stress, which can lead to a "perceptual shift" and resulting loss of confidence in leadership and erosion in unit cohesion.

## **SIGNS AND SYMPTOMS OF CSR**

The type of early interventions and responses by combat unit personnel to members exhibiting combat stress reactions following engagement can prove critical to a stress casualty's recovery. Because of this, possessing some basic knowledge as to what to look out for in terms of CSR signs and symptoms is considered essential.

Combat stress reactions can take the form of multiple symptoms, which in turn can vary somewhat from individual to individual. Despite this complexity, a useful guideline to CSR identification is to:

> ... beware of a noticeable change in the unit member's normal mode of behaviour or emotional state (for example, a notable and lingering decline in attitude, mood, general behaviour or performance). That is, a member "... who shows persistent, progressive behaviour that deviates from his baseline behaviour may be demonstrating the early warning signs and symptoms of a combat stress reaction."[4]

Keeping this general guideline to CSR identification in mind, a number of specific signs and symptoms can be direct indications of combat stress. The severity of individual reactions and symptom intensity can vary considerably from mild to extreme. A large sample of common signs and symptoms are provided and organized under three categories: thoughts, feelings, and behaviours.

* <u>Thoughts and Cognitive Processes</u>: difficulty speaking, communicating; disruption in logical thinking; difficulty making decisions; inability to concentrate or focus; forgetfulness; suspicion; decreased alertness to extreme overvigilance; recurring thoughts or nightmares related to aspects of the combat experience;

* <u>Emotions and Feelings</u>: rapid emotional shifts; inappropriate emotions (to the situation); fear; anger; self-blame and guilt (e.g., over acts committed or perceived failures associated with inaction); hatred; apathy (e.g., including apparent indifference to danger); feeling overwhelmed or out of control; and,

* <u>Behavioural and Physical Signs</u>: social withdrawal; change in communications/interactions with others; outbursts; verbal blaming/contempt toward others and/or self; tremors; excessive silence; anti-social acts; accident-prone; sleep disturbances; shakes; crying; vision problems; profuse sweating; nausea to vomiting; dehydration; insomnia; immobility.

## **COMBAT GUILT, SHAME, AND GRIEF REACTIONS**

The soldier's grief helps us comprehend the powerful bond that arises between men in combat. This bond may be so intense as to blot out the distinction between self and other, leading each to value the other's life above his own. But now the other is dead; the survivor still lives. "It should've been me!" is the cry of guilt that goes up in the midst of grief from a survivor condemned by his very survival.[5]

Combat-related guilt can take on many forms. Self-blame can result, for example, from acts of negligence, wrongdoing or error, inaction or perceived failure to protect others from harm, or from personal thoughts and feelings related to the combat experience. Research in this area has noted that "most military veterans experience some level of guilt related to their involvement in combat."[6] Survivor guilt, as reflected in the quote above, is said to be characterized by confusion over having lived while others perished, and the resulting personal struggle with the meaning of this survival.[7] Guilt can, for some combat veterans, serve as a way of honouring the dead and in a sense make those that were lost present.[8]

It is important to recognize that those surviving combat where special comrades were killed may experience any number of intense emotions and reactions, from anxiety and outrage to tension and deep sadness. Such responses should be recognized for what they are: natural (albeit painful) reactions to having experienced a significant loss. Where a soldier seems to be taking on total responsibility for another's death, they must come to appreciate that such an extreme position is neither appropriate nor rational. "A war veteran, failing to comprehend that, will blame himself for the death of a friend, failing to realize that the enemy was the killer. The war should be blamed, not those who lived through it."[9] It is stressed, however, that for many combat veterans, such rational reasoning may necessitate experiencing and processing the guilt first.[10]

## **GRIEF (REACTION TO LOSS)**

The nature of grieving is a very personal process and not every soldier will respond in a similar way to the same loss. Factors such as past experience with loss and death, cultural conditioning, and perception of preventability, can each shape the grief response.[11] A degree of appreciation for what others are going through can help to ensure realistic expectations and appropriate support and patience on the part of unit peers and immediate leaders. The symptoms associated with grief can be organized under three broad categories:[12]

* Avoidance: initial shock and disbelief; numbness is quite common; confused and dazed; this phase can serve as a buffer against becoming overwhelmed by the event;

* Confrontation: a period of intense grief; irritability, tension, bitterness; a time of "angry sadness"; feels indifferent to others; loss of faith; may experience some anger toward the deceased; angry at others seemingly less affected; possible "survivor guilt" coloured by self-reproach and/or a sense of worthlessness; intense yearning for those taken is common; and

* Re-establishment: learning to live with the loss; some emotional energy is reinvested in new persons, things, ideas; those lost are not forgotten but put in a special place/memory; guilt may accompany early reestablishment efforts as those remaining learn they can live on in spite of the significant loss; may coexist with earlier reactions and symptoms.

The grieving process is complex and symptoms associated with it can fade and then resurface over time. While the general phases described are valid, individuals will not necessarily move on from one stage to the next in a predictable, step-by-step manner. The duration as well varies from individual to individual. The grief response should be respected as a natural response to loss and is an important dimension of healing.

## **LEADER ACTIONS AND FORWARD TREATMENT OF CSR**

> On the battlefield, leaders, especially junior officers and noncommissioned officers who had direct personal contact with the men, exercised a great influence on them. Good leadership and intensified comradeship produced good morale, which in turn led to good performance and a low psychiatric casualty rate.[13]

The time-honoured military leadership principles of fostering unit cohesion, providing high realism training, caring for soldiers, and upholding professional norms are well understood by Army leaders. Efforts in line with these principles help foster psychological resilience and can serve as a buffer or lessen the effects of combat stress.

> Historical evidence suggests that the combat soldier's ability and willingness to fight depends not so much on the enemy as on his relations with those around him. Confidence in the ability and willingness of peers and leaders to protect one in combat and a feeling of obligation to do the same for them are at the heart of unit cohesion.[14]

As stated at the outset, combat stress reactions have often been misunderstood and responses to a soldier's behaviour varied, often undermining recovery and subsequent return to duty. Appropriate expectations and actions of peers and tactical leaders responding to unit members experiencing CSR are critical but not complicated.

Guiding principles specific to CSR treatment and management have been identified and proven effective in helping lessen the likelihood of soldiers becoming permanently disabled by combat stress.[15] These include establishing a positive expectancy toward recovery, treating CSR casualties in safe proximity to their units, and addressing the symptoms of combat stress as soon as practicable (immediacy):

## COMBAT STRESS REACTIONS

1) <u>Expectancy</u>: Soldiers experiencing CSR need to be reassured from the onset of signs and symptoms that they are experiencing a normal, temporary reaction to an abnormal situation, and that after a period of respite will rejoin their peers and resume their military functions (combat-related duties). Tactical leaders should take steps to:

   a. minimize the soldier's identification as a patient and maintain their military image;

   b. provide the soldier with rest, replenishment and a chance to clean up;

   c. ensure the soldier has an opportunity to talk of his experiences with an understanding person or select group in an atmosphere of supportive, non-judgmental listening;

   d. focus on the individual successes and accomplishments of the unit to help ensure realistic expectations and a balanced perspective;

   e. establish a degree of structure and routine (assign manageable military tasks and a physical exercise regime); and,

   f. supervise personally and select buddies to assist where appropriate and ensure they understand the return to duty goal.

Leaders must keep in mind that expressing "over-concern" for a soldier afflicted by CSR can undermine the expectancy aspect and negatively reinforce identification as a "patient."

2) Proximity: The principle of "proximity" implies looking after and treating soldiers experiencing CSR as close to the combat unit and area as practical, i.e., soldiers can perceive they are safe from attack. The use of and proximity to "natural" unit supports (immediate leaders and peers) is invaluable. It has been found that "the immediate stage of CSR (immediate hours or days after combat trauma) is most successfully treated in proximity to the unit. The farther a casualty is evacuated to the rear unnecessarily, the greater is the risk of chronic PTSD."[16] In addition to the recommended leader actions stated above, proximity affords tactical leaders the opportunity to:

   a. keep soldiers temporarily afflicted by CSR informed as to the unit's operations and the status of fellow soldiers;

   b. reinforce the soldier's ongoing identification as a valued unit member despite being temporarily out of action. This again reinforces a positive expectancy of returning to their sub-unit and duty in the near future, generally within a few days; and,

   c. organize visits and contact by select peers/friends and respected leaders who can meet with the soldier during a lull in operations.

3) <u>Immediacy</u>: Soldiers who exhibit CSR signs and symptoms to the point that they are unable to function normally need to be dealt with as soon as the tactical situation permits. Respect for the CSR treatment principle of immediacy or early intervention helps address the danger of symptoms becoming embedded, which can lead to a lack of receptiveness toward unit support and leader influence.

## SUMMARY

Although the focus of this chapter has been on tactical leader interventions in the early stages of CSR casualty management, there is no intent here to underemphasize or negate the need for in-theatre medical/psychological support expertise and consultation.

The critical points here are that unit members in general and tactical leaders in particular are vital to helping ensure effective front-line CSR recognition, response, and recovery.

## NOTES

1. Larry H. Ingraham and Frederick J. Manning, "Psychiatric Battle Casualties," *Military Review*, Vol. 60 (August 1980): 19–29.
2. M. MacPherson, *Long Time Passing: Vietnam and the Haunted Generation* (New York: Doubleday, 1984).
3. S. Noy, "Combat Stress Reactions," in *Handbook of Military Psychology*, eds. D.A. Mangelsdorff and R. Gal (London: John Wiley, 1991), 507–530.
4. Department of Defense, *Combat Stress* (Department of the Army, FM 6-22.5, June 23, 2000), in T.C. Helmus and R.W. Glenn, *Steeling*

*the Mind: Combat Stress Reactions and Their Implications for Urban Warfare*, RAND Corp, 2005.
5. J. Shay, *Achilles in Vietnam: Combat Trauma and the Undoing of Character* (New York: Simon & Schuster, 1995), 69.
6. K.R. Henning and B.C. Frueh, "Combat Guilt and its Relationship to PTSD Symptoms," *Journal of Clinical Psychology*, Vol. 53, No.8 (1997): 806.
7. T. Williams, ed., "Diagnosis and Treatment of Survivor Guilt — The Bad Penny," in *Post-Traumatic Stress Disorders: A Handbook for Clinicians* (Cincinnati, OH: Disabled American Veterans, 1987), 79.
8. Shay, 73.
9. Williams, 79.
10. R.E. Opp and A.Y. Samson, "Taxonomy of Guilt for Combat Veterans," *Professional Psychology: Research and Practice*, Vol. 20, No. 3, (1989): 163.
11. T.A. Rando, *Grief, Dying and Death* (Champaign, IL: Research Press Company, 1984), Chapter 3.
12. *Ibid.*, 28–36.
13. H. Spiegel, "Psychiatric Observations in the Tunisian Campaign," *American Journal of Orthopsychiatry* 14 (1944): 381–385.
14. F.J. Manning, "Moral, Cohesion and Esprit de Corps," in *Handbook of Military Psychology*, eds. D.A. Mangelsdorff & R. Gal (London: John Wiley, 1991), 468.
15. S. Noy, "Combat Stress Reactions," in *Handbook of Military Psychology*, eds. D.A. Mangelsdorff and R. Gal (London: John Wiley, 1991), 522.
16. *Ibid.*, 522.

## SELECTED READINGS

Binneveld, H. *From Shellshock to Combat Stress — A Comparative History of Military Psychiatry*. Amsterdam: Amsterdam University Press, 1997.

Department of Defense (DoD). *Combat Stress*. United States Department of the Army, Field Manual (FM) 6-22.5, June 2000.

Dispatches (Lessons Learned for Soldiers). *Stress Injury and Operational Deployments*, Vol. 10, No. 1, February 2004.

Helmus, T.C. and R.W. Glenn. *Steeling the Mind: Combat Stress Reactions and Their Implications for Urban Warfare*. Santa Monica, CA: RAND Corporation, 2005.

Horn, B. "Revisiting the Worm: An Examination of Fear and Courage." *Canadian Military Journal*, Vol. 5, No. 2 (Summer 2004): 5–16.

"Leadership in Land Combat," *Military Training*, Vol. 15, National Defence B-GL-318-015/PT-001, 1988.

NATO Task Group HFM 081/RTG, *A Leader's Guide to Psychological Support Across the Deployment Cycle*, 2007.

Noy, S. "Combat Stress Reactions," in *Handbook of Military Psychology*, edited by D.A. Mangelsdorff and R. Gal. London: John Wiley, 1991.

Rando, T.A. *Grief, Dying and Death*. Champaign, IL: Research Press Company, 1984.

Shay, J. *Achilles in Vietnam — Combat Trauma and the Undoing of Character*. Touchstone: Simon and Schuster, New York, 1995.

# 7
# COMMAND
## by Bernd Horn

There is nothing more sought after, nor any greater responsibility in the military, than command. Command is a very personal experience and each person approaches it in different ways, depending on their experience, circumstances, and personality. How an individual commands speaks more to the character and personality of the respective person than it does to the concept of command itself. In essence, command is far more an art than it is a science. That is why there is such a wide variance between commanders — some who reach legendary status and others who fade into ignominy.

Command, however, is not an arbitrary activity. It can only be exercised by those who are formally appointed to positions of command. The necessity to prepare individuals for command is not surprising. First, of course, is the heavy responsibility that command entails — namely, the lives of others. Second, commanders must accomplish their tasks, but they must do so with the minimum cost in lives and within their allocated resource envelope. This is often an enormous challenge. "A commander," explained legendary strategist B.H. Liddell Hart, "should have a profound understanding of human nature, the knack of smoothing out troubles, the power of winning affection while communicating energy, and the capacity for ruthless determination where required by circumstances." Hart added, "He needs to generate an electrifying current, and to keep a cool head in applying it."

## COMMAND DEFINED

Command is the vested authority an individual lawfully exercises over subordinates by virtue of their rank and assignment. The NATO-accepted definition, which has been adopted by Canada, defines command as "the authority vested in an individual of the armed forces for the direction, co-ordination, and control of military forces."[1] The essence of command is the expression of human will — an idea that is captured in the concept of commander's intent as part of the philosophy of mission command. In sum, command is the purposeful exercise of authority over structures, resources, people, and activities.

## MISSION COMMAND

Mission Command is a command philosophy that promotes decentralized and timely decision-making, freedom, and speed of action, as well as initiative that is responsive to superior direction. It entails three enduring tenets: the importance of understanding a superior commander's intent, a clear responsibility to fulfil that intent, and timely decision-making. At its core, the fundamentals of mission command are: unity of effort, decentralized authority, trust, mutual understanding, and timely and effective decision-making.

Simply put, mission command translates to a commander issuing his/her orders in a clear and detailed manner that ensures their subordinates fully understand the commander's intentions, their specific assigned missions, as well as the significance of their own missions within the context of the larger plan/framework. In essence, subordinates are given the effect they are to achieve and the reason why they must achieve it. Importantly, subordinates are also allocated the appropriate resources to achieve success. In addition, a commander should impose a minimum of control measures to ensure that s/he does not unnecessarily limit the initiative or freedom of action of their subordinates. This allows subordinates to decide within their respective freedom of action how best to achieve their assigned mission.

It must be understood that mission command is situational. It does not apply to all people or situations. While micro-management and/or rigid, superfluous direction may cause resentment and stagnation of creativity and initiative, a lack of direction can produce little effect. Subordinates must be well-trained and possess the ability and skill to be able to execute decentralized tasks. Junior, inexperienced subordinates provided by some coalition forces may not be capable of exercising mission command. Equally, some situations, such as immediate crises, which demand decisive, quick action may also not lend themselves to mission command.

## COMMANDER'S INTENT

Commander's intent is the critical component of mission command. The commander's entire effort (as well as that of his staff and subordinates), whether in planning, directing, allocating resources, supervising, motivating, and/or leading — is driven and governed by the commander's vision, goal, or mission, and the will to realize or attain that vision, goal, or mission. The Commander's intent is the commander's personal expression of why an operation is being conducted and what s/he hopes to achieve. It is a clear and concise statement of the desired end state and acceptable risk. Its strength is the fact that it allows subordinates to exercise initiative in the absence of orders, or when unexpected opportunities arise, or when the original concept of operations no longer applies.

## UNITY OF COMMAND

Unity of command is a concept that revolves around a single, clearly identified commander who is appointed for any given operation and who is accountable to only one superior. This ensures clarity and unity of effort, promotes timely and effective decision-making, and avoids conflict in orders and instructions. It is characterized by a clear chain of command, where command at each level is focused on one commander.

## COMMAND CLIMATE

A commander exercises command through the force of his or her personality and leadership to influence the attitude, direction, and motivation of the staff and subordinate commanders. A commander's ability to create an effective and positive command climate has a direct impact on the morale and level of performance of the personnel within the organization. Positive leadership, sincerity, and compassion by the commander stimulates subordinate confidence, enthusiasm, mutual trust, and teamwork. In addition, encouragement to think independently, use initiative and accept risk, as well as the inclusion of staff and subordinates in the decision-making process all assist in creating an effective command climate.

For mission command to be successful, the command climate must encourage subordinate commanders at all levels to think independently, take the initiative and not be risk averse. After all, the strength of mission command is the ability of commanders at all levels to react quickly to developing situations in an ambiguous, complex, fluid, and chaotic battlespace. In such an environment, it is critical to gain and maintain the initiative. Delay in decision-making can have serious consequences. As a result, subordinate commanders, acting within the commander's intent, must make decisions and take action.

This entails trust. The superior commander must trust his/her subordinates to act; to act in accordance with the commander's intent; and to make reasonable decisions regardless of the circumstances in which the subordinate commander finds himself/herself (which in an ambiguous and chaotic security environment may not necessarily be those the superior commander originally envisaged). This trust is critical since for mission command to function, the superior must minimize control mechanisms and allow the subordinate the necessary freedom of action and initiative to achieve the necessary effect.

Similarly, the subordinate must have confidence that s/he has not been given an unachievable task. Moreover, the subordinate must trust his/her superior to provide the necessary direction, guidance, and resources to successfully achieve the assigned mission. An important factor is the subordinate's ability to fully exercise initiative and accept the necessary degree of risk; s/he must be able to trust the superior to provide the necessary support should errors (i.e., neither malicious, nor due to negligence) occur.

## COMMAND RESPONSIBILITY

Every military professional, as an individual, is responsible for his/her actions and the direct consequences of those actions. Nonetheless, commanders are responsible to make decisions, issue orders, and supervise the conduct of their personnel. The legal authority vested in commanders necessitates subordinates to adhere to the lawful commands of their superiors. However, by the same token, the commander who gives the orders must accept responsibility for the consequences that flow from the execution of his/her orders. They are responsible for the actions that they knew, or ought to have known, were carried out in response to their direction.

## THE COMMAND — LEADERSHIP — MANAGEMENT NEXUS

Command, however, is not a uni-dimensional concept. The all-encompassing scope of command is why it consists of three, often reinforcing, components: authority, management, and leadership. Paradoxically, these terms are often seen as synonymous, or mutually exclusive. But each component is an integral and often interrelated element of command. Each can achieve a distinct effect. None are necessarily mutually exclusive — and when used judicially in accordance with prevailing circumstances and situational factors, combine to provide maximum effectiveness and success.

*Authority*

The first component is authority. Commanders can always rely on their authority to implement their will. Authority, which encompasses a legal and constitutional component (e.g. National Defence Act), is always derived from a higher or superior entity. It gives commanders the right to make decisions, transmit their intentions to subordinates, and impose their will on others. It is military authority, namely by virtue of a service person's unlimited liability and the commander's vested authority to send individuals into harm's way, complete with the support of substantial penalties for non-conformance, that differentiates military command from civilian positions/appointments of power. Although authority is a powerful tool for commanders — reliance on rank and position will never build a cohesive, effective unit that will withstand the test of crisis. At best, it may present a chimera of an efficient organization, but even this is doubtful.

Notwithstanding that, at times, such as in crisis and/or in the face of individual or group intransigence to necessary change, it can provide the necessary hammer required to clear the path to renewal or survival. In some circumstances and occasions, authority must be the tool of choice.

*Management*

The second component of command is management. Management is designed to control complexity and increase group effectiveness and efficiency. It is primarily concerned with the allocation and control of resources (i.e. human, financial, and material) to achieve objectives. Its focus is staff action such as allocating resources, budgeting, coordinating, controlling, organizing, planning, prioritizing, problem-solving, supervising, and ensuring adherence to policy and timelines.

Management is also based on formal organizational authority and it is unequivocally results-oriented. Its emphasis is on the correct and efficient execution of organizational processes. However, this is not to say that effective managers do not use leadership to increase their effectiveness in accomplishing their goals.

Clearly, management is of great importance to commanders and institutional leaders. Management skills and practices allow them to ensure that subordinates receive the necessary direction, guidance, and resources — on time and where required — to achieve the mission in accordance with the commander's intent. As such, management is a critical and necessary component of command. However, it is not leadership, but then, neither should it be. It serves a distinct and vital purpose necessary to command effectiveness and success. It neither replaces, nor substitutes for leadership. Rather, it is complimentary. It is but one of three instruments, designed to perform a specific function, in the command "tool belt."

*Leadership*

The third component of command is leadership. It is the "human" side of command, but it can also be exercised outside of the concept of command. It deals with the purpose of the organization — "doing the right thing" versus "doing it right" [management]. In accordance with CF doctrine, leadership is defined as "directing, motivating and enabling others to accomplish the mission professionally and ethically, while developing or improving capabilities that contribute to mission success."[2] Whereas management is based on authority and position, leadership relies on influence, either direct or indirect.

In the end, the leadership component of command is about influencing people to achieve some objective that is important to the leader, the group, and the organization. It is the human element — leading, motivating, and inspiring, particularly during times of crisis, chaos, and complexity when directives, policy statements, and communiqués have little effect on cold, exhausted, and stressed subordinates. Strong leadership will encourage subordinates to go beyond the obligation to obey and commit to the mission in a way that maximizes their potential. It is the very individualistic, yet powerful component that allows commanders and leaders at all levels to shape and/or alter the environment or system in which people function and thereby influence attitudes, behaviour, and the actions of others.

It is within this powerful realm of influence and potential change that leadership best demonstrates the fundamental difference between it and the concept of command. Too often the terms *command* and *leadership* are interchanged or seen as synonymous. But they are not. Leadership can, and should, be a component of command. After all, to be an effective commander the formal authority that comes with rank and position must be reinforced and supplemented with personal qualities and skills — the human side. Nonetheless, as discussed earlier, command is based on vested authority and assigned position and/or rank. It may only be exercised downward in the chain of command through the structures and processes of control. Conversely, leadership is not constrained by the limits of formal authority. Individuals anywhere in the chain of command may, given the ability and motivation, influence peers and even superiors. This clearly differentiates leadership from command. In sum, although leadership is an essential role requirement for successful commanders, it is neither command, nor management.

## **SUMMARY**

In the end, command is the vested authority assigned to an individual in a specific appointment. This authority empowers that individual, the commander, to provide the necessary vision, direction, and purpose to accomplish his/her assigned mission. To assist the commander with his/her task, most advanced militaries have evolved to a mission command philosophy, which promotes decentralized and timely decision-making, freedom, and speed of action and initiative that is responsive to superior direction. It entails three enduring tenets: the importance of understanding a superior commander's intent, a clear responsibility to fulfil that intent, and timely decision-making. At its core lie five fundamental principles: unity of effort, decentralized authority, trust, mutual understanding, and timely and effective decision-making.

Command, however, is made up of three often-reinforcing components: authority, management, and leadership. As command is a very personal experience, each commander adopts a unique approach and places a different emphasis on each of these components. In the end, what matters is the ability of a commander to make sound and timely decisions, as well as to motivate and direct subordinates in the accomplishment of assigned missions. This requires a vision of the desired end state, an understanding of military art and science, as well as an ability to assess personnel, situations and risk, and to act accordingly.[3]

## NOTES

1. Canada, *Command* (Ottawa: DND, 1997), 4.
2. Canada, *Leadership in the Canadian Forces: Conceptual Foundations* (Kingston: DND, 2005).
3. This chapter is based on an amalgamation of material drawn from sources listed in the "Selected Readings."

## SELECTED READINGS

Australian Army. *Command, Leadership and Management.* Puckapunyal: Land Warfare Development Centre, 2003.

Canada. *Leadership in the Canadian Forces: Leading the Institution.* Kingston: DND, 2007.

Canada. *Leadership in the Canadian Forces: Conceptual Foundations.* Kingston: DND, 2005.

Canada. *Command.* Kingston: Canadian Land Forces Command and Staff College, 1997.

Granatstein, Jack. *The Generals: The Canadian Army's Senior Commanders in the Second World War.* Toronto: Stoddart, 1995.

Horn, Bernd, *The Buck Stops Here: Senior Commanders on Operations.* Kingston: CDA Press, 2007.

Melvin, Brigadier R.A.M.S. "Mission Command." *British Army Review*, No. 130 (Autumn 2002): 4–9.

Slim, Field Marshal Viscount. *Defeat into Victory.* London: Pan Books, Reprint 1999.

Storr, Major Jim. "A Command Philosophy for the Information Age: The Continuing Relevance of Mission Command." *Defence Studies*, Vol. 3, No. 3 (Autumn 2003): 119–129.

Sullivan, Gordon R. and Michael V. Harper. *Hope is Not a Method.* New York: Random House, 1996.

# 8
# COMMUNICATION
by Allister MacIntyre and Danielle Charbonneau

There are countless definitions of leadership in existence. Leadership models continue to be developed, and the ongoing debate flourishes with respect to the particular attributes that might best characterize the qualities of effective leaders. Yet there is a single common denominator, influence, which threads its way throughout the various meditations about leadership. No matter how one chooses to think about leadership itself, influence of leaders over their followers will always be a constant factor. For example, the recently published Canadian Forces (CF) doctrine on leadership offers a value-neutral definition of this concept as "directly or indirectly influencing others, by means of formal authority or personal attributes, to act in accordance with one's intent or a shared purpose."[1] Without even taking into consideration whether the leadership is good or bad, effective or ineffective, the influence aspect is firmly rooted in the definition. However, influence cannot possibly exist if leaders do not communicate their intentions. In short, without communication, there can be no leadership.

## COMMUNICATION DEFINED

Everyone knows that communication takes place when information of some sort is transferred between the person who is sending the message and the person, or persons, on the receiving end of that message. But it is not often that people stop to think about the process itself, and how communication functions.[2] Before messages can be transmitted, senders have to first think about the content of what they wish to communicate. The message is then encoded into the appropriate words and the subject matter is conveyed to the receiver (face to face, letter, e-mail, etc.). Those on the receiving end must first perceive the message, and then decode the content to extract meaning, and finally reach an understanding of what they believe was meant by the message. Although this sounds like a simple process, there are many opportunities for the message to become derailed. How often do people hear others say things like: "I don't care what you heard from John, that's not what I said" or "Why did you do that? That's not what I asked you to do"? This happens because, despite best efforts, the communication process often fails. The various sources of message error, and barriers to effective communication, will be discussed later in this chapter.

## **ROUTES TO PERSUASION**

It is easy to accept that communication is important if leaders expect to be able to influence followers. In a military environment, leaders can influence behaviour by falling back on their legitimate authority and issuing orders. However, if leaders want their followers to shift attitudes and internalize messages, they will need to use influence and persuasion. Several approaches to understanding persuasion can be found in the literature, but the Elaboration Likelihood Model (ELM) of persuasion, as developed by social psychologists Richard Petty and John Cacioppo, is one of the most widely researched and accepted approaches.[3] The ELM is a comprehensive attempt to integrate the many assorted elements of persuasion research that were in existence prior to its conception. This 1986 model is light years ahead of anything that preceded, or has succeeded, its inception.

This chapter is not the place for a comprehensive presentation of the ELM; however, there are some points worth addressing. The concept of two routes to persuasion, a central route and a peripheral route, seems to capture most of the relevant properties of the persuasion process. In simple terms, this means that one will either take the time to think about a persuasive message (central) or one will be compelled to action without a great deal of thought (peripheral). Those who follow the central route are typically motivated to consciously think about the message and pay attention to the quality of the arguments. Furthermore, when persuasive messages entice message recipients to expend a greater amount of time elaborating, the content of the message will more likely become ingrained and resistant to change. When persuasion occurs through a peripheral route, people tend to spend more time attending to superficial aspects like the number of arguments or the attractiveness or likableness of the communicator, and less time paying attention to the specific details of the message.

The implication for leaders is, if they simply want followers to comply, then issuing an order is sufficient. However, if they want colleagues, including superiors, peers, subordinates, and "followers," to internalize the content of the persuasive message, then they must not only provide sufficient rationale and arguments, they have to allow the time necessary for message recipients to process what they have heard. Whether the persuasive route followed is central or peripheral, there are four critical aspects contributing to the effectiveness of the communication. These factors are associated with the characteristics of the source, the message itself, the channel selected, and the audience.[4]

## THE SOURCE

The source characteristics that come into play during the communication process are primarily associated with the credibility and trustworthiness of the message sender. Credibility ties into the person's[5] credentials and expertise, particularly with respect to the topic of the communication. Hence, some people may be credible in one situation, but not another. A physician will be a credible source if the topic is associated with a health issue, but most people will want to talk to an accountant if they need advice regarding income tax returns. In a military context, members are more likely to grant credibility to a combat arms officer than a logistics officer when a subject matter like battlefield tactics is the issue under discussion. While credibility is most closely linked to a communicator's expertise, trustworthiness has more to do with honesty and integrity. If message recipients suspect that message senders are being less than candid, or have a motive to mislead, or if they sense deceit through body language or other cues, then they will be less prone to accept the message content.

Naturally, any assessment of source credibility is incredibly subjective. In this regard, recipients can be influenced by how well they know the message senders and by how much they like, admire, or respect them. At times, recipients will even be influenced by how similar message senders are to them (e.g., gender, race, religion, socio-economic status, age, professional qualifications). The way people dress, their body language, and whether message recipients find them attractive may also have an impact.

## **THE MESSAGE**

Paying attention to the characteristics of the message is always important, but it becomes especially crucial if there is any possibility that source issues are in question. The ideas that a communicator wishes to convey must be encoded in a manner that is easy for the receiver to interpret. Long, rambling messages will not normally be well received and the content has to remain relevant, targeted, and coherent.

Phillip Clampitt, a University of Wisconsin professor of organizational communication and information sciences, provides a model that can help communicators to choose the right message characteristics and appropriate channel, given the objective of the message and the personal attributes of the receiver.[6] For example, the message choices include factors such as terminology, facts, arguments, evidence, tone, emotionality, complexity, sequencing, and formality. Canadian psychologists Alan Auerbach and Shimon Dolan add that a "… message that conveys even a hint of a threat tends to have the opposite of the desired effect on the person being persuaded."[7] However, messages that contain fear appeals can be effective for topics like health-related issues.

It will almost always be preferable to state conclusions explicitly rather than allowing receivers to reach their own conclusions. In this light, communication and management specialist Lani Arredondo says that a good approach when giving a presentation is to tell recipients what they are going to be told, then tell them, and then wrap up by telling them what they have been told.[8]

## **THE AUDIENCE**

Message senders are less likely to be successful if they do not know their audience. It is important to understand that a message transmission will not be complete until receivers have decoded the content. Receivers will decode messages by converting the content into something that they, the receivers, will understand. When senders and receivers share a culture and language, there is a greater likelihood that the messages will be decoded accurately.[9] Noise, a term that is used to refer to anything that impedes the communication process, can interfere with message decoding. This can include actual noise, interruptions, distractions, cultural differences, and language barriers.

As with source characteristics, the range of possible attributes that will exist in an audience is virtually endless. To decide how much effort should be spent in developing and presenting arguments of good quality or whether a more superficial message would suffice, a communicator should factor in aspects like intelligence, education, occupation, and knowledge of the subject matter. Is it possible that the receivers are already committed to a position on the issue? Do they have a reason to be interested in the topic? Are they motivated to listen and understand? What sort of beliefs and values do they have? Is there a difference in socio-economic status between the senders and receivers? Even aspects like personality, gender, age, and religious orientation can play a significant role in the communication process.

Obviously, there is a great deal that senders must take into consideration. However, it does not end with knowing as much of these characteristics as possible. Effective communicators must appreciate that different messages will have a different impact on different audiences. If they are armed with this awareness, and knowledge of their audience, these communicators will be better equipped to tailor their messages for maximum impact.

## **THE CHANNEL**

Communicators may have no doubts about the message to be transmitted, and be aware of audience characteristics, but they still must take into consideration the most effective medium for the communication itself. There are a host of possibilities in this regard. The available channels of communication include oral (one on one, phone, voice message, making a presentation to an audience), written, e-mail, meetings, video conference, fax, web pages, and bulletin boards. Organizational psychologist Richard Field presents a continuum where the media richness of a communication channel goes from the lowest (e.g., impersonal static media: flyers, bulletins) to the highest (e.g., physical presence: face to face).[10] It is argued that communication will be most effective when rich media are used for non-routine organizational messages and lean media for routine matters. If a lean approach is used for a non-routine subject, the complexity of the message will be lost and the use of a rich medium for a routine topic may lead to information overload and confusion.

When the subject matter of a message is important, and the channel selected is face-to-face, it is always advisable to use a written medium as a follow-up to the conversation with a summary of the key points. This will solidify the aspects of agreement, ingrain any timelines or deliverables, and allow an opportunity for clarification if necessary. It is when communicators move away from the written form that peripheral cues become increasingly salient. When something in put in writing, it will more likely result in elaboration because the reader controls the presentation pace, sections can be read again and again, and the subject matter is more focused.

## **TARGETING THE MESSAGE TO DIFFERENT GROUPS**

One of the biggest mistakes that a communicator may make is to assume that all audiences are the same. There are three important questions to be answered regarding the target that will influence a communication strategy. First, how well does the target (or audience) know the topic? Second, what is the importance level of the topic to the target? And, third, what is the target's pre-existing opinion on the topic? Harry Mills[11] has identified six possible types of audience, each requiring a different communication approach.

*The Hostile Audience*

Being against the message initially, the hostile audience is likely to be indifferent to the communicator's credibility. A communicator will first need to win the target's attention, which may be done by using humour or a little story. Then, it is useful to identify points of agreement, such as goals, objectives, or disadvantages of the status quo. This may be followed by a presentation of the arguments in favour of the message. A realistic goal for a hostile audience may be to neutralize an opinion, rather than a radical change in opinion.

*The Neutral Audience*

The neutral audience needs to be given a reason to listen to the message. Hence, they have to be shown how the message will directly affect them, positively and negatively, before they will become interested in listening.

*The Uninterested Audience*

For the uninterested audience, the message sender needs to catch their attention at the beginning of the message with a story, a news item, or an unusual fact. Then, the communicator needs to explain how the message concerns these uninterested audience members, and needs to give them some reasons for listening.

*The Uninformed Audience*

For the uninformed audience, credibility matters, so the background and qualifications of the communicator should be highlighted. The presentation should start with two or three good arguments in support of the message and provide concrete or easy to understand examples. Questions should be encouraged.

*The Supportive Audience*

With a supportive audience, the communicator could use stories and testimonies to revive the audience's current enthusiasm. This will emphasize the similarities between the speaker and the audience. To strengthen their pre-existing opinion, the presentation could even include arguments against the message along with rebuttals to refute these arguments.

*The Mixed Audience*

The last group, the mixed audience, initially presents a variety of opinions and knowledge. In this case, there is a need to identify the important sub-groups and tailor the message toward the largest groups. That is, different parts of the message will address different sub-groups. Care must be taken to avoid contradictions when trying to win the approval of these sub-groups.

## **NON-VERBAL COMMUNICATION**

Humans do not rely just on the written and spoken word to communicate their messages. When people have the opportunity to view a communicator, they can learn a great deal about the meaning of the message from non-verbal cues such as a communicator's stance, facial expressions, gestures, proximity to the receiver, and eye behaviour. Message recipients often will listen to a stream of words as they emanate from someone's mouth, yet have a sensation that the real message is different from what is heard. Recipients may have a vague feeling that words are somehow disconnected from body language. For example, when a person of higher status, possibly a "boss," says that s/he understands an employee's work situation, while an almost imperceptible head-shaking that says "no" accompanies these words, the words will be perceived as lacking sincerity.

One of the more obvious non-verbal cues is facial expression. Research has discovered two important aspects of facial expressions. First, even though the use of gestures can vary dramatically from culture to culture, facial expressions are fairly universal. Second, facial expressions are really just a combination of six different emotions: happiness, sadness, surprise, anger, fear, and disgust.[12] Smiles can be powerful because smiling people will be viewed as being happy and will be rated as being more attractive. They will stimulate others to smile.[13] However, a smile of real enjoyment is different than other smiles because it is the only type of smile that also involves the contraction of muscles around the eyes. Message recipients have all encountered people who seem to have a smile plastered on their face while their eyes betray their lack of genuine enjoyment. Recipients respond differently to these smiles.

Additional non-verbal cues include the amount and type of eye contact, the manner in which people use their personal space, their rate of speech, the tone of their voice, and their posture. People with higher status will appear to be more relaxed than those of lower status, people who maintain eye contact are assessed as being more competent. Recipients even use non-verbal cues to help them detect lies and deception. Cues associated with lying include lack of spontaneity, less smiling, dilation of pupils, and hesitation in speech. Similarly, indicators of deception include less sustained eye contact, more posture shifts, longer response times, more speech hesitations, and higher pitch.[14]

## EFFECTIVE LISTENING

Effective communication is not a one-way street. The process also involves being on the receiving end of a message, and this demands effective listening. Thomas E. Harris and John Sherblom, two professors of communications studies, argue that "Active, effective listening is hard work."[15] They offer several barriers to effective listening,[16] briefly summarized as:

*Lack of Interest*

If recipients are uninterested in a subject, or judge it to be too difficult, they will become bored, impatient, and likely daydream or become preoccupied with something else. Although it may require some effort, it is critical that recipients stay focused on what is being communicated; they must listen actively.

*Distracting Delivery*

Recipients have a tendency to judge a speaker's personal characteristics. If communicators fidget, seem disorganized, speak with an accent, or are inefficient in their delivery, then recipients will concentrate more on the communicators' mannerisms than on the message. It is important for recipients to watch for the warning signs that this is happening and to force their attention back to the subject matter. If necessary, visual attention should be shifted away from the speaker to allow attention to be maintained on absorbing the content of the message.

*Noise*

Any noise that either masks the content of the message, or distracts message recipients' attention, will impair listening. If the noise is a temporary interruption (e.g., train, plane, sirens) recipients should signal the communicator to wait for the interruption to pass. If the noise is of an enduring nature, recipients could either dampen the sound (e.g., close doors or windows) or move to a quieter location.

*Arrogance and Disrespect*

When communicators display arrogance (e.g., a know-it-all attitude) or disrespect, recipients will respond on an emotional level and not listen as effectively. This will always be challenging for recipients to address because it is difficult to find a rational solution after having an emotional reaction. Once again, message recipients need to be alert for the warning signs and do what they can to keep their emotions in check. Alternatively, recipients can postpone the communication until after each party has had the chance to calm down.

*Pre-programmed Emotional Responses*

If the communicator touches on a subject for which recipients have a strong emotional reaction, the message content will be filtered through these pre-programmed responses. As above, the best that recipients can do is to watch for the warning signals and maintain an emotional balance. Recipients need to make an effort to turn the filters off without pre-judging the message. Once again, they can also take a break from the conversation or communication to calm down.

*Listening for Facts*

If message recipients only listen to the facts in a message, they may fail to understand the overall point being made by the communicator. Listening for the facts is admirable, but recipients need to take in the complete picture. They need to pay attention to things like body language, tone of voice, sense of urgency, and they need to try to understand the reasons behind the message content.

*Faking Attention*

Because most people have been faking attention since grade school, message recipients have become proficient at appearing to be interested rather than truly listening. People have to avoid this trap at all costs. Not only does the communicator deserve our attention, it is a terrible interpersonal faux pas to get caught while feigning interest. Recipients' personal credibility will be damaged and the communication process will cease. When communicators take the time to speak, recipients, in most cases, should show them respect by listening.

*Thought Speed*

Humans have an ability to think three or four times faster than the average person can talk. As a consequence, recipients may be filling this free time by thinking about how they will respond to what they think is being said, or even by thinking about unrelated subjects. It is difficult to avoid this trap because, once people's minds shift to other thoughts, people will normally not even be aware that they have strayed away from the topic. The best advice, once again, is to stay focused.

Additional barriers to effective communication include the Mum Effect (an unwillingness to communicate bad news up the chain of command, the absence of "truth to power," so to speak), fatigue, time pressures, status differences, value judgments, inappropriate language or jargon, mixed messages, inadequate information, the use of an improper channel, and information overload.[17]

An effort should be made to overcome the above barriers. A person who is successful at active listening can enable the process further by paraphrasing what they have heard, by expressing their understanding, and by asking questions. This will not only signal interest and attention, it will provide a communicator with the opportunity to clarify the content of a message.

## SUMMARY

Leaders must master the skills associated with effective communication if they expect to be able to influence their followers and achieve missions and goals. They need to appreciate that there is more than one route to persuasion and they have to keep in mind that communication is not simply a one-way exercise of position power. Leaders must become credible and trustworthy in the eyes of their superiors, peers, followers, and other interested people, and it is crucial that they learn to appreciate the characteristics of their audience. By knowing themselves and their message recipients, leaders will be able to appropriately tailor their communication for maximum impact and understanding. They also will be better equipped to select the best channel for delivering the message.

The most successful leaders will learn to appreciate the nuances associated with non-verbal behaviours and will learn to listen actively. Finally, they will take steps to overcome the barriers to effective communication and learn to identify their dominant communication style as well as the style of others. Furthermore, they will strive to develop a more expressive communication style.

Renowned communication specialist Frank Luntz argues that what people hear is far more important than what people say. His sentiment can be summed up in these powerful words:

> You can have the best message in the world, but the person on the receiving end will always understand it through the prism of his or her own emotions, pre-conceptions, prejudices, and pre-existing beliefs. It's not enough to be correct or reasonable or even brilliant. The key to successful communication is to take the imaginative leap of stuffing yourself right into your listener's shoes to know what they are thinking and feeling in the

deepest recesses of their mind and heart. How that person perceives what you say is even more *real*, at least in a practical sense, than how you perceive yourself.[18]

**NOTES**

1. Canada. *Leadership in the Canadian Forces: Doctrine* (Kingston: DND, 2005).
2. See R.H.G. Field, *Human Behaviour in Organizations: A Canadian Perspective*, 2nd ed. (Scarborough, ON: Prentice Hall Canada, 1998); and G. Johns and A.M. Saks, *Organizational Behaviour: Understanding and Managing Life at Work*, 6th ed. (Toronto: Pearson Education Canada, 2005).
3. R.E. Petty and J.T. Cacioppo, *The Elaboration Likelihood Model of Persuasion* (New York: Academic Press, 1986).
4. See D. O'Hair, G.W. Friedrich, and L.D. Dixon, *Strategic Communication in Business and the Professions* (Boston: Houghton Mifflin Company, 1992).
5. The source does not have to be an individual. It can also be a group, organization, or a form of media.
6. P.G. Clampitt, *Communicating for Managerial Effectiveness*, 2nd ed. (Thousand Oaks, CA: Sage Publications, 2001).
7. A.J. Auerbach and S.L. Dolan, *Fundamentals of Organizational Behaviour: The Canadian Context* (Toronto: International Thomson Publishing, 1997), 42.
8. L. Arredondo, *Communicating Effectively* (New York: McGraw-Hill, 2000).
9. Auerbach and Dolan.
10. Field.
11. H. Mills, *Artful Persuasion: How to Command Attention, Change Minds, and Influence People* (New York: American Management Association, 2000).
12. P.B. Paulus, C.E. Seta, and R.A. Baron, *Effective Human Relations: A*

*Guide to People at Work*, 3rd ed. (Boston: Allyn & Bacon, 1996).
13. *Ibid.*
14. *Ibid.*
15. T.E. Harris and J.C. Sherblom, *Small Group and Team Communication* (Boston: Allyn & Bacon, 1999), 111.
16. *Ibid.*, 111–113.
17. See Field; and see O'Hair and Friedrich.
18. F. Luntz, *Words That Work: It's Not What You Say, It's What People Hear* (New York: Hyperion, 2007), xiii.

## SELECTED READINGS

Arredondo, L. *Communicating Effectively.* New York: McGraw-Hill, 2000.

Auerbach, A.J. and S.L. Dolan. *Fundamentals of Organizational Behaviour: The Canadian Context.* Toronto: International Thomson Publishing, 1997.

Clampitt, P.G. *Communicating for Managerial Effectiveness*, 2nd edition. Thousand Oaks CA: Sage Publications Inc., 2001.

Luntz, F. *Words That Work: It's Not What You Say, It's What People Hear.* New York: Hyperion, 2007.

Mills, H. *Artful Persuasion: How to Command Attention, Change Minds, and Influence People.* New York: American Management Association, 2000.

O'Hair, D., G.W. Friedrich, and L. Dixon. *Strategic Communication in Business and the Professions.* Boston: Houghton Mifflin Company, 1992.

Paulus, P.B., C.E. Seta, and R.A. Baron. *Effective Human Relations: A Guide to People at Work*, 3rd edition. Boston: Allyn & Bacon, 1996.

# 9
# CONFLICT MANAGEMENT
## by Bradley Coates

Interpersonal conflict is a common aspect of human social interaction. Although not inherently negative, when unresolved or poorly handled, such conflict can have negative repercussions for both individuals and organizations. As a result, the management of interpersonal conflict remains an important leadership competency. This chapter discusses conflict, its causes and impacts, and an overview of common dispute resolution approaches.[1]

## INTERPERSONAL CONFLICT DEFINED

The pervasiveness of interpersonal conflict is readily apparent in virtually all aspects of life, including the family, the community, and the work environment.[2] Interpersonal conflict has been defined as "… an expressed struggle between at least two interdependent parties who perceive incompatible goals, scarce resources, and interference from others in achieving their goals."[3] Thus, disputes are often framed as apparently competing and mutually exclusive positions or desired outcomes. In the workplace, interpersonal disputes are manifest in various ways, such as disagreement, anger, unwillingness to cooperate, avoidance, and gossip. It is important to note that conflict itself is not necessarily negative. In fact, the process of successfully working through conflict can have positive benefits: it can trigger creativity, strengthen relationships, and lead to increased productivity.[4] When not effectively addressed, however, conflict can have significant adverse consequences.

## CAUSES AND CHARACTERISTICS OF CONFLICT

To better manage workplace conflict, it is important to briefly consider its causes and characteristics. With regard to causes, the following is an overview of common categories of workplace conflict and some of their sub-components.[5]

Facts and Data: Insufficient information, misinformation, and different methods of information assessment.

Relational: Strong negative emotions, misperceptions, stereotypes, and poor communication.

Values: A perceived incompatibility of belief systems, attempts to impose values or to claim exclusive value systems that do not allow for divergent beliefs.

Structural: Limited physical resources, insufficient decision-making authority, geographical factors, time restraints, and organizational structure.

Interests: A perceived or actual incompatibility of needs. These are often framed in a manner wherein the concerns or interests of only one party can be met.

In addition to an appreciation of potential causes of conflict, it is also important to consider some of its characteristics. In this regard, two points are particularly noteworthy; first, the role that assumptions can play in conflict, and, second, the nature of conflict escalation. In this context, assumptions are understood as the perceptions a person holds regarding the intentions and motivations of other individuals. Among conflict resolution professionals, assumptions are frequently cited as a leading factor in workplace conflict.[6] When interpreting their personal actions, individuals tend to do so through the lens of their intentions — that is, what were their objectives? When interpreting the actions of others, however, this subjective perspective is not available. That is to say, we judge ourselves based on our intentions while we judge others by their actions.[7] Making these assumptions and inferring intent through factors such as body language, intonation, and situational context is a normal part of social interaction. In unfamiliar or tense situations, however, these assumptions, when left unchecked, can be misleading and trigger or exacerbate ongoing conflict.[8]

Closely related to assumptions is conflict escalation. While each dispute scenario is unique, the progression of many conflicts often shares common characteristics. For instance, one of the first casualties of conflict frequently is effective communication between the parties. Once this link has been severed, the number of assumptions tends to increase and others in the work environment are often drawn into the conflict. If this continues unchecked, over time these negative assumptions can deepen and lead to destructive behaviour (e.g. blaming and sabotage). As conflict becomes entrenched, it can take on a life of its own and become more about the dispute itself than the actual precipitating event. Eventually, if allowed to continue to escalate, conflict can lead the affected parties to see themselves as incompatible and unable to work together.

## **IMPACT OF CONFLICT**

Poorly managed workplace conflict has significant negative repercussions for both individuals and organizations. From the perspective of the individual, disputes can impact work performance, spill over into family and personal matters, and trigger health concerns. Over time, these factors can result in decreased productivity, increased absenteeism, and, in certain cases, can lead to staff turnover.

While this impact on the individual is important, from a Canadian Forces (CF) perspective, arguably the most significant aspect of conflict is the impact it can have on work units. Paramount among these consequences is the effect on unit morale and cohesiveness. This is particularly important in the CF, where work-unit membership tends to be comparatively long-standing and teamwork is highly valued. A key aspect of the CF's organizational culture is the view that people, both individually and as members of a combat unit, are critical to success on the battlefield.[9] Although adequate resources are essential for a military to accomplish its objectives, from an armed forces perspective, ultimate success or failure is viewed in large part as contingent upon the ability and motivation of its personnel.[10] Although the precise role that organizational culture fulfils within armed forces continues to be discussed, the view that group cohesiveness is crucial in combat remains widely accepted.[11] In this context, the provision of timely and effective conflict-resolution mechanisms can be an important part of maintaining a strong, cohesive, and productive workplace.

## CONFLICT RESOLUTION APPROACHES

Within the CF there are both formal and informal conflict-resolution approaches. Within the formal category, there are two long-standing dispute-resolution approaches (power-based and rights-based); while in the informal category, there is what is generally known as interest-based conflict resolution. In discussing these different approaches, it is important to recognize that no single approach is optimized to deal with all conflict scenarios. Accordingly, when addressing workplace conflict, it is essential to consider situational specifics and objectives. The following paragraphs provide an overview of the different resolution methodologies and highlight some of their respective strengths and weaknesses. In reality, there is often overlap between these different approaches, but for ease of analysis they are examined separately.[12]

## POWER-BASED RESOLUTION

Power-based conflict resolution can be understood as any process wherein a party in a position of formal authority uses this organizational power to influence the outcome of a dispute.[13] Within an organizational setting, power-based conflict resolution is commonly understood to be those settlements effected by the chain of command.[14] As such, it is essentially an adjudicative process wherein the organizational hierarchy considers disputes and renders decisions.

The appeal to authority as a means of dispute resolution is a time-honoured approach within armed forces. Indeed, formal authority in the CF is often employed as a means of resolving conflict.[15] Within an operational environment where risk is high and timeliness crucial, a decisive, authority-driven leadership style is often necessary. The use of this hierarchical approach, however, is not solely limited to operational settings. For example, with respect to more mundane day-to-day disagreements, supervisors are often perceived to be particularly well placed to understand the nuances of a dispute, the participants, and broader organizational considerations.

This problem-solving method does, however, have potential limitations from both chain-of-command and member perspectives. From an organizational viewpoint, a power-based approach may not always be seen as appropriate. In certain situations, the chain of command may feel that, given the nature of the dispute, it is unable to adjudicate (or to be perceived as adjudicating) in an unbiased fashion. In other circumstances, supervisors may lack the necessary time, training, or skill to effectively address a specific conflict. There is also the concern that overuse of this approach absolves individuals from taking responsibility for solving their own disputes and normalizes dependency on the chain of command.[16]

From the individual's perspective, conflict resolution effected by the chain of command may raise concerns regarding objectivity as well as process control.[17] With respect to objectivity, the member may feel that, given unit history and/or workplace relationships, they will not receive fair and non-prejudicial treatment. In terms of the actual process, as noted above, power-based conflict resolution is largely an adjudicative process wherein a solution is developed and rendered by the chain of command.[18] From the member's perspective, a superficial application of this approach can at times result in seemingly arbitrary decisions that fail to adequately address key underlying issues.

## RIGHTS-BASED RESOLUTION

The rights-based approach also appeals to third-party adjudication for conflict resolution. The key difference is that, in this model, the role of the third party is guided by the applicable body of governing policies, procedures, and regulations. As noted by American conflict commentator William Ury, "… this method seeks to appeal to some independent standard such as organizational policy, law or commonly accepted social norms to determine who is right."[19] Within the CF context, such processes include administrative investigations, harassment investigations, grievances, summary trials, and courts martial.

Rights-based approaches are solidly established and well used within the CF.[20] These processes are inherently formal and positional, limited primarily to addressing to the particular symptoms of a given dispute rather than any underlying concerns or issues that may exist. These characteristics of rights-based processes may be less than ideal for situations where unit cohesiveness and morale are paramount considerations. Despite these limitations, however, rights-based approaches have an important role within the conflict-management continuum.[21] For instance, in certain types of policy issues where a formal and public precedent is sought, utilization of a rights-based process may be the appropriate resolution mechanism. This approach also provides an essential option for members who feel uncomfortable with or have exhausted other avenues of resolution.

## INTEREST-BASED RESOLUTION

Interest-based conflict resolution attempts to break the adjudicative and adversarial paradigm of the more formal conflict resolution approaches by shifting the onus for solution development from interveners to participants and from one-sided to joint outcomes.[22] Interest-based resolution is aimed at early and informal resolution wherein the parties can deal with conflict before it escalates or spreads to the broader work environment. Although interest-based problem-solving can be assisted by a third party, as in the case of mediation or facilitation, at its core this approach is intended as a tool to allow individuals, and their local chain of command, to more actively participate in, and take ownership of, conflict resolution. The interest-based model is focused on identifying and addressing underlying needs and concerns as a means of resolving conflict. To accomplish this task, it seeks to shift discussion from the level of "positions" (what outcome is desired) to "interests" (why these outcomes are important) in order to develop understanding and build mutually satisfactory solutions that meet the needs of both parties.

Similar to the other models, the interest-based process has both strengths and limitations. The interest-based model's emphasis on understanding and consensual agreement can be particularly beneficial when addressing workplace scenarios wherein a key objective is the restoration or enhancement of unit cohesion and morale. Within a classic positional discussion, there are a limited number of resolution options; namely, those of the two disputants and a compromise between their viewpoints. Through exploration of underlying concerns, the interest-based model attempts to expand the menu of potential outcomes and increase the likelihood of achieving a mutually satisfactory outcome.[23]

Despite its potential benefits, it may not be the appropriate method in situations where individuals are unwilling to engage in the process. Collaborative problem-solving is a demanding process and participants who are forced into participating in such a process may lack a sufficient level of commitment to make it work. This approach may also not be ideal in circumstances where organization-wide policy interpretations pertaining to issues such as rights and benefits are being determined, or where the dispute is an aspect of a larger systemic problem.[24]

## **SELECTION OF AN APPROACH**

In examining the different models it is apparent that, though they each have strengths, none of the approaches is optimized for all situations.

The lack of a universally applicable model highlights the requirement for a combination of approaches within a comprehensive conflict-management system. When determining how to best deal with conflict, it can be helpful to use assessment criteria that consider both the situational specifics of the dispute and the respective resolution approaches.

Examples of such criteria are: cost in time and money; participant satisfaction with outcome; durability; and effect on relationship.[25] In comparing the different approaches, a noteworthy difference is their level of process formality. Moving from interest-based, to power-based, to rights-based, the methodologies become increasingly structured and formal. This progression in bureaucratic structure and formality has two notable effects: the resolution process becomes more time-consuming and costly as institutional procedures and regulations increasingly come into play; and, decision-making power is transferred from the disputants to a third party along this same continuum.

## SUMMARY

From time immemorial, conflict management has been a critical organizational and leadership challenge. This chapter has provided an overview of interpersonal conflict and explored different resolution approaches. Although conflicts often share important similarities, it is essential to recognize the unique challenges of each situation.

## NOTES

1. Although the focus of this chapter is workplace conflict, much of the discussion is generalizable to a broader social context.
2. Christopher Moore, *The Mediation Process*, 3rd ed. (San Francisco: Jossey-Bass Publishers, 2003), 3.
3. Joyce Hocker and William Wilmot, *Interpersonal Conflict*, 4th ed. (Madison: WCB Brown and Benchmark, 1995), 21.

4. Lewis Coser, *The Functions of Social Conflict* (New York: The Free Press of Glencoe, 1964), 33–38.
5. Christopher Moore, *The Mediation Process*, 64–66.
6. In informal discussions with conflict resolution practitioners, assumptions are often cited as a leading cause of interpersonal disputes.
7. Chris Argyris, *Knowledge for Action* (San Francisco: Jossey-Bass Publishers, 1993), 57.
8. Peter Senge, Art Kleiner, Charlotte Roberts, Richard Ross, and Bryan Smith, eds., *The Fifth Discipline Fieldbook* (New York: Doubleday, 1994), 242–252.
9. For a discussion of unit cohesion and its importance to the military see Christopher Straub, *The Unit First* (Washington: National Defence University Press, 1988).
10. The primacy of personnel in the success of military operations is widespread throughout military literature. This emphasis in the Canadian context can be seen in Department of National Defence, *Duty with Honour: The Profession of Arms in Canada* (Ottawa: CFP A-PA-005-000 AP-001, 2003).
11. For an overview of literature regarding the importance of group loyalties and organizational culture with the military, see Donna Winslow, "Misplaced Loyalties" in *The Human in Command*, eds. Carol McCann and Ross Pigeau (New York: Kluwer Academic/Plenum Publishers 2000), 293–307.
12. In the CF, leaders are expected to use an informal interest-based approach when appropriate. Department of National Defence. *CANFORGEN064/03 ADMHRMIL 022 May 03*.
13. For a broader view of the relationship between power and conflict, see Bernard Mayer, *The Dynamics of Conflict Resolution* (San Francisco: Jossey-Bass Publishers, 2000), 50–71.
14. Although individual concerns will be considered in many power-based processes, they are not by definition integral to this approach.
15. For an overview of CF leadership philosophy, see Canada, *Leadership in the Canadian Forces: Conceptual Foundations* (Kingston: DND, 2005), http://www.cda-acd.forces.gc.ca.

16. Members may become accustomed to this style and develop an increased dependency on the chain of command for conflict resolution. This lack of solution ownership can also result in less motivation by the disputants to make the settlement work or to direct blame at the decision-maker.
17. In a "classic" sense, a power-based resolution entails the transfer of control from participants to the supervisor.
18. This could take the form of a decision rendered by a third party, such as the supervisor, or by the disputant with the greater organizational authority.
19. William Ury, Jeanne Brett, and Stephen Goldberg, *Getting Disputes Resolved* (San Francisco: Jossey-Bass Publishers, 1998), 7.
20. In a hierarchical organization such as the CF, rights-based approaches can be perceived as the only recourse against perceptions of power-based abuse.
21. Mary Rowe, "People Who Feel Harassed Need a Complaint System With Both Formal and Informal Options," *Negotiation Journal* (a), Vol. 6, No. 2 (1990 ): 164–165.
22. The term *interest-based conflict resolution* is sometimes used synonymously with other terms such as *informal resolution* or Alternative Dispute Resolution (ADR).
23. This approach seeks to be integrative rather than distributive, i.e., to increase options and value rather than merely distribute that which has already been identified.
24. In theory, there is no limit to the application of interest-based conflict resolution. In organizational settings, however, it is employed primarily as a tool to handle interpersonal dispute.
25. William Ury et al., *Getting Disputes Resolved*, 11–12.

## SELECTED READINGS

Bush, B. and J. Folger. *The Promise of Mediation*. San Francisco: Jossey-Bass, 1994.

Canada. Department of National Defence. *CANFORGEN064/03*

Canada. Department of National Defence. *ADMHRMIL 022 May 03*.

Canada. *Leadership in the Canadian Forces: Conceptual Foundations*. Kingston: DND, 2005.

Coates, B. "Alternative Dispute Resolution in the Canadian Forces." *Canadian Military Journal*, Vol. 7, No. 2 (2006): 39–46.

Deutsch, M. *The Resolution of Conflict*. New Haven, CT: Yale University Press, 1973.

Fisher, R. and W. Ury. *Getting to Yes*. Boston: Houghton Mifflin, 1981.

Mayer, B. *The Dynamics of Conflict Resolution*. San Francisco: Jossey-Bass, 2000.

Moore, C. *The Mediation Process*. San Francisco: Jossey-Bass, 2003.

Senge, P., A. Kleiner, C. Roberts, R. Ross, and B. Smith, eds. *The Fifth Discipline Fieldbook*. New York: Doubleday, 1994.

# 10
# COUNSELLING[1]
## by Paul Pellerin

When subordinates or colleagues are facing difficulties either at work or at home, it is often difficult to acknowledge the need for help. They know that they do not cope as well as they used to be able to. They also know that they do not want to appear weak and vulnerable or simply jeopardize their careers by being unable to cope. Sometimes, it is easier to say: "*I'm fine. There is nothing wrong with me!*" than to face what needs to be done in order to resolve those difficulties.

However, leaders, in addition to their daily responsibilities such as assignment of work, performance evaluation, and the imposition of discipline, are required to look after their people by being vigilant in noticing, evaluating, and providing counsel and advice to a subordinate in need.[2] This chapter provides a better understanding of the counselling process and offers guidelines for achieving full benefits of that process.

## WHAT IS COUNSELLING?

A leader often is called upon to communicate with subordinates and to deal with myriad professional and personal difficulties. Counselling is an important tool in assisting subordinates and followers. The main purpose of counselling is to "discover an employee's principal problem and to find a way of handling it."[3]

In a work environment such as the military, a leader counsels subordinates to:[4]

* praise and reward good performance;
* develop teamwork;
* inform on how well or how poorly they are performing;
* assist a subordinate to reach required standards;
* cause subordinates to set personal and professional goals; and
* help resolve personal problems.

## THE IMPORTANCE OF COUNSELLING[5]

It is the leader's responsibility to ensure that subordinates are well trained and developed both professionally and personally. Unit readiness and mission accomplishment depend on every member's ability to adhere to established standards. Therefore, a leader must be prepared to be a mentor, a teacher, a coach, and a counsellor. The counsellor's role implies that a leader will:

* identify weaknesses;
* set goals;
* develop and implement a plan of action; and
* provide oversight and motivation throughout the process.

More importantly, the leader's main concern is the well-being of the subordinate: "When counselling an individual, talk about the behaviour, not the person. Talking about behaviour is easier for the individual to handle. Do not criticize the individual; criticize the behaviour he or she is exhibiting."[6]

Subordinates often perceive counselling as an adverse action. Effective leaders who counsel properly can change that perception. Leaders conduct counselling to help subordinates become better members of the team, maintain or improve performance, and prepare for the future. There are no obvious answers for exactly what to do in all counselling situations, just as there are no easy answers for exactly what to do in all leadership situations. However, to conduct effective counselling, leaders should develop a counselling style that encompasses these characteristics:

* Purpose: Clearly define the purpose of the counselling session;

* Flexibility: Adapt the counselling style to the character of each subordinate. Individuals are different and the counselling style should be in concordance with the individual needs;

* Respect: View subordinates as unique, complex individuals, each with their own sets of values, beliefs, and attitudes;

* Communication: Establish open, two-way communication with a particular attention both to verbal and non-verbal language (gesture), and body language. Effective counselors listen more than they speak;

* Support: Encourage subordinates through actions while guiding them through their problems;

* Motivation: Get every subordinate to actively participate in counselling and understand its value.

## THE "HOW TO" FOR COUNSELLORS

Leaders must demonstrate certain qualities to be effective counsellors. These qualities include respect for subordinates, self-awareness and cultural awareness, empathy, and credibility.[7]

*Respect for Subordinates*

Leaders show respect for subordinates when they allow them to take responsibility for their own ideas and actions. Respecting subordinates helps create mutual respect in the leader–subordinate relationship. Mutual respect improves the chances of changing (or maintaining) behaviour and achieving goals.

*Self-Awareness and Cultural Awareness*

Leaders must be fully aware of their own values, needs, and biases prior to counselling subordinates. Cultural awareness is a mental attribute. Leaders need to understand there are similarities and differences between individuals of different cultural backgrounds and how these factors may influence behaviours. Leaders should not let unfamiliarity with cultural backgrounds hinder them in addressing cultural issues, especially if they generate concerns within the unit or hinder team-building. Cultural awareness (see also the chapter in this handbook on cultural intelligence) enhances a leader's ability to display empathy.

*Empathy*

Empathy is the action of being understanding of and sensitive to the feelings, thoughts, and experiences of another person to the point that a counsellor can almost feel or experience them personally. Leaders with empathy can put themselves in their subordinate's shoes; they can see a situation from the other person's perspective. By understanding the subordinate's position, the empathetic leader can help a subordinate develop a plan of action that fits the subordinate's personality and needs, one that works for the subordinate. If a leader does not fully comprehend the situation from the subordinate's point of view, the leader may guide or influence the subordinate into a direction that is not suited for them.

*Credibility*

Leaders achieve credibility by being honest and consistent in their statements and actions. Credible leaders use a straightforward style with their subordinates by repeatedly demonstrating their willingness to assist a subordinate and being consistent in what they say and do. They behave in a manner that subordinates respect and trust. Leaders who lack credibility with their subordinates will find it difficult to influence them.

## SKILLS OF A COUNSELLOR[8]

One challenging aspect of counselling is selecting the proper approach to a specific situation. Effective counselling techniques must suit the situation, the leader's capability, and the subordinate's expectations. In some cases, a leader may only need to listen or simply give information in the form of a brief word of praise. Other situations may require structured counselling followed by recommended actions.

All leaders should seek to develop and improve their own counselling abilities. Everyone can improve their counselling techniques by studying human behaviour, learning what usually concerns subordinates, and developing interpersonal skills. The techniques needed to provide effective counselling will vary from person to person and session to session. However, general skills that one will need in almost every situation include active listening, responding, and questioning.

*Active Listening*

During counselling, leaders must actively listen to their subordinates. When actively listening, leaders communicate verbally and nonverbally that they have received the subordinate's message. To fully understand a subordinate's message, a leader must listen to the words and observe the subordinate's manners. Elements of active listening that counsellors should attend to include:

- Eye Contact: Maintaining eye contact without staring helps show sincere interest. Occasional breaks of contact are normal and acceptable. Subordinates may perceive excessive breaks of eye contact, paper shuffling, and clock-watching as a lack of interest or concern. Based on cultural background, participants in a particular counselling session may have different ideas about what proper eye contact is.

- Body Posture: Being relaxed and comfortable will help put the subordinate at ease. However, a too-relaxed position or slouching may be interpreted as a lack of interest.

- Head Nods: Occasionally nodding the head shows that attention is being paid, and encourages the subordinate to continue.

- Facial Expressions: Facial expressions need to be kept natural and relaxed. A blank look or fixed expression may disturb the subordinate. Smiling too much or frowning may discourage the subordinate from continuing.

- Verbal Expressions: Leaders need to refrain from talking too much and avoid interrupting. Let the subordinate do the talking while keeping the discussion on the counselling subject. Speaking only when necessary reinforces the importance of what the subordinate is saying and encourages the subordinate to continue. Silence also can do this, but with caution. Occasional silences may indicate to the subordinate that it is fine to continue talking, but a long silence sometimes can be distracting and make the subordinate feel uncomfortable.

* <u>Common Themes:</u> Active listening also implies listening carefully to the way a subordinate uses common themes, or expresses himself/herself with opening and closing statements as well as recurring references. These all may indicate his/her priorities. Inconsistencies and gaps may suggest avoidance of the real issue. This confusion and uncertainty could require additional questions.

While actively listening, leaders must pay attention to gestures. Sometimes, these gestures could be indications of the feelings behind the words. Not all gestures are proof of a subordinate's feelings, but they should be taken into consideration. Leaders must note differences between what the subordinate says and does. Nonverbal indicators of a subordinate's attitude may include:

* <u>Boredom:</u> Drumming on the table, doodling, clicking a ballpoint pen, or resting the head in the palm of the hand;

* <u>Self-Confidence:</u> Standing tall, leaning back with hands behind the head, and maintaining steady eye contact;

* <u>Defensiveness:</u> Pushing deeply into a chair, glaring at the leader, and making sarcastic comments as well as crossing or folding arms;

* <u>Frustration:</u> Rubbing eyes, pulling on an ear, taking short breaths, wringing the hands, or frequently changing total body position;

* <u>Interest, friendliness, and openness:</u> Moving toward the leader while sitting;

* <u>Openness or anxiety:</u> Sitting on the edge of the chair with arms uncrossed and hands open.

Leaders/counsellors must consider these indicators carefully. Although each indicator may show something about the subordinate, meaning cannot be assumed. Counsellors need to ask the subordinate about the indicator in order to better understand the behaviour and allow the subordinate to take responsibility for it.

*Responding*

Active listening skills are followed by responding skills. A leader's response is a form of understanding of the situation presented by the subordinate. It is important to respond to subordinates both verbally and nonverbally. From time to time, counsellors should check their understanding: clarify and confirm what has been said. Verbal responses consist of summarizing, interpreting, and clarifying the subordinate's message. Nonverbal responses include eye contact and occasional gestures such as a head nod.

*Questioning*

Although a necessary skill, questioning must be used with caution. Too many questions can place the subordinate in a passive mode. The subordinate may also react to excessive questioning, which may be perceived as an intrusion of privacy, and s/he may become defensive. Counsellors need to ask questions to obtain information or to get the subordinate to think about a particular situation. Generally, the questions should be open-ended, requiring more than a yes or no answer. Such questions may help to verify understanding, encourage further explanation, or help the subordinate advance in the stages of the counselling session.

## COUNSELLING ERRORS[9]

Effective leaders avoid common counselling mistakes. Dominating the counselling session by talking too much, giving unnecessary or inappropriate "advice," not truly listening, and projecting personal likes, dislikes, biases, and prejudices all interfere with effective counselling. Leaders should also avoid other common mistakes such as rash judgments, stereotypes, loss of emotional control, and improper follow-up.

## THE LEADER'S LIMITATIONS[10]

Leaders must realize that they cannot help everyone in every situation. Therefore, they must recognize their limitations and, when necessary, refer a subordinate to a professional. Although it is generally in an individual's best interest to seek help first from his/her immediate supervisor, leaders must always respect an individual's right to contact most of these professional agencies on his/her own.

## APPROACHES TO COUNSELLING[11]

An effective leader approaches each subordinate as a unique individual. Moreover, situations and difficulties differ from one individual subordinate to another and approaches to resolve the situations should reflect these differences. The approaches have different techniques, but they all maintain the overall purpose and definition of counselling. The three approaches to counselling usually are referred to as non-directive, directive, and combined. A major difference is the degree in which the subordinate participates and interacts during the counselling session.

## Non-directive

The non-directive approach to counselling is preferred for most counselling sessions. Leaders can use their own experience and judgment to assist subordinates in developing solutions. The leader partially structures this type of counselling by telling the subordinate about the counselling process and explaining what is expected.

During the counselling session, counsellors listen rather than make decisions or give advice. They clarify what has been said, causing the subordinate to bring out important points, so as to better understand the situation. When appropriate, the discussion is summarized. Leaders avoid providing solutions or rendering opinions; instead, they maintain a focus on individual and organizational goals and objectives, ensuring the subordinate's plan of action supports those goals and objectives.

## Directive

The directive approach works best to correct a simple problem, make on-the-spot corrections, and correct aspects of duty performance. The leader using the directive style does most of the talking and tells the subordinate what to do and when to do it. In contrast to the non-directive approach, the leader directs a course of action for the subordinate.

Leaders usually choose this directive approach when time is short, when the leader alone knows what to do, or if a subordinate has limited problem-solving skills. It is also an appropriate approach when a subordinate needs guidance, is immature insecure.

## Combined

In the combined approach, the leader uses techniques from both the directive and non-directive approaches, adjusting them to articulate what is best for the subordinate. The combined approach emphasizes the subordinate's planning and decision-making responsibilities.

With the leader's assistance, the subordinate develops his or her own plan of action. Leaders should listen, suggest possible courses of action, and help analyze each possible solution to determine its good and bad points. Counsellors should then help the subordinate to fully understand all aspects of the situation and encourage the subordinate to decide which solution is best.

## **THE COUNSELLING PROCESS**

Effective leaders use the four-stage counselling process in identifying the need for counselling, preparing for counselling, conducting the counselling, and following up.[12]

Stage 1 — Identify the Need for Counselling: Quite often, organizational policies, such as counselling associated with an evaluation or counselling required by the command, may be the focus on a counselling session. However, a leader may conduct developmental counselling whenever the need arises. Developing subordinates consists of observing the subordinate's performance, comparing it to the standard, and then providing feedback to the subordinate in the form of counselling.

Stage 2 — Prepare for Counselling: Successful counselling requires preparation. To prepare for a counselling session, the following points should be taken in consideration: selecting a suitable place, scheduling the time, notifying the subordinate well in advance, organizing the information, outlining the counselling session components, planning the counselling strategy, and establishing the right atmosphere.

* *Select a Suitable Place:* Schedule counselling in an environment that minimizes interruptions and is free from distracting sights and sounds.

* *Schedule the Time:* When possible, counsel a subordinate during normal working hours. Counselling

after working hours may be rushed or perceived as unfavourable. The length of time required for counselling depends on the complexity of the issue. Generally a counselling session should last less than an hour. If more time is needed, schedule a second session. Additionally, select a time free from competition with other activities and consider what has been planned after the counselling session. Important events can distract a subordinate from concentrating on the counselling.

* *Notify the Subordinate Well in Advance:* For a counselling session to be a subordinate-centred, a two-person effort, the subordinate must have time to prepare for it. The subordinate should know why, where, and when the counselling will take place. In a case of performance or professional-development counselling, the subordinates may need a week or more to prepare or review specific information/documents, such as personal forms or previous counselling records.

* *Organize Information:* Solid preparation is essential to effective counselling. Review all pertinent information. This includes the purpose of the counselling, facts and observations about the subordinate, identification of possible problems, main points of discussion, and the development of a plan of action. Focus on specific and objective behaviours that the subordinate must maintain or improve as well as a plan of action with clear, obtainable goals.

* *Outline the Components of the Counselling Session:* Using the information obtained, determine what to

discuss during the counselling session. Note what prompted the counselling, what the aim to achieve is, and what the role of counsellor is. Identify possible comments or questions to help keep the counselling session subordinate-centred and help the subordinate progress through its stages. Although no one will ever know what a subordinate will say or do during counselling, a written outline helps organize the session and enhances the chance of positive results.

* *Plan Counselling Strategy:* As there are many approaches to counselling, a leader should learn to approach each situation with the proper strategy. The directive, non-directive, and combined approaches to counselling were addressed earlier. Use a strategy that suits subordinates and the situation.

* *Establish the Right Atmosphere:* The right atmosphere promotes two-way communication between a leader and subordinate. To establish a relaxed atmosphere, the counsellor may offer the subordinate a seat or a cup of coffee. In most cases, it is preferred to sit in a chair facing the subordinate, since sitting behind a desk can act as a barrier.

<u>Stage 3 — Conduct the Counselling</u>: Be flexible when conducting a counselling session. Often, counselling for a specific incident occurs spontaneously as leaders encounter subordinates in their daily activities. Such counselling can occur in the field, in the office, in barracks, wherever subordinates perform their duties. Good leaders take advantage of naturally occurring events to provide subordinates with feedback.

Even when a counsellor has prepared for formal counselling, s/he should address the four basic components of a counselling session. Their purpose is to guide effective counselling rather than mandate a series of rigid steps. Ideally, a counselling session results in a subordinate's commitment to a plan of action, and to the assessment of that plan of action, which becomes the starting point for follow-up counselling. Counselling sessions consist of opening the session, discussing the issues, developing the plan of action, and recording and closing the session.

* *Opening the Session:* In the session opening, counsellors state the purpose of the session and establish a subordinate-centred setting. They establish the preferred setting early in the session by inviting the subordinate to speak. The best way to open a counselling session is to clearly state its purpose. If applicable, leaders start the counselling session by reviewing the status of the previous plan of action.

* *Discussing the Issues:* The counsellor and the subordinate should attempt to develop a mutual understanding of the issues, using active listening, responses, and questions without dominating the conversation. Counsellors aim to help the subordinate better understand the subject of the counselling, for example, performance or a problem situation and its impact, or potential areas for growth.

    Both the counsellor and the subordinate should provide examples or cite specific observations to reduce the perception that either is unnecessarily biased or judgmental. However, when the issue is substandard performance, the leader should make clear how the performance did not meet the standard.

* *Developing a Plan of Action:* A plan of action identifies a method for achieving a desired result. It specifies what the subordinate must do to reach the goals set during the counselling session. The plan of action must be specific: it should show the subordinate

how to modify or maintain his behaviour. The plan must use concrete and direct terms. A specific and achievable plan of action sets the stage for successful development. For example, a counsellor might say: "Next week you will attend the map reading class with 1st Platoon. After the class, Sgt Bloggins will coach you through the land navigation course. He will help you develop your skill with the compass. I will observe you going through the course with the Sgt, and then I will talk to you again and determine where and if you still need additional training."

* *Recording and Closing the Session:* Although requirements to record counselling sessions vary, a leader always benefits by documenting the main points of a counselling session. Documentation serves as a reference to the agreed plan of action and the subordinate's accomplishments, improvements, personal preferences, or problems. A complete record of counselling aids in making recommendations for professional development, schools, promotions, and evaluation reports.

   To close the session, leaders summarize its key points and ask if the subordinate understands the plan of action. The subordinate is invited to review the plan of action, and also what is expected of the leader. With the subordinate, a leader establishes any follow-up measures necessary to support the successful implementation of the plan of action. These may include providing the subordinate with resources and time, periodically assessing the plan, and following through on referrals. Any future meetings are scheduled, at least tentatively, before dismissing the subordinate.

Stage 4 — Follow-up: For the leader with his/her responsibilities, the counselling process does not end with the counselling session. It continues through implementation of the plan of action and evaluation of results. After counselling, the leader and the counsellor must support subordinates as they implement their plans of action. Leaders must observe and assess this process and possibly modify the plan to meet its goals. Appropriate measures after counselling include follow-up counselling, making referrals, informing the chain of command, and taking corrective measures.

As a crucial part of follow-up, the leader also must continue to assess the plan of action. The purpose of counselling is to develop subordinates who are better able to achieve personal, professional, and organizational goals. During the assessment process, a leader periodically reviews the plan of action with the subordinate to determine if the desired results are being achieved. The leader and the subordinate should determine the dates for this assessment process during the initial counselling session. The assessments of the plan of action provide useful information for fine-tuning the counseling in future follow-up sessions.

## **SUMMARY**

Effective counselling is one of the most important processes when leading people. Generally speaking, counselling is employed to keep small problems from becoming larger. For leaders, counselling provides insight into group situations that reflect the bigger picture. For subordinates, proper counselling will not just alleviate performance degradation, but will develop subordinates positively, and can also improve group performance.

The functions of counselling are to provide advice, reassurance, communication, release of emotional tension, clarified thinking, and reorientation, with an overall goal of reducing problems for the individual, the team or group, or the organization. As summarized in this chapter, numerous guidelines should be considered to assist the leader to help a person to benefit to the maximum from a counselling session.

Counselling can be employed in correcting disciplinary problems or personal problems, with the counsellor using a variety of techniques. Regardless of the technique selected, good listening is of fundamental importance to the counsellor. Listening involves not only hearing what the person says, but also observing what the person avoids saying, what is unable to be said without help, the use of voice intonations and physical gestures and expressions, all to be observed in order for the counsellor to achieve the most accurate interpretation possible.

In times of stress, leaders, counsellors, peers, subordinates, and followers all find it more difficult to see clearly for themselves, and to benefit from the counsel of others. Importantly, counselling can offer a safe place to confide and resolve problems and issues while minimizing career-limiting consequences. Counselling has the potential to help everyone to manage more effectively the everyday pressures of life and work.[13]

## NOTES

1. This chapter is based on FM 6-22 (U.S.) *Army Leadership Manual*, Appendix B. 2006. Counsellor/counselor, counselling/counseling, all are acceptable variations of these key words.
2. Impact Factory, Leadership training, available on Impact Factory website: http://www.impactfactory.com/p/counselling_skills_training/snacks_1643-2103-62450.html.
3. Douglas A. Benton, *Applied Human Relations*, 6th ed. (Upper Saddle River, NJ: Prentice Hall, 1998), 110.
4. Definition from Northeastern Illinois University website: http://www.neiu.edu/~dberhrlic/hrd408/glossary.html (accessed 28 February 2007).
5. This sub-section is based on FM 6-22 (U.S.) *Army Leadership Manual*, Appendix B.
6. Canada, *Leadership in the Canadian Forces: Leading People* (Kingston: DND, 2007), 82.
7. This sub-section is based on FM 6-22 (U.S.) *Army Leadership Manual*, Appendix B.

8. *Ibid.*
9. *Ibid.*
10. *Ibid.*
11. *Ibid.*
12. This sub-section is taken from FM 6-22 (U.S.) *Army Leadership Manual*, Appendix B.
13. Canada, A-PD-131-002/PT-001, *Leadership: The Professional Officer*, Vol. 2 (Ottawa: DND, 1973).

**SELECTED READINGS**

Benton, Douglas A. *Applied Human Relations*, 6th ed. Upper Saddle River, NJ: Prentice Hall, 1998.

Canada. A-PD-131-002/PT-001. *Leadership: The Professional Officer*, Vol. 2. Kingston: DND, 1973.

Fritz, M. Susan, Joyce Povlacs Lunde, William Brown, and Elizabeth A. Banset. *Interpersonal Skills for Leadership*, 2nd ed. New York: Prentice Hall, 2004.

McLeod, John. *An Introduction to Counselling*, 3rd ed. Maidenhead, Berkshire: Open University Press, 2003.

Milne, Aileen. *Teach Yourself Counselling*, 2nd revised ed. Toronto: McGraw-Hill, 2003.

Mynatt, R. Clifford and Michael E. Doherty. *Understanding Human Behavior*, 2nd ed. Boston: Allyn and Brown, 2001.

United States. FM 6-22 *Army Leadership Manual*. Washington, DC: Department of the Army, 2006.

# 11
# COURAGE
by Bernd Horn

Courage has often been cited as a virtue and is perhaps the most admired and respected of human qualities. Quite simply, everyone wants to be courageous. "I do not believe," extolled Field Marshal William Slim, "that there is any man who would not rather be called brave than have any other virtue attributed to him."[1] Combat veterans overwhelmingly characterized the best combat soldiers as those who demonstrated fearless behaviour, which is commonly associated with courage.[2] Within the context of the military, courage, both physical and moral, is the foundation upon which fighting spirit and success in operations depends. It is a quality needed by all military personnel, and particularly by leaders.

## COURAGE DEFINED

The great writer Ernest Hemingway defined courage as "grace under pressure."[3] In the traditional military context courage is often related to "masculine," "military," or "warrior" values and virtues. In many ways it is an accepted standard of conduct in action, if not almost a code to be followed. In essence, it is often equated to fear overcome. Clearly, some ambiguity exists and the concept requires greater clarity and definition.

Courage is often seen in two lights: one, as an act or action such as a single desperate act (e.g. the storming of a pillbox or falling on a grenade to save comrades); the second, as the ancient philosopher Socrates extolled, as a very noble quality. The two themes, however, are not mutually exclusive. Israeli psychologist Ben Shalit offered, "Courage is the right act at the right time and in the right place." He added, "It must be an act that is perceived to be outstanding in a setting it can dramatically affect."[4]

In more lay terms, courage is normally broken down into physical and moral courage. The *Canadian Oxford Dictionary* defines courage as "the ability to disregard fear; bravery" (i.e. [brave] "able or ready to face and endure danger, pain, adversity, etc.").[5] Similarly, the American *Standard Dictionary* describes courage as "the quality of mind which meets danger or opposition with intrepidity, calmness and firmness; the quality of being fearless; bravery." It further states, "The brave man combines confidence with firm resolution in the face of danger. Courageous is more than brave, adding the moral element. The courageous man steadily encounters perils to which he may be keenly sensitive, at the call of duty."[6] The British *Chambers Dictionary* explains courage as "the quality that enables men to meet danger without giving way to fear; bravery (courage, heroism; to be brave — to meet boldly, to defy, to face, spirit)."[7] Lastly, the Israeli *Ben Shushan* dictionary depicts courage as "strength, power, might."[8] In all cases, the underlying theme of the definition underscore that courage is clearly a human trait or quality.

This motif is further developed by scholars, researchers, and veterans. Samuel Stouffer, in his seminal study of the American combat soldier in the Second World War, noted there was an internal struggle between an individual's impulses toward personal safety and comfort and the social compulsions that drove them into danger and discomfort.[9] Research has revealed that "courage is not fearlessness; it is being able to do the job even when afraid."[10] For instance, Professor S.J. Rachman formulated that "true courage" was a quality of "those people who are willing and able to approach a fearful situation despite the presence of subjective fear and psychophysiological disturbances."[11] Similarly, S.L.A. Marshall, the renowned soldier/scholar who studied combat behaviour in the Second World War, considered courage to be more than an innate quality — courage and cowardice to him were alternative free choices that came to every man and woman. Professor William Miller broadly defined courage as fortitude — "a certain firmness of mind."[12] Finally, one military researcher explained courage as:

> [A] state of mind which leads people to disregard their own safety, to make all kinds of sacrifice, and to press forward toward their goals with single-minded determination. Courage takes three forms. It can be acquired, be instinctive, or be generated by following the example of others. Acquired courage is born out of intelligence and adherence to an ideology. Once instilled in a person, this form of courage is the least likely to fail when put to the test. Instinctive courage stems from the natural desire for self-preservation; the concept of "self" here includes home and family. The third form of courage is achieved by following the example of a courageous leader. It is gained through intensive training, strict discipline and complete confidence in their leader by the led.[13]

These views accord well with that of Lord (Sir Charles Wilson) Moran in his classic work, *The Anatomy of Courage*. Moran theorized that courage is "a moral quality" and "not a chance gift of nature." He asserts that "it is the cold choice between two alternatives, it is the fixed resolve not to quit, an act of renunciation which must be made not once but many times by the power of will." Moran concluded, "Courage is willpower."[14]

The definitions to this point have largely emphasized the physical component of courage. However, it is important to recognize that moral courage is equally critical to the success of operations and particularly to leadership in general. "Moral courage means taking personal responsibility," explained eminent scholar Michael Ignatieff. "Moral behaviour is always individual behaviour."[15] It is the courage to do the right thing even when it may be unpopular, involve risk of censure, ridicule, or danger. It is the insistence of maintaining the highest standards of decency and integrity at all times and under all circumstances regardless of consequence. It is for this reason that the ancient philosophers Aristotle and Socrates held a conception of courage that required an individual to have a noble goal and to use practical reasoning to obtain it. For them, a courageous person was one who engaged in thoughtful, practical deliberations and acted coolly in a way that was calculated to most successfully achieve the noble purpose in a particular situation.[16]

## **HOW TO OPTIMIZE COURAGE**

The mechanisms and strategies to optimize courage are very similar, if not the same, as those measures used to control fear (see Chapter 20 on Fear). In essence, the core components are: confidence, education and training, experience and group cohesion, and leadership. All are interrelated and mutually supporting.

Confidence — in self, in others, in the equipment one uses, as well as in the cause for which one is fighting, is an important catalyst toward courage. Research has revealed that confidence is perhaps the greatest source of emotional strength that military personnel can draw upon. "With it," asserted behavioural expert Bernard Bass, "he willingly faces the enemy and withstands deprivations, minor setbacks, and extreme stresses, knowing that he and his unit are capable of succeeding."[17] As such, confidence generates capability and mental toughness — a belief that an individual or group can overcome any obstacle. Israeli research concluded that "heroes are not born but become heroes through force of circumstance."[18]

In essence, research has repeatedly demonstrated that troops who expressed a high degree of self-confidence before combat were more likely to perform with relatively little fear during battle and were more likely to demonstrate courageous behaviour. Confidence, in turn, can be achieved in self through training, education, and fitness. It can also be generated through sound leadership, team cohesion, and dependable equipment.

As mentioned earlier, education and training are also key components of courage.[19] Flavius Renatus asserted in 378 AD, "The courage of the soldier is heightened by the knowledge of his profession." Knowledge is the key in fuelling confidence — not only in self, but in one's comrades, equipment, and tactics. It provides the perception, as well as ability, to respond effectively to potential or real threats in the environment preemptively or reactively. Importantly, knowledge assists in removing ambiguity and uncertainty in regards to the future, particularly self-doubt as to whether or not an individual will be capable of withstanding the challenges that await them.

This level of confidence is also achieved through realistic training, as well as a complete understanding of the realm of conflict. In addition, realistic training reduces the element of the unknown and assists individuals with a realization of what to expect. For example, realistic training (e.g. battle simulation, full combat loads, non-templated enemy action, intense tempo, imposed stress, physical exertion, and fatigue) can create reasonable expectations of how far an individual/group/unit can go, how long they can fight, and how well they can cope. In addition, realistic training is also valuable to the extent it inculcates in military personnel a realization that they are capable of far more than they originally believed and that they can survive on the battlefield. Major John Masters, a Second World War commander, explains that "easy it is to be brave when a little experience has taught you that there is nothing to be afraid of."[20]

The value of training is also derived from its ability to create an element of habit, routine (i.e., developing instinctive reactions) and discipline. Drill, for instance, is utilized to teach the instinctive reaction of a body of troops to commands. "What is learnt in training," insisted commando commander Lord Lovat, "is done instinctively in action — almost without thinking — down to the last man."[21] Moreover, Field Marshal Bernard Law Montgomery extolled, "Discipline strengthens the mind so that it becomes impervious to the corroding influence of fear … it instills the habit of self control."[22]

Although education and training are important factors in engendering courage, so too is experience. Experience is the application and fusion of all education, training, and personal development that an individual has conducted in actual operational and exercise settings. As such, it emboldens individuals with the knowledge and confidence of their own capability, an awareness of realistic threat and risk, possible solutions or courses of action, as well as an understanding and confidence in the capabilities of their equipment, personnel, and allies. Experience rips away the shroud of uncertainty and the unexpected. It reinforces a sense of control and a mindset that the individual is the master of his or her own destiny. All told, experience feeds courage.

Another mechanism to optimize courage in individuals is strong group cohesion. It has long been documented and consistently reinforced that one of the greatest motivators for military personnel is the desire not to let down their comrades. Pride in unit and regiment, not to mention personnel reputation, is a powerful catalyst to courageous behaviour.[23] Veteran paratrooper John Agnew explained, "Being able to depend on each other makes individuals courageous regardless of fear, don't let your comrades down."[24] S.L.A. Marshall asserted, "I hold it to be one of the simplest truths of war that the thing which enables an infantry soldier to keep going with his weapons is the near presence or the presumed presence of a comrade."[25] Many military commanders and researchers maintain that an individual behaves as a hero or coward according to the expectations of others. Marshall insisted, "no matter how lowly his rank, any man who controls himself automatically contributes to the control of others." He added, "Fear is contagious but courage is not less so."[26] American Congressional Medal of Honor recipient Captain Roger Donlon acknowledged, "That was the core that allowed us to do things beyond what we thought we'd be capable of. When we saw what the other members of the team were doing, we gained more strength."[27]

The final mechanism to be discussed to optimize courage is leadership. "The commander strolled across the battlefield issuing orders in an icily calm voice," wrote historian Daniel Dancocks. "By setting such a courageous example [Lieutenant-Colonel A.A. 'Bert'] Kennedy stiffened the determination of his regiment. The riflemen advanced again into the face of the enemy fire."[28] The example of others, particularly that of the leader, has endured throughout conflict as one of the prime reasons for courageous behaviour. Samuel Stouffer found that "cool and aggressive leadership was especially important "in pressing troops forward in dangerous and fearful situations such as storming across a beach raked by fire.[29] "A brave captain," affirmed

Sir Philip Sidney, "is as a root, out of which, as branches, the courage of his soldiers doth spring."[30]

This is not surprising. Research has shown that courage is contagious — simply, role-modelling has an extremely important influence on a person's reaction to threatening situations. "Courage, when it spreads," revealed Marshall, "is not so much caught as mimicked."[31] In this regard, leadership is key. Veterans explained that the generation of courageous behaviour was directly linked to "leadership from in front."[32] Most research has reinforced the intuitive deduction that "men like to follow an experienced man ... [he] knows how to accomplish objectives with a minimum of risk. He sets an example of coolness and efficiency which impels similar behaviour in others." In this regard, the presence of strong thoughtful leadership creates "a force which helps resist fear."[33] For example, Field Marshal Slim's "remarkable calmness in crisis, despite his own inner fears and anxieties, contributed significantly to a lessening of the storm of panic which erupted at every new and unexpected Japanese move."[34] As such, leadership from the front, setting the example, sharing risk and hardship, and ensuring that subordinates understand that the leader would never expect them to do something that he or she would not do, all acts to fuel courage.

## **SUMMARY**

In conclusion, courage, whether physical or moral, is critical to military leadership. In the simplest of terms, it is the bedrock upon which fighting spirit and success in operations depends. Courage is central to the military ethos and it is a quality required of all military personnel, but especially those who have been imparted with the authority to lead others into harm's way.

**NOTES**

1. Quoted in Major-General F.M. Richardson, *Fighting Spirit: A Study of Psychological Factors in War* (London: Leo Cooper, 1978), 67.
2. S.J. Rachman, *Fear and Courage* (San Francisco: W.H. Freeman and Company, 1978), 235.
3. Quoted in Douglas N. Walton, *Courage: A Philosophical Investigation* (Los Angeles: University of California Press, 1986), 31.
4. Ben Shalit, *The Psychology of Conflict and Combat* (New York: Praeger, 1988), 97.
5. Katherine Barber, ed., *The Canadian Oxford Dictionary* (Don Mills, ON: Oxford University Press, 1998), 323, 170.
6. Quoted in Shalit, 97.
7. *Ibid.*
8. *Ibid.*
9. Samuel A. Stouffer, *The American Soldier: Combat and Its Aftermath. Vol. II* (Princeton, NJ: Princeton University Press, 1949), 84.
10. John Dollard, *Fear in Battle* (Westport, CT: Greenwood Press, Publishers, 1944), 57; Walton, 32; and Anthony Kellett, *Combat Motivation: Operational Research and Analysis Establishment ORAE Report No. R77* (Ottawa: DND, 1980), 293.
11. Rachman, 25.
12. William Ian Miller, *The Mystery of Courage* (Cambridge, MA: Harvard University Press, 2000), 5.
13. Major P.B. Deb, "The Anatomy of Courage," *Army Quarterly*, Vol. 127, No. 4 (October 1997): 403.
14. Sir Charles Wilson (Lord) Moran, *The Anatomy of Courage* (New York: Avery Publishing Group Inc., 1987), 61.
15. Michael Ignatieff, *Virtual War: Ethical Challenges* (Annapolis: United States Naval Academy, March 2001), 17. He added, "The responsibilities we're talking about in ethical life are individual ones, and they have to be shouldered by each of you. Therefore, moral responsibility is a habit of the heart, and it's a habit of the mind."
16. Douglas N. Walton, *Courage: A Philosophical Investigation.* (Los Angeles: University of California Press, 1986), 33–34.

17. B.M. Bass, *Leadership and Performance beyond Expectations* (New York: Free Press, 1985), 69.
18. Colonel Ian Palmer, "The Emotion That Dare Not Speak Its Name?" *British Army Review*, No. 132 (Summer 2003): 32.
19. Education is defined as "the reasoned response to an unpredictable situation — critical thinking in the face of the unknown" as opposed to training, which is defined as "a predictable response to a predictable situation." Professor Ronald Haycock, "Clio and Mars in Canada: The Need for Military Education," presentation to the Canadian Club, Kingston, Ontario, 11 November 1999.
20. John Masters, *The Road Past Mandalay* (London: Cassell, 2003 reprint), 271.
21. Quoted in Will Fowler, *The Commandos at Dieppe: Rehearsal for D-Day* (London: HarperCollins, 2002), 55.
22. Field Marshall Bernard L. Montgomery, "Discipline from Morale in Battle: Analysis," in Canada, *The Officer: A Manual of Leadership for Officers in the Canadian Forces* (Ottawa: DND, 1978), 66.
23. Some would argue shame is a vital factor in cultivating courage.
24. Captain T.M. Chacho, "Why Did They Fight? American Airborne Units in World War II," *Defence Studies,* Vol. 1, No. 3 (Autumn 2001): 81.
25. Miller, 214.
26. *Ibid.*, 209.
27. Roger Donlon, "Army Values — Personal Courage," *Special Warfare*, Spring 1999, 25.
28. Daniel G. Dancocks, *The D-Day Dodgers* (Toronto: McClelland & Stewart, 1991), 182.
29. Stouffer, 68.
30. Quoted in Lieutenant-Colonel Dave Grossman, O*n Killing* (New York: Little, Brown and Company, 1996), 85.
31. Quoted in Miller, 209.
32. Kellett, *ORAE Report No. R77*, 299.
33. Dollard, 44.
34. Robert Lyman, *Slim, Master of War* (London: Constable, 2004), 108.

## SELECTED READINGS

Becker, Selwyn W. and Alice H. Eagly. "The Heroism of Women and Men." *American Psychologist*, Vol. 59, No. 3, (2004): 163-178.

Deb, Major P.B. "The Anatomy of Courage." *Army Quarterly*, Vol. 127, No. 4 (October 1997): 403-406.

Dinter, Elmar. *Hero or Coward*. London: Frank Cass, 1985.

Horn, Colonel Bernd. "Revisiting the Worm: An Examination of Fear and Courage." *Canadian Military Journal*, Vol. 5, No. 2 (Summer 2004): 5-16.

Kellett, Anthony. *Combat Motivation — Operational Research and Analysis Establishment ORAE Report No. R77*. Ottawa: DND, 1980.

Miller, William Ian. *The Mystery of Courage*. Cambridge, MA: Harvard University Press, 2000.

Moran, Sir Charles Wilson (Lord) Moran. *The Anatomy of Courage*. New York: Avery Publishing Group Inc., 1987.

O'Connell, Robert L. "Courage." *Military History Quarterly*, Vol. 3, No. 1 (Autumn 1990): 62-67.

Rachman, S.J. *Fear and Courage*. San Francisco: W.H. Freeman and Company, 1978.

Shalit, B. *The Psychology of Conflict and Combat*. New York: Praeger, 1988.

Walton, Douglas N. *Courage: A Philosophical Investigation*. Los Angeles: University of California Press, 1986.

# 12
# CREATIVITY
by Robert W. Walker

How creatively might one describe "creativity"? Why is creativity so difficult to grasp while also being, increasingly, so important in the 21st century? How does creative thinking differ from an informed analysis, from disciplined, linear, cause-and-effect thinking that can effectively solve many problems and support important decisions? Is creativity the purview of senior people, leaving analytic thinking in general to junior people? And, what relevance does creativity have to leadership? Does it have particular relevance to military institutions, for which leadership often is stereotyped as authoritative, impersonal, bureaucratic, punitive, and even authoritarian? Challenging questions, with this as a complex chapter of answers.

Creativity, generally, is associated with adjectives like imaginative, original, inventive, innovative, even fertile or prolific. One English-language dictionary describes creativity as being characterized by originality of thought, displaying imagination, a sophisticated bending of the rules or conventions, and a consequence of situations where disagreement or discord is encouraged and ultimately gives rise to better ideas or outcomes.[1] The Public Service of Canada, particularly its subcomponent in the Department of National Defence (DND), has defined creativity as, "continually seeking and implementing new and innovative ways to accomplish objectives and improve the status quo."[2] Creativity has been perceived as the antithesis of analysis, the latter being associated with phrases like examining in detail, investigating components, subdividing and isolating into essential features, and separating into constituent parts in order to determine relationships and relative values. "Analytic philosophy" serves as an example of the school of formal logic that flourished in the first half of the 20th century but, since then, generally has fallen into disuse.[3]

And so, with this somewhat tenuous grip on the concept of creativity, a comparatively easier challenge would be to articulate just why creativity is relevant and important in today's work world. In a substantial study of military and non-military leadership, American academic and applied researcher Stephen Zaccaro summarized the challenges, both conceptual and behavioural, facing leaders of today.[4] Leaders, particularly senior leaders, he stated, function in increasingly complex environments characterized by rapid information processing systems, volatile and ambiguous challenges, and novel human resource and institutional problems. With society's changing nature, the disarray in the global security environment, and expanding demands on senior military leaders, the requisite capacities for leaders, Zaccaro emphasized, included the need to creatively develop new ideas and outcomes that respond to these environmental changes, successive institutional transformations, and the shaping and implementation of change.[5]

Zaccaro's synthesis of the leader literature also established that this creative conceptual complexity could not be a stand-alone metacompetency, but needed to be intertwined and interdependent with flexible, challenging, and "risk-propensity" behavioural complexity, as well as with what Zaccaro labelled "informational complexity" and "social complexity." One could conclude that these complexities constitute the collective challenge for which creativity is the necessary solution! Zaccaro put forward Robert E. Quinn's competing-values performance model as an excellent example of complex, conflicting, and competing demands on institutional leaders, and the consequent numerous (and creative) roles such leaders are required to balance.[6]

So, creativity in leadership is indeed important in the working environment and, as Zaccaro and other leadership gurus have stated, is equally important for institutions with professions, such as the profession of arms, embedded in them. Now, since creativity is important for military institutions, do creative and effective military leaders already exist? If not, can sufficient numbers of them be professionally developed to function appropriately as creative military leaders?

## **CREATIVITY — A BRIEF MILITARY HISTORY**

The Canadian Defence Academy refers to a Professional Development Framework (PDF) in its oversight of training and development.[7] (See the PDF chapter in this handbook.) This PDF provides the structure over which leader levels and requisite leader capacities are distributed. It consists of five metacompetencies — Expertise, Professional Ideology, Cognitive Capacities, Social Capacities, and Change Capacities — cross-tabulated with four leader levels: junior, intermediate, advanced, and senior. Within the Cognitive Capacities continuum, with its four leader levels, the requisite capacities for the first, junior, leader level are seen to be the analytical skills:

Analytical thinking and reasoning is at the theorems-and-practical-rules stage where the intentions are to utilize task procedures using simple theorems and simple scientific principles or laws. These principles, intertwined with relevant expertise, would not be dissimilar from a "cookbook" approach for practical troubleshooting or problem-solving, using limited innovation.

At the other end of the continuum of leader levels, the senior level, the leader appropriately is engaged in cognitive capacities that involve creative, abstract, innovative skills.

New-knowledge-creation capabilities are engaged to generate, organize, and manage the theory-based body of knowledge of a profession, and to function beyond the analytic capacities to work through the conceptual complexity of strategic expertise and professional ideology via systems thinking, cognitive mapping, and "metacognition," all of these being conceptual activities similar, perhaps, to knowledge creation at a post-graduate level.

Richard Hughes and Katherine Beatty, American researchers of military leadership, have emphasized that, unfortunately, strong operational leaders, having fought alligators daily at more junior and intermediate levels, can experience great difficulty shifting gears, finding focus, and bringing forward the cognitive skills required to address long-term, larger-picture, swamp-draining, institutional, strategic and inter-institutional challenges at advanced and senior leader levels. Hughes and Beatty saw these institutional leaders as very challenged to "let go of the day-to-day issues, even if they are potentially in conflict with the long-term issues."[8] They emphasized that, for complex challenges, analytical skills needed to be substantially supplemented by creative thinking skills:

* synthesis (creating a complex whole) in addition to analysis;
* nonlinear (disruptive change thinking) as well as linear;
* visual (imaging) as well as verbal;
* implicit (patterns and relationships) as well as explicit; and
* engaging the heart (committed, inspired) as well as the head.[9]

In 2002, then-Major-General Rick Hillier, subsequently the Canadian Forces Chief of Defence Staff (CDS), made a more direct but related observation, stating that the more junior Canadian military officers, in general, seemed less likely to seize the broader-scope initiatives than did their international peers, particularly the British and Dutch.[10] While these younger Canadian officers were good platoon commanders, they did not demonstrate the cognitive attributes to think beyond the immediate task, or about the bigger-picture elements. Hillier wanted all of his leaders, for example, to effectively synchronize the efforts of the CF, those of non-governmental and international organizations, and those of civil organizations with which the military operated, all in order to ensure broad national mission success, i.e., to function creatively at the more engaging and complex levels of cognition. As CDS, Hillier continued his CF transformation and change.

## **EVIDENCE OF STYMIED CREATIVITY**

Are these perceptions of military leaders and their conceptual capacities supported by applicable research? Would such studies identify the challenges of military members to acquire a creative capability and to address the potential professional development for such creativity? Fortunately, the research does provide enlightenment but, unfortunately, the research also indicates that much remains to be done for the development of creative military leaders. There are American, Canadian, and Australian military research examples that reflect this.

American Army lieutenant-colonels and colonels, over 250 successful career leaders engaged as executive students at the U.S. Army War College, underwent assessments as part of their professional development.[11] Results indicated the propensity for logical, analytical, instrumental, objective, cause-and-effect thinking, and judgment that was impersonal, decisive, sought closure, and reflected a high need for control. While these officers were less open with other people, they wanted people to be open with them. In a Canadian study, 70 military officers were participants in master of arts in leadership, or executive leadership, programs at a Canadian university.[12] They underwent testing related to their thinking styles and the impact of such on their leadership styles, styles that could have been in alignment with, and therefore reinforced by, the surrounding leadership culture. For this group, the preference in thinking styles was for logical and conservative thinking. As the authors stated, these results tended to support the popular belief with respect to hierarchical and structured organizations that foster a reliance on specific, rational, efficient, and impersonal thinking. These thinking skills are seen to leave little room for personal discretion or flexibility in problem-solving and, according to the researchers, could create organizational stagnation, a task-centred culture, and a lack of synthesis of separate ideas, elements, and concepts into something new. One of the least preferred work elements for this university-student military group proved to be "creative." The researchers concluded that, in order for such military leaders to increase the likelihood for successful institutional and cultural change, stronger relationship building, creation of a shared vision, and motivation for effective transformation, the CF's professional development programs for upwardly-mobile senior leaders would need to incorporate more attention to enhancing conceptual and creative skills.

Eric Stevenson, researcher of leadership effectiveness in the Australian Defence Force, worked with over 900 leaders at junior, middle, and senior levels who were in residence for 1–3 years of professional development.[13] One of the themes of this development was the generation of a sense of value for "intellectual capital" appropriate to operating effectively in joint and integrated military-civil environments. He investigated how the 907 leaders constructed and categorized their ideas on leadership and how these ideas, in turn, affected the efficiency of their decision-making and their leadership of subordinates. His results indicated that junior officers link transformational leader behaviour with effectiveness and satisfaction, while middle and senior leaders had become increasingly task-oriented, addressing only followers' basic needs and not their needs for intellectual stimulation, potentially generating followers' dissatisfaction with senior leadership. Stevenson has speculated that, in a military environment with task-focused responsibilities and established routines, ambitious officers would be less inclined to take risks, show initiative, or be creative. Promotions to higher rank and selection for senior courses would appear to reward such leader behaviours, perhaps generating what Stevenson termed a "command presence," while junior leaders in this particular research were of an impressionable age and still somewhat idealistic about their leader models. Stevenson expressed concerns that this significant sampling of senior officers, with their well-ordered perspectives of leadership, might recognize only some of the factors important in leadership and decision-making, whereas those with less structured, more transformational, more creative thinking styles would be capable of considering more or most of the important factors appropriate to their levels of experience, thereby generating a greater overall leader-effectiveness in mission success and a higher level of follower satisfaction.

## OVERCOMING STYMIED CREATIVITY

As Bernd Horn, Canadian military officer and leadership academic, articulated in 2000, the post–Cold War Canadian military comprehension of leadership was limited, archaic, ill-defined, and embedded in authoritative, hierarchical, top-down mindsets.[14] He emphasized that the traditional, experience-based approach to learning leadership, devoid of theory and effective professional development, was inadequate. Horn's work prophesized, and reflected the urgency for a 21st-century shift of the role of leadership from this earlier, narrow focus on task or mission experience to one with expansive leadership concepts — concepts like transformational, values-based, institutional, distributed, learning-organization-anchored, creative, and professional leadership — that reflected a broad responsibility for the alignment of values in support of a collective effectiveness. As Canadian leadership researcher Alan Okros has stated, such a conceptual shift has led to a significant amendment to CF leadership doctrine, including the adoption of a values-based leadership philosophy and the development of an institutional effectiveness model.[15] (The recent American, Canadian, and Australian research cited above, regrettably, would seem to indicate that the more senior leaders of these evolving and increasingly effective military institutions must first catch up and then stay ahead.) Okros stipulated that values-based leadership draws on the perspective that values represent what is centrally important in guiding decisions and actions. This, he states, requires new thinking about thinking (metacognition), increasingly possible when senior leaders and senior stewards of the profession of arms of any nation focus on redefining their own cognitive talents, as well as those of the next generation of leaders, rather than obligating themselves and peers, as well as new members, to comply with a predetermined mindset that no longer effectively and creatively addresses military requirements.

So, as Horn and Okros have well articulated, "It needs to be 'out with the old, in with the new.'" But how? What solutions exist or are needed beyond broad support for, and vague ideas about, military organizations becoming learning organizations[16] instead of remaining just knowing organizations, or beyond expansive adoption of transformational leadership throughout the military institutions, or beyond encouraging group or team interactivity in hopes of increasing the probability of creativity?[17] These may be foundational requirements in support of a culture of creativity and innovation and insight, but a more applied, focused attention to educating individual military leaders about creativity, and to enlighten them about creative processes and procedures, is necessary.

David Garvin, a professor at the Harvard Business School, addressed the dedication to and educating of creativity in his co-authored book, *Education for Judgment*.[18] He identified the best leader traits to be developed and to be encouraged in teams or among workers:

* to work at acquiring an extraordinary ability to adapt;
* to perform a balancing act [with competing values and demands];
* to avoid blacks and whites and to embrace greys in challenges and circumstances;
* to simultaneously consider both short-term and long-term goals;
* to simultaneously encourage creativity without losing control; and
* to support a culture that permits dissent in order to arrive at better decisions.

Garvin is well known for his use in classes of the Harvard Case Method for developing creativity, although it must be emphasized that his, or similar methods, are not exclusive to academic settings or higher levels of professional development. Methods in any work setting can include:

* orchestrating group participation to reach sound decisions with limited known facts at the time;
* having challenging multi-day strategic exercises (with "spanners" thrown in periodically to shake up the flow of creativity);

* encouraging decision-making processes that include conflict, debate, and minority views;
* supporting open discussions that can mitigate risk, "groupthink" and worst-case consequences; and
* leading the learning by asking questions and determining how participants came to think of their solutions.

All of this should appeal to upwardly mobile and/or conscientious senior military leaders. However, extracting meaning from a Harvard Business School graduate-level classroom and applying it to leaders in diverse military learning situations is not straightforward. On the positive side, efforts to do so extend over the last 10-plus years. Robert Murphy of the U.S. Army War College, at an international conference in 1998, advised how professional development of senior military executives already was moving away from normal pedagogical styles of traditional classroom learning to the andragogical styles appropriate to adult learners, utilizing case-method learning.[19] He identified activities that encouraged critical and creative thinking — experienced students providing classroom briefings, facilitators and not lecturers influencing the learning process, multi-day experiential exercises and simulations, subject matter experts as observers asking difficult questions in a wide open forum, and diverse cultural participants and advisors from around the globe, all of this making the case method realistic, challenging, and founded on the recognition of the commitment of experienced learners to stretch their expertise through creativity.

The "What Leaders Need to Know" article by Shalley and Gilson in *The Leadership Quarterly*, identified above, supports leaders at any level by providing specific guidance for them in their fostering of creativity in their teams and groups, and even with their peers and superiors:[20]

* Use selection criteria for members who favour higher dispositions to be creative among members in the team, group, platoon, etc., thereby facilitating team members' willingness and mutual support

to be creative. Address job–person congruence, i.e., when placing members in different jobs, consciously factor in whether individuals will fit well with the job and with the creativity level desired. Provide professional guidance in creative problem-solving, systems thinking, and content-specific skills needed to be more creative. Determine how to handle negative feedback to team members without harming a positive relationship of creativity;

* Generate a creative, supportive work context by exemplifying leader behaviour that lets individuals feel that environment and observe behaviour between leaders and employees, among coworkers and team members, and with others outside of the working situations. Creative and supportive expectations can be exemplified by sufficient resources being made available, by evaluations and rewards and recognition, and by a supportive climate that is perceived as fair;

* Recognize the closest factors to the day-to-day, job level, and (meaningful) work as supporting creativity, or not, such as the complexity and demands of the outcomes/outputs, while also recognizing that organization-level aspects (pay systems, holidays, health benefits) are unrelated to creativity. Recognize, as well, that, increasingly, work is more mobile, changing jobs is more frequent, and creativity, unfortunately, may be lost or jeopardized by turnover, postings, and novice team members;

* Communicate to workers that creativity is desired, that the setting of goals or stating of role requirements reflects an expectation of creative outcomes.

Modelling by leaders of creative types of behaviour (group activities, unfettered brainstorming, flattened organizational structures, group dynamics, and rewarding creative contributions) can send powerful messages;

* Support workers, subordinates, peers to be in contact with people of diverse skills, interests, specialties, cultures to interact with different functional groups across the organization, and to interact with all or most of their team members as well as members outside their own team. Try to generate a climate of creativity, or enhance the one in existence, across a broader spectrum, from team to department to directorate/regiment to the broader organizational environment.

With respect to military leadership applications, military researcher David Schmidtchen explored the development and application of creativity by incorporating it into the Australian Army's practice of Mission Command, a philosophy for overcoming inherent problems in the dynamic environment of warfighting.[21] Decision-making was to be dispersed maximally to the lowest levels with acknowledgement that a "rule-book" approach was inadequate and creativity was expected. Schmidtchen emphasized that this philosophy needed to be reflected in the culture and practices of the organization at every level and in every undertaking. He summarized the underlying principles as mutual trust and understanding, and acceptance of responsibility and risk, each being foundational in support of creativity, and he emphasized that a supportive culture would include decentralized decision-making, discretion about work-completion practices among subordinates, a flat organizational structure of work teams, fluid and flexible job descriptions, and authority correlated with individual skills and abilities instead of positional power. Schmidtchen emphasized that the responsibility rested with the leader(s) to generate the supporting culture and environment for creativity by recognizing those military members committed and motivated to such creativity and innovation, to encourage flexibility and adaptability in an unfamiliar context, to teach and practise creativity, and to accept errors, mistakes, and risks as part of creativity in progress. Leaders needed to be role models and to exemplify core values. Finally, as Schmidtchen emphasized, initiative in creative and innovative ways was as much a way of thinking as it was a way of doing.

## **REQUISITE LEADER CREATIVITY — THE BOTTOM LINE**

The chapter so far has been about complex leader challenges that need to be responded to through creative leadership. For many experienced leaders, be they senior or not, creative leadership would represent a transformation for which an elegantly simple, step-by-step methodology would be appreciated, even needed, by them in order to, literally, transform their conceptual processes toward greater creativity. For such leaders to do so, two issues are crucial: What action now is to be taken by both leaders and other team members? How might creative talent be developed and/or enhanced in leaders or members?

The first question is most substantially addressed by a Center for Creative Leadership book published specifically to address creative leader capacities or behaviours,[22] while the second question is addressed convincingly by Canadian Navy Commodore Jennifer J. Bennett in her work on (post-pedagogical) andragogical (adult-led) learning strategies and methodologies for professional development, in this case, military leader capacities required for complex and creative leadership.[23]

In the Center for Creative Leadership book, Charles Palus and David Horth identified the following behaviours for leaders, team members, followers, etc., at any level in institutions. They categorized these behaviours as "complex challenge competencies":[24]

> \*   Paying Attention: using multiple modes of perception, seeing or sensing things others might miss, looking for conflicting components, recognizing own and others' bias, concentrating on details;

* Personalizing: tapping into team members' unique life experiences and personal passions, such as hobbies, activities, personal quests, and identity, to gain insight and create energy;

* Imaging: making sense of complex information, communicating effectively using images, pictures, stories, metaphors to reduce that complexity, displaying data clearly;

* Serious Play: generating knowledge through free exploration, levity, play, experimentation, possibly using puzzles, a sport or game worth playing, bending rules;

* Co-inquiry: collaborative dialoguing across language, culture, functions in order to exercise critical thinking, building new perspectives and finding answers; and

* Crafting: synthesizing issues, events, actions, orchestrating the parts of work and life into integrated and meaningful wholes that appear complete, original, appealing, and creative.

For leaders committed to evolving toward creative leadership, such can commence through focused and structured courses attended in groups, by leader-led initiatives in team-based working groups, or by individual self-development. Any of these approaches would require emphasis on the professional development of the attributes, capacities, and competencies provided above. Bennett emphasized that leadership in successful organizations needed to be adaptable, innovative, and knowledgeable, requisite characteristics that would call into question the effectiveness of past professional development practices and programs.[25]

Bennett also emphasized that andragogical learning is necessary for highly motivated, broadly experienced, self-directed, and systems-thinking learners comfortable with, even insistent upon, active learning and increasing responsibility for their lives, careers, and development. Learner-centric, interactive, process-focused, critical-thinking learning involving case studies, role-playing, simulations, coaching, interactive mentoring, syndicate discussions, self-evaluation, and reflection are all effective components of andragogical learning. These broad learning strategies, in turn, can be integrated, intertwined, and totally focused on the complex creative leadership competencies identified above by the Center for Creative Leadership, as an intertwining of content and process. Leaders should recognize and extract the most crucial learning aspects of members' previous experiences that were successes or failures, for these personal learning aspects can be integrated with the active learning processes and complex challenge competencies listed above. Bottom line: this integration of relevant learning content with the most interactive learning methods constitutes a significant foundational aspect of creative leadership. It represents a must-use active-learning approach for finding "real but creative solutions to real problems."[26]

**SUMMARY**

American military researcher Dr. Leonard Wong, currently with the Strategic Studies Institute of the U.S. Army War College, employing the U.S. Army motto applicable to leadership and professionalism, wrote this, "A strategic leader must 'Be, Know, Do' just about everything."[27] By default, the argument then would be that, to be able to be, know, and do this "everything," a leader's requisite capacities would need to include expertise, social capacities, change capacities, a professional ideology, and, of course, a very broad and complex sophistication with cognitive capacities. This conceptual wherewithal would need to include creativity, permitting that leader to display creative leadership. This relationship, however, is not obvious or straightforward, as this chapter has emphasized.

It needs to be recognized that opportunities for creativity in a military organization will occur more in one segment of the junior–senior leader continuum than another. There is a predominant relationship between junior leadership and analytical cognitive capacities on one side, and between senior leadership and creative and innovative cognitive capacities on the other side. Accordingly, the business of creative leadership does in fact appear to be the purview of senior, executive, and institutional leaders. Not exclusively, of course because, for special circumstances and certain projects, in hierarchical teams or multicultural groups, or for military staff functions, as examples, participants could contribute to and welcome creativity in these problem-solving roles, rather than leaving it all to senior and institutional leaders.

## NOTES

1. *Collins English Dictionary,* 6th ed. (Glasgow: HarperCollins, 2003), 393.
2. DND Executive and Manager Competency Leadership Continuum (Ottawa: Department of National Defence ADM [Human Resources-Civilian], 2002), 6.
3. *Collins English Dictionary,* 57
4. Stephen J. Zaccaro, *Models and Theories of Executive Leadership: A Conceptual/Empirical Review and Integration* (Alexandria, VA: United States Army Research Institute, 1996); Stephen J. Zaccaro, *The Nature of Executive Leadership: A Conceptual and Empirical Analysis of Success* (Washington, DC: American Psychological Association, 2001).
5. Stephen J. Zaccaro, *The Nature of Executive Leadership,* 136.
6. Robert E. Quinn, *Beyond Rational Management: Mastering the Paradoxes and Competing Demands of High Performance* (San Francisco: Jossey-Bass, 1988); and Kim Cameron and Robert E. Quinn, *Diagnosing & Changing Organizational Culture* (New York: Addison-Wesley, 1999). *Leadership in the Canadian Forces:*

*Conceptual Foundations*, Chapter 2, Annex A (see Figure 2A-1) explains Quinn's framework and how a CF-specific institutional effectiveness model, as an organizational adaptation of Quinn's generic model, was created. Available at http://www.cda-acd.forces.gc.ca/cfli.

7. Robert W. Walker, *The Professional Development Framework: Generating Effectiveness in Canadian Forces Leadership*, CFLI Technical Report 2006-01 (Kingston: Canadian Forces Leadership Institute, 2006), 29–30, http://www.cda-acd.forces.gc.ca/cfli.

8. Richard L. Hughes and Katherine C. Beatty, *Becoming a Strategic Leader* (San Francisco: Jossey-Bass and the Center for Creative Leadership, 2005), 17.

9. Ibid., 45.

10. Major-General Rick J. Hillier, *Leadership in the Canadian Army in the 21st Century*. Presentation at the Seminar "Canadian Army Leadership in the 21st Century," Kingston, 6–7 February 2002.

11. Herbert F. Barber, "Some Personality Characteristics of Senior Military Officers" in *Measures of Leadership* (West Orange, NJ: Leadership Library of America Inc., 1990).

12. Guy Nasmyth, Anne Schultz, and Tony Williams, "Thinking Skills and Their Impact on Military Leadership Practices," research paper completed for the Canadian Forces Leadership Institute, 2002, http://www.cda-acd.forces.gc.ca/cfli/researchpapers/.

13. Eric J. Stevenson, *Command Presence: Australian Military Officers' Mental Models of Effective Leaders* (Canberra: Australian Defence Force Academy, 2003), http://www.defence.gov.au/adc/cdclms/; and Eric J. Stevenson and James R. Warn, "Effective Leadership Development: Creating Better Mental Models." Presentation at Annual Conference of the International Military Testing Association, Canberra, Australia, 2001.

14. Bernd Horn, "Wrestling with an Enigma: Executive Leadership," in *Contemporary Issues in Officership: A Canadian Perspective*, ed. Bernd Horn (Toronto: Canadian Institute of Strategic Studies, 2000), 123–139.

15. Alan Okros, "Applying Complexity Science to Military

Professionalism: Recent Advances in Canadian Military Doctrine," (Kingston: Royal Military College Department of Military Psychology and Leadership, 2005), 6. Unpublished paper. See also Canada, *Leadership in the Canadian Forces: Conceptual Foundations* (Kingston: DND, 2005), http://www.cda-acd.forces.gc.ca/cfli.

16. Almost 15 years ago, the Harvard Business Review published articles on learning organizations. David Garvin defined a learning organization as an organization skilled at creating, acquiring, and transferring knowledge, and at modifying its behaviour to reflect new knowledge and insights. David A. Garvin, "Building a Learning Organization," *Harvard Business Review* Vol. 71, No. 4, (1993): 78–91.

17. Christina E. Shalley and Lucy L. Gilson, "What Leaders Need to Know: A Review of Social and Contextual Factors That Can Foster or Hinder Creativity," *The Leadership Quarterly* 15 (2004): 33–53. (During 2003 and 2004, *The Leadership Quarterly* published numerous articles that focused on creativity and leadership.)

18. See the Toronto *Globe and Mail* column by Diane Francis, "Best Leaders Are More like Mentors than Moguls," 6 October 2005. Original source is C. Roland Christensen, David A. Garvin, and Ann Sweet, eds., *Education for Judgment: The Artistry of Discussion Leadership* (Cambridge, MA: Harvard Business School, 1991).

19. Robert Murphy, "Educating U.S. Senior Military Leaders: Case Method Teaching in Action." Presentation at the World Association for Case Method Research and Application Worldwide Conference, Edinburgh, 1998.

20. Christina E. Shalley and Lucy L. Gilson, 47–48.

21. David Schmidtchen, "Developing Creativity and Innovation through the Practice of Mission Command," *Australian Defence Force Journal* 146 (January/February 2001): 11–17.

22. Charles J. Palus and David M. Horth, *The Leader's Edge: Six Creative Competencies for Navigating Complex Challenges* (San Francisco: Jossey-Bass and the Center for Creative Leadership, 2002).

23. Jennifer J. Bennett, "Effective Professional Development Strategies

for Institutional Leaders," in *Institutional Leadership in the Canadian Forces: Contemporary Issues,* ed. Robert W. Walker (Kingston: Canadian Defence Academy Press, 2007), 169–194.

24. Charles J. Palus and David M. Horth, most specifically 3, 187, 223–226.
25. Jennifer J. Bennett, 169–174.
26. David Giber, Louis Carter, and Marshall Goldsmith. *Best Practices in Leadership Development Handbook* (San Francisco,: Linkage Inc., Jossey-Bass/Pfeiffer, 2000), xvi.
27. Leonard Wong, Stephen Gerras, William Kidd, Robert Pricone, Richard Swengros, *Strategic Leadership Competencies* (Carlisle Barracks, PA: U.S. Army War College, Strategic Studies Institute, 2003). Available at http://www.carlisle.army.mil/research_and_pubs/research_and_publications.shtml.

## SELECTED READINGS

Canada. *Leadership in the Canadian Forces: Conceptual Foundations.* Kingston: DND, 2005. http://www.cda-acd.forces.gc.ca/cfli.

Hughes, R.L., and K.C. Beatty. *Becoming a Strategic Leader.* San Francisco: Jossey-Bass and the Center for Creative Leadership, 2005.

Matthew, L.J., ed. *The Future of the Army Profession.* New York: McGraw-Hill, 2002.

Palus, C.J., and D.M. Horth. *The Leader's Edge: Six Creative Competencies for Navigating Complex Challenges.* San Francisco: Jossey-Bass and the Center for Creative Leadership, 2002.

Quinn, R.E. *Beyond Rational Management: Mastering the Paradoxes and Competing Demands of High Performance.* San Francisco: Jossey-Bass, 1988.

Shalley, C.E. and L.L. Gilson. "What Leaders Need to Know: A Review of Social and Contextual Factors That Can Foster or Hinder Creativity." *The Leadership Quarterly* 15 (2004): 33–53. (From 2003 to 2004, *The Leadership Quarterly* published numerous articles on creativity, leadership, creative leadership, and professional development in support of creative leadership.)

Walker, Robert W., ed. *Institutional Leadership in the Canadian Forces: Contemporary Issues*. Kingston: Canadian Defence Academy Press, 2007.

Wong, L., S. Gerras, W. Kidd, R. Pricone, and R. Swengros. *Strategic Leadership Competencies*. Carlisle Barracks, PA: U.S. Army War College, Strategic Studies Institute, 2003. http://www.carlisle.army.mil/research_and_pubs/research_and_publications.shtml.

Zaccaro, S.J. *The Nature of Executive Leadership: A Conceptual and Empirical Analysis of Success*. Washington, DC: American Psychological Association, 2001.

# 13
# CULTURAL INTELLIGENCE
## by Emily Spencer

Over ten years ago, an American veteran of several foreign interventions observed of the U.S. military, "What we need is cultural intelligence." He continued, "What I [as a soldier] need to understand is how these societies function. What makes them tick? Who makes the decisions? What is it about their society that is so remarkably different in their values, in the way they think compared to my values and the way I think?"[1]

Indeed, the desire for cultural knowledge of one's enemy is as old as war itself. Yet, even in war, it is not only cultural knowledge of our enemies that we wish to have in order to maximize support for the war effort. In addition to knowing what makes the enemy "tick," it is also important to appreciate cultural nuances in the national, international, and host nation domains.

Cultural Intelligence (CQ) is basically the ability to recognize the shared beliefs, values, attitudes, and behaviours of a group of people and to apply this knowledge toward a specific goal. When applied concurrently in the national, international, host nation, and enemy domains, CQ can be an effective force multiplier. Importantly, demonstrating a lack of CQ in any of these domains can negatively impact a mission. It is thus important to understand what CQ is, how it applies to each of the four domains and what this means for leadership in the Canadian Forces (CF).

## DEFINING CULTURAL INTELLIGENCE

The literature on CQ can be divided along roughly civilian/military lines. The civilian literature, which is largely academic and management-focused, views CQ in terms of facilitating cross-cultural boundaries within and between businesses with profit as the major incentive. For obvious reasons, the military's use of CQ diverges from the civilian. Militaries generally view CQ as a force multiplier, and while both civilians and militaries use CQ to advance their particular needs, militaries may resort to a more pragmatic application. The U.S. Marine Center for Advanced Operational Culture Learning (CAOCL) is clear in its mandate: "CAOCL does not teach culture for its own sake, or for a non-directed appreciation of or sensitivity towards foreign peoples. CAOCL executes operationally focused training and education in individual training, PME, [professional military education] and pre-deployment phases, reflecting current and likely contingencies and functions, to ensure Marines and leaders deploy a grasp of culture and indigenous dynamics for use as a force multiplier."[2] Nonetheless, the two schools of thought are inextricably linked and each requires further explanation.

One of the leading authors on CQ in the civilian domain is scholar P. Christopher Earley. Earley and Elaine Mosakowski, in a 2004 *Harvard Business Review* article, describe CQ as an outsider's "… ability to interpret someone's unfamiliar and ambiguous gestures in just the way that person's compatriots and colleagues would, even to mirror them." They continue, "A person with high cultural intelligence can somehow tease out of a person's or group's behaviour those features that would be true of all people and all groups, those peculiar to this person or this group, and those that are neither universal nor idiosyncratic. The vast realm that lies between those two poles is culture."[3]

In a more complex analysis of CQ, Earley and Soon Ang define CQ as "a person's capability to adapt effectively to new cultural contexts." While slightly vague, they further explain that CQ has both process and content features that comprise cognitive, motivational, and behavioural elements.[4] Earley and Randall S. Peterson elaborate on this concept. They surmise, "At its core, CQ consists of three fundamental elements:

metacognition and cognition (thinking, learning, and strategizing); motivation (efficacy and confidence, persistence, value congruence, and affect for the new culture); and behavior (social mimicry and behavioral repertoire)."[5]

Other researchers have also explored the idea of CQ being composed of cognitive, motivational, and behavioural domains or similar variations of this triplex system. For instance, James Johnson et al. define CQ in terms of attitude, skills, and knowledge and David C. Thomas emphasizes knowledge, skills, and mindfulness.[6]

Most of this literature, however, prioritizes CQ as pertaining to other cultures and notably not one's own. Earley and Ang are clear when they state, "CQ reflects a person's adaptation to new cultural settings and capability to deal effectively with other people with whom the person does not share a common cultural background."[7] Indeed, they even goes so far as to suggest that individuals who are part of their own cultural in-group will find it particularly difficult to adjust to a new cultural setting, as it may be one of the first times that they experience alienation from the in-group and lessons learned in one culture may not be useful in another.[8]

This argument, however, ignores the support of, and reactions from, the home population. This is something that may work for businesses, but is not acceptable for militaries that serve democratic nations. The ability of an individual to understand the behavioural patterns, beliefs, values, and attitudes of their own society must remain an important aspect of the definition of CQ as it applies to the CF. Certainly, most of the military literature that discusses CQ recognizes this fact.

Seen collectively, the military literature underscores the need for CQ at the national, international, host nation, and enemy levels while also attempting to answer what it is about culture that we need to know for CQ to be an effective force multiplier. The current conflicts in Afghanistan (2001–) and Iraq (2003–) underscore the necessity of some form of cultural intelligence/awareness among military personnel serving overseas. The ongoing belligerence and soldiers' requests for more cultural knowledge is important to appreciate because it frames how definitions of CQ have been adapted to match militaries' needs and understandings.

A common theme that surfaces in accounts by soldiers serving in these areas is the need for a deeper understanding of host nation (HN) peoples. "The pitfalls presented by a different culture and an ill-defined, poorly functioning (or non-existent) local judicial, administrative, and political system," extols U.S. Marine Corps Major P.M. Zeman, "are enormous."[9] U.S. Army, Major-General (Ret) Robert H. Scales Jr. echoes these sentiments while describing the vital "cultural" phase of the war where "intimate knowledge of the enemy's motivation, intent, will, tactical method, and cultural environment has proved to be far more important for success than the deployment of smart bombs, unmanned aircraft, and expansive bandwidth."[10] Lorenzo Puertas, U.S. Naval Reserve, aptly notes, "Every war is a war of persuasion ... we must destroy the enemy's will to fight." He continues, "Persuasion always is culturally sensitive. You cannot persuade someone if you do not understand his language, motivations, fears, and desires."[11]

Additionally, in this global age of media, decisions by soldiers in remote areas can have far-reaching consequences for home and host populations, thereby highlighting the importance of CQ to mission success. Puertas illustrates this point by describing the potential consequences of one corporal and his decisions after being fired on in an alley in Iraq. "Without cultural training, his reaction will be a product of his personal experiences and beliefs," Puertas asserts. He adds, "He might have cultural misunderstandings that lead to serious errors in judgment. He might fail in his mission — and he might find himself despised by one poor neighborhood, or by a billion horrified TV viewers." He cautions, "Cultural knowledge of the battlespace should not be left to on-the-job training."[12]

To help mitigate this problem, western militaries are starting to define CQ and underscore important aspects about culture that can contribute to mission success. For example, the U.S. Center for Advanced Defense Studies defines cultural intelligence (CULTINT) as "the ability to engage in a set of behaviors that use language, interpersonal skills and qualities appropriately tuned to the culture-based values and attitudes of the people with whom one interacts." Culture is clarified as comprising "equivocal layers based on language, society, customs, economy, religion, history and many other factors."[13] Leonard Wong et al. describe cultural savvy, or in our terms, CQ, for their report to the U.S. Army War College as enabling "an officer [to] see perspectives outside his or her own boundaries." They explain, "It does not imply, however, that the officer abandons the Army or U.S. culture in pursuit of a relativistic worldview. Instead, the future strategic leader is grounded in National and Army values, but is also able to anticipate and understand the values, assumptions, and norms of other groups, organizations, and nations."[14]

A more thorough definition is provided by John P. Coles, Commander U.S. Navy. Coles defines CQ as "analyzed social, political, economic, and other demographic information that provides understanding of a people or nation's history, institutions, psychology, beliefs (such as religion), and behaviours." He explains, "It helps provide understanding as to why a people act as they do and what they think. Cultural intelligence provides a baseline for education and designing successful strategies to interact with foreign peoples whether they are allies, neutrals, people of an occupied territory, or enemy." Coles emphasizes, "Cultural intelligence is more than demographics. It provides understanding of not only how other groups act but why."[15]

In the end, an analysis of the literature concludes that CQ is the ability to recognize the shared beliefs, values, attitudes, and behaviours of a group of people and to apply this knowledge toward a specific goal. More specifically, CQ refers to the cognitive, motivational, and behavioural capacities to understand and effectively respond to the beliefs, values, attitudes, and behaviours of individuals and institutions of their own and other groups, societies, and cultures under complex and changing circumstances in order to effect a desired change. For the CF, in particular, CQ must be applied at the national, international, host nation, and enemy levels and the focus must be on its ability to be a force multiplier. The focus for the CF must be on providing the skill sets and knowledge to individuals so that they can exploit the benefits of CQ in order to achieve mission success.

Notably, CQ is different than cultural awareness. Learning limited language and social skills that pertain to a host nation's population may facilitate operating in an unfamiliar setting, yet cultural awareness alone does not provide sufficient skills to operate successfully in today's complex security environment. CQ, on the other hand, while encompassing aspects of cultural awareness, additionally provides some of the framework for dealing with the often competing demands placed upon soldiers in theatre by their home government (and general and flag officers) and the operational demands of their mission, their coalition partners, the social, political, and environmental challenges that they face while serving overseas, and the enemy. Notably, many CF members recognize that while cultural awareness pertaining to mission-specific regions is fundamental to military operations in foreign environments, there is a need for a template regarding CQ to which one can apply mission specific information.[16]

In essence, CQ provides a framework in which cultural-specific training can operate. CQ should be a process that is continuously developed within military personnel and prepares them for pre-deployment cultural specific training.

## **UNDERSTANDING THE FOUR CQ DOMAINS**

CQ empowers individuals to see "reality" through the eyes of another culture, specifically the one with which they are interacting. This ability, in turn, provides individuals the skills to be able to adapt their attitudes and behaviours in order to better influence the target audience to achieve specific aims. For the CF, CQ requires an appreciation of the role of the CF within the broader spectrum of Canadian society, the role the CF plays in multinational alliances, the complexities that may arise when operating in an overseas environment, particularly with host nation institutions and populations, as well as an in-depth understanding of the "enemy." Additionally, interactions between these domains must be recognized and understood. Indeed, CQ demands that all four domains are properly balanced.

*National Domain: Winning and Keeping the "Hearts and Minds" of Canadians*

For CF members, understanding the beliefs, values, and customs that comprise Canadian culture is important because the CF both represents and serves this culture. Applying this knowledge to maximize support for a mission from the Canadian public and within the CF is imperative for mission success. A military that serves a democratic nation cannot be fully successful if the will of the home population does not support the mission. The media in the 21st century exacerbates this point. Academic Robert W. Walker stresses that within the CF there is the need for postmodern military personnel to be warriors and technicians as well as scholars and diplomats.[17] As nation builders, soldiers act as ambassadors to their host nation, but, an often neglected point, and one that is perhaps more important, is that through this role, soldiers are also ambassadors to their home nation. CQ requires an appreciation of these connections.

*International Domain: Playing with Others — Military Coalitions, Inter-Governmental Organizations, Non-Governmental Organizations, and Host Nation Partners*

Another important domain in which CQ can be used as a force multiplier is in dealing with international organizations that desire the same general end state. Increasingly, the CF is called on to operate within military coalitions, Inter-Governmental Organizations (IGOs), such as the United Nations (UN) and the North Atlantic Treaty Organization (NATO), and to work in cooperation with other Canadian governmental departments and domestic and international Non-Governmental Organizations (NGOs). Additionally, CF members deployed overseas must often help train and work with HN partners. Unity of command is rarely possible in complex scenarios involving multiple players. Unity of effort, however, can and should be achieved. Understanding other organizations' cultural beliefs, values, attitudes, and behaviours, and appreciating how others see yours — thus CQ — facilitates the achievement of unity of effort.

*Host Nation Domain: In the Land of OZ — Applying CQ in an Unfamiliar Environment*

Another domain in which CQ is particularly valuable is that of the Host Nation. Aside from HN military forces, which are grouped together with other IGOs with which the CF deals intimately and on a regular basis, the HN domain incorporates all other HN groups including political, civilian, and other HN agencies. Notably, these groups themselves, while normally sharing a common national culture, will likely exhibit unique subcultures. The decision to group HN military forces with other IGOs is an example of the degree to which a subculture might supersede national culture. Of course, this relationship is unlikely to be black and white, existing rather in varying shades of grey. This ambiguity should be noted and dealt with on a case-to-case basis.

*Enemy Domain: Knowing the Enemy*

In war, it is important to know your enemy. As in each of the three other CQ domains, the overarching cultural grouping within the enemy domain can be further divided into subcultures (e.g. HN insurgents, foreign fighters, mercenaries, criminals). It is important to recognize that the enemy domain does not necessarily adhere to national borders. Yet, different national identities and cultures do contribute to the cultural composition of the enemy. Generally, however, the culture of the enemy is more than simply the sum of its component members. For example, suicide bombings have become a hallmark of the war in Afghanistan. Afghans, however, do not believe in suicide. Yet the Taliban is still able to use this tactic even though it is comprised of many Afghans.

It is also important to recognize that the enemy is not necessarily reflective of the HN. Appreciating that there is a difference between many HN members and the enemy, and that each group abides by a separate set of beliefs, values, attitudes, and behaviours is important for remaining focused on the task of winning the "hearts and minds" of local populations.

*Balancing the Balls: Interactions between the Four CQ Domains*

Balancing the four CQ domains so that CQ can be an effective force multiplier is the most important phase. This does not mean that people should be cultural chameleons as they jump between each domain; rather, individuals need to balance the knowledge that they acquire of each domain and apply it in a manner that allows them to further their goals, which, in the case of the CF, ultimately align with those of the Canadian government and population and are reflective of Canadian cultural values.

Balancing each of the four CQ domains helps to align the national discourse on war with the reality of war. Minimizing the gap between how Canadians perceive war and what is actually occurring maximizes the likelihood of being successful. Narrowing the distance between the discourse and the reality of war can be done, through a series of modifiers, in one of two ways: either the national discourse on war can change, or the reality of war can change. More often than not, each of these two components will occur, sometimes enlarging the gulf and sometimes reducing it.[18] Applying CQ in each of the four domains helps to assure that the distance between the Canadian discourse on war and the reality that CF members face when engaged in belligerence is kept to a minimum.

CQ requires that individuals know their audience so that they may exhibit appropriate behaviour in order to further their cause. In the end, their basic set of beliefs and values does not need to change but they may alter how they represent themselves and the emphasis that they place on certain behaviours. Balancing the four CQ domains is of critical importance because behaving appropriately in each cultural domain is essential for mission success.

## **IMPLICATIONS FOR CF LEADERSHIP**

The ability for CF leaders to be able to recognize what CQ is and to apply this knowledge as a force multiplier has many implications for mission success. CQ facilitates winning the "hearts and minds" of Canadians and HN populations, as well as the cooperation of military allies, IGOs, and NGOs. Moreover, it can also help retain the support of CF members. For example, a way that CQ has been applied effectively in Afghanistan is the attention provided to the wounded and the repatriation of those who have fallen. These acts let soldiers know that their leaders and their country value them. This recognition of sacrifice is important because it expresses Canadian cultural values associated with life and death and assures that all Canadian soldiers will be returned home with respect and dignity.

CQ can also be applied at the tactical, operational, and strategic levels. It is important when planning at any of these levels to be aware of the four CQ domains — national, international, host nation, and enemy — and to appreciate how they interact and contribute to mission success. Different levels of leadership may need to prioritize the attention paid to specific domains; however, a balance between all four is always necessary.

The CQ domain paradigm allows leaders to consciously address cultural gaps in knowledge with specific information concerning the various cultures they may face during operations. This can be done through a combination of strategies and methodologies, such as programmed cultural-awareness training; designated reading lists tapping scholarly studies, travel books, sociological and anthropological studies and literary works; discussions among peers and veterans with specific country experience; and through role-playing. Notably, when learning cultural specific information, it is important to try to see the world through the eyes of the group that is being examined. This skill will help with appropriate decisions and will contribute to an ability to shift others to another way of thinking.

## SUMMARY

In sum, CQ can be viewed as the skeleton upon which cultural-specific information can be hung. The more cultural-specific knowledge that one places on this framework, the more developed it will be. The more developed the structure, the more useful it will be as a force multiplier. Importantly, the four CQ domains — national, international, host nation, and enemy — should always be balanced.

## NOTES

1. Frank G. Hoffman, "Principles for the Savage Wars of Peace," in *Rethinking the Principles of War,* ed. Anthony McIvor (Annapolis: Naval Institute Press, 2005), 304.
2. U.S. Marines, CAOCL home page, http://www.tecom.usmc.mil/caocl/ (accessed 19 February 2007).
3. P. Christopher Earley and Elaine Mosakowski, "Cultural Intelligence," *Harvard Business Review* (October 2004): 139–140.
4. P. Christopher Earley and Soon Ang, *Cultural Intelligence: Individual Interactions Across Cultures* (Stanford: Stanford University Press, 2003), 59, 67.
5. P. Christopher Earley and Randall S. Peterson, "The Elusive Cultural Chameleon: Cultural Intelligence as a New Approach to Intercultural Training for the Global Manager," *Academy of Management Learning and Education* 3 (2004): 105.
6. James P. Johnson, Thomasz Lenartowicz, and Salvador Apud, "Cross-Cultural Competence in International Business: Toward a Definition and a Model," *Journal of International Business Studies* 4 (2006): 525–544; David C. Thomas, "Domains and Development of Cultural Intelligence: The Importance of Mindfulness," *Group and Organization Management* 1 (2006): 78–96.
7. Earley and Ang, *Cultural Intelligence*, 12.
8. *Ibid.*, 94.
9. P.M. Zeman, "Goat-Grab: Diplomacy in Iraq," *Proceedings*, November 2005, 20.
10. Robert H. Scales Jr., "Culture-Centric Warfare," *Proceedings*, October 2004, 32.

11. Lorenzo Puertas, "Corporal Jones and the Moment of Truth," *Proceedings,* November 2004, 44.
12. Puertas, "Corporal Jones and the Moment of Truth," 43.
13. Center for Advanced Defense Studies Staff, "Cultural Intelligence and the United States Military," in *Defense Concepts Series* (Washington, DC, July 2006), 1.
14. Leonard Wong, Stephen Gerras, William Kidd, Robert Pricone, and Richard Swengros, "Strategic Leadership Competencies," *Report,* U.S. Department of the Army, 7.
15. John P. Coles, "Incorporating Cultural Intelligence Into Joint Doctrine," *IOSphere: Joint Information Operation Center,* Spring 2006, 7.
16. Interviews conducted by Colonel Bernd Horn and Dr. Emily Spencer at CFB Edmonton, January 2007.
17. Robert W. Walker, *The Professional Development Framework: Generating Effectiveness in Canadian Forces Leadership* (Canadian Defence Academy: CFLI Technical Report, September 2006), 21.
18. This model is adapted from John A. Lynn, *Battle: A History of Combat and Culture* (Boulder: Westview Press, 2003), Appendix: The Discourse and the Reality of War: A Cultural Model.

**SELECTED READINGS**

Coles, John P. "Incorporating Cultural Intelligence into Joint Doctrine," *IOSphere: Joint Information Operation Center*, Spring 2006, 7–13. Also available online at http://www.au.af.mil/info-ops/iosphere/iosphere_spring06_coles.pdf.

Earley, P. Christopher and Soon Ang. *Cultural Intelligence: Individual Interactions Across Cultures.* Stanford: Stanford University Press, 2003.

Hernandez, Prisco R. "Developing Cultural Understanding in Stability Operations: A Three Step Approach," *Field Artillery,* January–February

2007, 5–10. Also available online at: http://sill-www.army.mil/famag/2007/JAN_FEB_2007/JAN_FEB_2007_PAGES_5_10.pdf.

McFarland, Maxie. "Military Cultural Education," *Military Review*, March–April 2005, 62–69. Also available online at: http://www.au.af.mil/au/awc/awcgate/milreview/mcfarland.pdf.

McIvor, Anthony, ed. *Rethinking the Principles of War*. Annapolis: Naval Institute Press, 2005.

Ng, Kok-Yee and P. Christopher Earley. "Culture + Intelligence: Old Constructs, New Frontiers," *Group and Organizational Management* 1 (2006): 4–19.

Pretraeus, David H. and James F. Amos. *Counterinsurgency*. Washington, DC: Headquarters, Department of the Army, December 2006, specifically, Chapter 2, "Unity of Effort: Integrating Civilian and Military Activities" and Chapter 3, "Intelligence in Counterinsurgency."

Salmoni, Barak A. "Advances in Predeployment Culture Training: The U.S. Marine Corps Approach," *Military Review*, November–December 2006, 79–88.

# 14

# CULTURE

by Karen D. Davis

In recent years, the relationship between culture and leadership has gained increasing significance along with the proliferation of research on organizational culture. Within organizational contexts, such as the military, culture has been held responsible for many shortcomings, including cover-ups, unethical decision-making, inability to adapt to changing circumstances, and discrimination against women and other under-represented groups in military organizations. Conversely, culture and culture change are held up as keys to the future. Organizations, including the military, strive toward numerous cultural goals including development as a learning culture, a fitness culture, an ethical culture, an adaptive culture, and a culture that values diversity. The potential implications for leaders are endless. The discussion below addresses the concept of culture and highlights some of those implications for leaders.

## **CULTURE DEFINED**

In recent years, culture has become, perhaps, one of the most frequently used and least understood concepts in the English language. Military historian Allan English notes that culture was first defined in the social sciences over 125 years ago and has since been defined in as many as 250 different ways.[1] In his history of human progress over 10,000 years, historical philosopher Ronald Wright defines culture from a technical anthropological perspective to include "… the whole of any society's knowledge, beliefs, and practices…" Culture, according to Wright, is everything:

> … from veganism to cannibalism; Beethoven, Botticelli, and body piercing; what you do in the bedroom, the bathroom, and the church of your choice (if your culture allows a choice); and all of technology from the split stone to the split atom. Civilizations are a specific kind of culture: large, complex societies based on the domestication of plants, animals, and human beings. Civilizations vary in their makeup but typically have towns, cities, governments, social classes, and specialized professions.[2]

Recent scholarship, for the most part, is not quite this ambitious or far-reaching. Many cultural projects focus on conceptions of culture at various levels of society, including national, community, and organizational/institutional domains, thus providing relatively well-defined units of analysis to facilitate understanding. However, culture is an intangible entity that permeates boundaries while being simultaneously influenced by entities both inside and outside of boundaries; that is, culture is not fully contained within defined structures and groups. For example, the military, as a national public institution, is embedded within the national "cultural" environment. In reinforcing the notion that "culture is the bedrock of military effectiveness," Allan English highlights the relationship between values, beliefs, attitudes, and behaviours in organizational culture.[3] However, such relationships can be quite complex as beliefs, attitudes, and behaviours vary between and within highly developed and differentiated organizations and societies. Societies contain multiple value and meaning systems associated with regions, ethnic groups, classes, occupational communities, and generations.[4] Given the complexities, it is important to acquire some understanding of culture beyond the boundaries of any particular structure or organization.

In addition, English argues that Canadian military culture is influenced by the inherent ambiguities of war and other military operations, the history, traditions, and experiences of Canada's Army, Navy, and Air Force, and American military values and ethics.[5] Extrapolating from this analysis it is fair to say that organizational cultures are influenced by the role or raison d'être of the organization, the history and traditions of the culture and its subcultures, and external significant organizations, including competition, allies, and other organizations representing the host nation.

In a cultural study of difference and identity in Canada, anthropologist Eva Mackey explores the meaning of national culture and the ways in which national identity and homogenous culture is encouraged through structure, narrative forms, symbols, community activities, national celebrations, and political processes. Rather than defining culture, she explores processes of nation-building and identity construction, concluding that although a "Canada first" core culture does exist in Canada, there is no core culture that all Canadians share. However, there is a shared *ideal,* in terms of both culture and identity, that spans multiple differences across cultures in Canada, and that contributes to a shared project within which various viewpoints and conflicts are negotiated and managed.[6] Importantly, Mackey's analysis highlights the importance of a shared ideal to the notion of a culture and an associated cultural identity.

Edgar H. Schein, a well-known expert on organizational culture, has warned practitioners against superficial understandings of culture and argued that in understanding organizational culture it is important to build upon the deeper, more complex, understandings that have developed within anthropology through the study of societies over the course of their history.[7] Keeping this in mind, he has noted that several major categories of overt phenomena have been associated with culture in organizations:

- observed behaviour regularities when people interact (language, customs and traditions, rituals);
- group norms (standards, values, what is fair);
- espoused values (publicly articulated principles and values);
- formal philosophy (policies, ideological principles);

* rules of the game (implicit ways of becoming an accepted member);
* climate (immediate surroundings, interactions with insiders and outsiders);
* embedded skills (special competencies, abilities not written down but passed from generation to generation);
* habits of thinking, mental models and/or linguistic paradigms (taught to new members as part of socialization process);
* shared meanings (created by group members); and
* root metaphors or integrating symbols (ideas, feelings, and images that group develops to characterize themselves).[8]

According to Schein, all of these phenomena relate to culture, but none of them *are* the culture of an organization or group.[9] In addition to various combinations of these phenomena, culture implies some level of stability in the group, as well as patterning or integration of the various elements of culture into a larger paradigm that lies at a deeper level;[10] that is, according to Schein, culture is deep, broad, and stable.[11]

Culture guides behaviour and action in ways that are not visible, in ways that are quite unconsciously taken for granted, and in ways that are frequently rooted in decades of historical practice and understanding. Understanding culture begins with an understanding of cultural artifacts: language (including jargon), ritual and ceremony, symbols, myths, and technology.[12] These attributes are of particular import because they comprise the patterns that set military culture apart from civilian society as well as other military cultures.[13] Cultural artifacts are representative of culture; however, as already discussed, many other factors and processes can influence culture. Borrowing from M. Alvesson and D. B. Billing[14] as well as Canadian Forces leadership doctrine,[15] culture, for the purposes of this handbook, is defined as: "A set of shared and relatively stable pattern of behaviours, values, and assumptions that have evolved over time, and are transmitted to new members of a group."

In general, there are three different ways to conceptualize culture and thus apply a definition of organizational culture. The integration perspective, which is the most common approach to organizational culture, places predominant focus on cultural homogeneity across the organization;[16] that is, focuses on the behaviours, values, and assumptions that are shared across an organization such as the Canadian Forces. The differentiation perspective places emphasis on the clarity and consensus that exists within the subcultures of the organization; however, the subcultures are considered to be only pieces of the whole.[17] Within the Canadian Forces, for example, the Army, Navy, and Air Force would comprise significant subcultures. The fragmentation perspective stresses multiplicity within the group or organization, and recognizes even small groups consisting of a few dozen people as a cultural entity.[18] For example, from a fragmentation perspective, the focus might be placed on various pilot cultures within the Air Force (i.e. fighter pilot, search and rescue pilot). Regardless of perspective, culture plays an important role in organizations.

**IMPORTANCE OF CULTURE**

As networks of human relationships, cultures provide stability and a sense of belonging or identity for its members. In fact, it has been noted that, "As a stabilizing force in human systems, culture is one of the most difficult aspects to manage in a climate of perpetual change."[19] Stability represents comfort in times of change; however, stability or the human need for stability can present a significant barrier to effective change. The greater the alignment between the various cultural influences embodied within an organization, the less likely that dominant cultural norms will create conflict. The historical stability of a culture will be particularly comforting for those who are strongly aligned with the culture in terms of the roots of their individual identities.[20] Indeed, as cultural sites in the construction of homogenous gendered identities,[21] Western military cultures have been particularly challenged in their attempts to integrate women and members of minority groups.

At a very general level, organizational culture unifies the institution and distinguishes it and its members from others. In this sense, culture is an important medium for developing pride and identity within an organization. Culture is the vehicle for the perpetuation of the organization and for the mitigation, synthesis, and rationalization of change. Much of the focus of the literature on organizational culture emphasizes its role as a tool for inclusion. Attention is paid to the cohesive aspects of culture "... as a defence against the unknown and a means of providing stability."[22] These elements of culture are valid and important. They provide parameters around the institution that allow people to operate effectively and comfortably within the system, socialize or regulate the absorption of new members, and maintain an identity over time. Simply put, without culture, it would be difficult to get things done. However, the other face of this aspect of culture, often ignored, is that it can function to exclude. The same processes and cultural elements that are brought to bear to achieve inclusion, cohesion, and team effectiveness can exclude others and thus impair team performance.

The bottom line is that organizational culture can be a very powerful tool for leaders. Military organizations, including the Canadian Forces, are extremely dependent on culture to fulfill many roles such as:

* defining boundaries;
* conveying a sense of identity for its members;
* creating commitment to something larger than individual self-interests;
* enhancing stability;
* providing the social glue that holds the organization together through setting appropriate standards for behaviour;
* acting as a control mechanism to guide and shape peoples' attitudes and behaviours; and
* helping members to make sense of the organization.[23]

However, it is also important to note that culture can be powerful in ways that are not aligned with organizational strategy and effectiveness outcomes, including the exclusion of valuable members and reinforcement of values, attitudes, and behaviours that are dysfunctional in changing environments. Leaders have a particular responsibility to step in and effect change when cultural practices and behaviours become dysfunctional.

## **LEADERSHIP AND CULTURE**

According to psychologist Edgar Schein:

> Organizational cultures are created in part by leaders, and one of the most decisive functions of leadership is the creation, the management, and sometimes even the destruction of culture ... If one wishes to distinguish leadership from management or administration, one can argue that leaders create and change cultures, while managers and administrators live within them.[24]

From Schein's perspective, leaders have the potential and the responsibility to manage culture, rather than simply perpetuate culture. As noted earlier, organizational cultures are also influenced by the broader cultures and environments in which they exist. Advocates of the environment perspective are somewhat skeptical about the potential influence of leaders, and instead believe that culture inside the organization is deeply connected to the outside world.[25] Anne Khademian, a political scientist specializing in public affairs and leadership, believes that the leadership and environment perspectives can be integrated; that is, "leaders can be influential in the development and change of culture, but they must look for, understand, and work with the environmental factors that influence and interact with the culture they seek to manage."[26]

Importantly, culture also determines the boundaries within which cross-cultural leadership development is possible.[27] The role of leaders is important in shaping, maintaining, and changing culture; developing future leaders of the culture; and influencing the relationship of the organization with other cultures and organizations, from both cross-cultural and interoperable perspectives. In the military context, for example, United States Major-General Robert H. Scales asserts that "… wars are won as much by creating alliances, leveraging non-military advantages, reading intentions, building trust, converting opinions, and managing perceptions — all tasks that demand an exceptional ability to understand people, their culture, and their motivation."[28]

The cultural competence required to be effective within and across cultures is dependent not only upon knowledge of other cultures, but also knowledge of yourself and the way in which those from other cultures will perceive your behaviours, including leadership behaviours.

In his discussion of cross-cultural issues in the development of leaders, organizational behaviour and leadership expert Michael H. Hoppe emphasizes the role of leadership both within and across cultures: "It is assumed that every society is engaged in leader development of some sort. However, what the notion means, how it is practiced, and where, when, and through whom it is done may differ significantly among cultures. Yet the need for leaders and their development is universal."[29]

In addition, Hoppe recognizes that the term *leader* itself — "or its equivalent in societies around the world — carries different historical, cultural, and political connotations."[30] Based on his analysis of leadership development in 10 countries, Hoppe has identified differences along six value dimensions:

* Individual/Collective: individual versus collective leadership development and responsibility;
* Same/Different: equality versus inequality as a variable informing who can lead and who can learn to lead;
* Tough/Tender: work to live versus live to work;
* Dynamic/Stable: acceptance of ambiguity and uncertainty versus need for stability and continuity;
* Active/Reflective: learn by doing versus learn through reflection and intellect; and

* Doing/Being: pursuing progress and improvement as a priority versus living in harmony with the universe as essential.[31]

Each of these dimensions will inform the priorities that leaders implement within organizations as well as the ways in which members of the organization learn to understand leadership in other cultures. The culture within an organization also determines the capacity of the organization to develop leaders within the organization across differences such as gender[32] and race.[33]

## **CLIMATE**

In terms of day-to-day direct influence, whether consciously or unconsciously, leaders play a critical role in setting the stage for cultural reinforcement and cultural change. As noted by Schein, the climate of the workplace, or, the immediate surroundings of the work environment, and the interactions between people is one phenomenon that is related to culture. Organizational climate can be altered in the short term, and is impacted by perceptions about reward and punishment, information flow, expectations about job performance, characteristics of the work, the fairness of the administrative system, and the example set by leaders.[34] In their study of army culture and climate in Canada, Colonel Mike Capstick, Lieutenant-Colonel Kelly Farley, Lieutenant-Colonel (ret'd) Bill Wild, and Lieutenant-Commander Mike Parkes noted that "culture determines how and why certain things are done in the organization," and *organizational climate* refers to: "… how people feel about their organization. Satisfaction with leaders, pay, working conditions, and co-workers are all aspects of climate. Oftentimes, climate is influenced by the underlying values and beliefs that comprise culture. Similarly, changes to climate can result in changes to the culture over time." [35]

Leader awareness of unit climate is an essential first step in deciding how best to influence climate. For example, unit climate data in the Canadian Forces has facilitated greater understanding of several essential components of military culture and climate: morale, task cohesion, social cohesion, confidence in leadership, experience of stress, and effective coping strategies.[36] In a discussion of identity and culture, Canadian Forces leadership doctrine establishes the relationship between military culture, climate, and identity, emphasizing:

> A wide range of customs and traditions associated with membership in the CF, including branch and environmental affiliations, form characteristics that bond members. These customs and traditions produce special social structures that contribute to a sense of unity and military identity. Leaders actively seek to build morale and cohesion by building on or establishing a unique identity for their unit. This identity is reflected in particular ways of doing things, distinct standard operating procedures, and sometimes variations in uniforms.[37]

Sub-group identity is important in reinforcing group morale and military ethos. On the other hand, excessive sub-group cultural identity in an organization can also serve to reinforce values, beliefs, and responses — to new and changing circumstances in particular — that are not aligned with the values of the organization and the environments within which it must operate.[38] Visible sub-group identifiers can represent the tip of the iceberg[39] in terms of the values, beliefs, and practices within a particular unit. Undue focus on unit distinctions may also create false cultural boundaries in terms of where and how culture is manifested within and across various parts of the organization. That is, as noted earlier, culture is not necessarily contained within defined structure, however tempting it may be to simplify cultural impact by linking various cultural characteristics to specific sub-groups within the organization.

It is important to understand this social construction of diversity within and among units in an organization, including a critical examination of the formation of underlying taken-for-granted assumptions, values, and practices that are considered legitimate.[40] This is central to understanding the extent to which subcultures reflect organizational values, and the potential impact on maintaining effective relationships within the organization, as well as with other organizations and cultures.[41] Leaders reinforce positive assumptions, values, and practices through leadership example and support of related activities in their workplace. On the other hand, leaders take decisive, visible action to put an end to practices based upon dysfunctional assumptions and values.

## SUMMARY

Organizations face multiple and complex cultural influences with varying relationships and implications for the organization. In fact, the inclusion of various cultural influences in an organization, and publicly funded organizations, particularly the military, is increasingly essential to its identity as a viable entity. This context, in addition to the increasingly complex relationships between ethnic, community, organizational, and national cultures in both the domestic and international contexts, including military operations, reinforces the importance for all leaders to understand culture, cultural processes, and cultural outcomes both within and external to their own organization.

## NOTES

1. Allan D. English, *Understanding Military Culture: A Canadian Perspective* (Montreal and Kingston: McGill-Queen's University Press, 2004), 15.
2. Ronald Wright, *A Short History of Progress* (Toronto: House of Anansi Press, 2004), 32.
3. English, 12.
4. See for example, W. Richard Scott, *Organizations: Rational, Natural, and Open Systems*, 3rd ed. (Englewood Cliffs, NJ: Prentice Hall, 1992), 124–149, for discussion of the environments within which organizations operate.

5. English, 55–70.
6. Eva Mackey, *The House of Difference: Cultural Politics and National Identity in Canada* (Toronto: University of Toronto Press), 2002.
7. Edgar H. Schein, *Organizational Culture and Leadership*, 2nd ed. (San Francisco: Jossey-Bass, 1992), 3.
8. *Ibid.*, 8–10.
9. *Ibid.*, 10.
10. *Ibid.*
11. Edgar H. Schein, *The Corporate Culture Survival Guide: Sense and Nonsense About Culture Change* (San Francisco: Jossey-Bass Publishers, 1999), 25–26.
12. Schema adapted from R.O. Parker, "The Influences of Organizational Culture on the Personnel Selection Process" (Ph.D. diss., York University, 1995), cited in Donna Winslow, *The Canadian Airborne Regiment in Somalia: A Sociocultural Inquiry* (Ottawa: Minister of Public Works and Government Services Canada, 1997).
13. Winslow, *The Canadian Airborne Regiment in Somalia*.
14. Cited in Joseph L. Soeters, Donna J. Winslow, and Alice Weibull, "Military Culture" in *Handbook of the Sociology of the Military*, ed. G. Caforio (New York: Kluwer Academic, 2003), 238.
15. Canada. *Leadership in the Canadian Forces: Conceptual Foundations* (Kingston: DND, 2005), 129.
16. Soeters, Winslow, and Weibull, 239.
17. *Ibid.*
18. *Ibid.*
19. Schein, *Organizational Culture and Leadership*.
20. Karen D. Davis and Brian McKee, "Culture in the Canadian Forces: Issues and Challenges for Institutional Leaders" in *Institutional Leadership in the Canadian Forces: Contemporary Issues* ed. Robert W. Walker (Kingston: Canadian Defence Academy Press, 2007).
21. See Iris Aaltio and Albert J. Mills, "Organizational Culture and Gendered Identities in Context," in *Gender, Identity and the Culture of Organizations,* eds. Iris Aaltio and Albert J. Mills (London and New York: Routledge, 2002), 3–18.

22. S. Rutherford, "Organizational Cultures, Women Managers, and Exclusion" *Women in Management Review* 16 (2001): 371–382.
23. This summary list of roles adapted from Stephen P. Robbins and Nancy Langton, *Organizational Behaviour: Concepts, Controversies, Applications,* 3rd Canadian ed. (Toronto: Pearson Prentice Hall, 2003), 349.
24. Edgar H. Schein, *Organizational Culture and Leadership.*
25. Anne M. Khademian, *Working with Culture: The Way the Job Gets Done in Public Programs* (Washington, DC: CQ Press, 2002).
26. *Ibid.,* 33.
27. Michael H. Hoppe, "Cross-Cultural Issues in the Development of Leaders," in *Handbook of Leadership Development,* 2nd ed., eds. Cynthia D. McCauley and Ellen Van Velsor (San Francisco: Jossey-Bass, 2004), 333.
28. Major-General Robert H. Scales, "Culture-Centric Warfare," *Proceedings,* October 2004, 33.
29. *Ibid.*
30. *Ibid.*
31. *Ibid.,* 336–337.
32. Marian N. Ruderman, "Leader Development Across Gender," in *Handbook of Leadership Development,* 2nd ed., eds. Cynthia D. McCauley and Ellen Van Velsor (San Francisco: Jossey-Bass Publishers, 2004), 271–303.
33. Ancella B. Livers and Keith A. Caver, "Leader Development across Race," in *Handbook of Leadership Development,* 2nd ed., eds. Cynthia D. McCauley and Ellen Van Velsor (San Francisco: Jossey-Bass Publishers, 2004).
34. Walter F. Ulmer, Jr., et al. *American Military Culture in the Twenty-First Century* (Washington, DC: CSIS Press, 2000), cited in English, *Understanding Military Culture: A Canadian Perspective,* 29–30.
35. Mike Capstick, Kelly Farley, Bill Wild, and Mike Parkes, *Canada's Soldiers: Military Ethos and Canadian Values in the 21st Century* (Ottawa: National Defence, Land Personnel Concepts and Policy, 2005).
36. Kelly M. J. Farley, *A Model of Unit Climate and Stress for Canadian*

*Soldiers on Operations* (Unpublished Doctoral Dissertation, Carleton University, 2002).

37. Canada, *Leadership in the Canadian Forces: Leading People* (Kingston: DND, 2007), 87.
38. See, for example, Donna Winslow, *The Canadian Airborne Regiment in Somalia: A Socio-Cultural Inquiry*. This was a study prepared for the Commission of Inquiry into the Deployment of Canadian Forces to Somalia.
39. An iceberg analogy is frequently used to describe the contrast between the most readily identifiable aspects of culture (e.g. symbols, language, rituals, stories) and those aspects that are hidden below the surface and thus much more difficult to identify (e.g. beliefs, values, assumptions). See, for example, Stephen P. Robbins and Nancy Langton, *Organizational Behaviour: Concepts, Controversies, Applications*, 3rd Canadian ed. (Toronto: Pearson Prentice Hall, 2003), 333.
40. See Stella M. Nkomo and Taylor Cox Jr., "Diverse Identities in Organizations" in *Handbook of Organization Studies,* eds. Stewart Clegg, Cynthia Hardy, Walter R. Nord (London: Sage Publications, 1996), 338–356.
41. Karen D. Davis, "Culture, Climate and Leadership in the Canadian Forces: Approaches to Measurement and Analysis," in *Dimensions of Military Leadership,* eds. Allister MacIntyre and Karen D. Davis (Kingston: Canadian Defence Academy Press, 2006), 311–336.

## SELECTED READINGS

Aaltio, Iris and Albert J. Mills. "Organizational Culture and Gendered Identities in Context." In *Gender, Identity and the Culture of Organizations,* edited by Iris Aaltio and Albert J. Mills. London and New York: Routledge, 2002, 3–18.

Capstick, Mike, Kelly Farley, Bill Wild, and Mike Parkes. *Canada's Soldiers: Military Ethos and Canadian Values in the 21st Century.* Ottawa: National Defence, Land Personnel Concepts and Policy, 2005.

Davis, Karen D. and Brian McKee. "Culture in the Canadian Forces: Issues and Challenges for Institutional Leaders." In *Institutional Leadership in the Canadian Forces: Contemporary Issues,* edited by Robert W. Walker. Kingston: Canadian Defence Academy Press, 2007.

English, Allan D. *Understanding Military Culture: A Canadian Perspective.* Montreal and Kingston: McGill-Queen's University Press, 2004.

Hoppe, Michael H. "Cross-Cultural Issues in the Development of Leaders." In *Handbook of Leadership Development,* 2nd ed., edited by Cynthia D. McCauley and Ellen Van Velsor. San Francisco: Jossey-Bass, 2004, 331–360.

Khademian, Anne M. *Working with Culture: The Way the Job Gets Done in Public Programs.* Washington, DC: CQ Press, 2002.

Mackey, Eva. *The House of Difference: Cultural Politics and National Identity in Canada.* Toronto: University of Toronto Press, 2002.

Nkomo, Stella M. and Taylor Cox, Jr. "Diverse Identities in Organizations." In *Handbook of Organization Studies,* edited by Stewart Clegg, Cynthia Hardy, and Walter R. Nord. London: Sage Publications, 1996, 338–356.

Schein, Edgar H. *Organizational Culture and Leadership,* 2nd ed. San Francisco: Jossey-Bass Publishers, 1992.

Soeters, Joseph L., Donna J. Winslow, and Alice Weibull. "Military Culture." In *Handbook of the Sociology of the Military,* edited by G. Caforio. New York: Kluwer Academic, 2003.

# 15
# DECISION-MAKING
by Bill Bentley

Doctrine teaches that decision-making is one of the military leader's primary responsibilities. In essence, the leader is charged with incredible responsibility. Not only does the profession of arms carry with it the "unlimited liability" associated with the possibility of laying down one's life, but leaders must also make decisions that could result in others paying the ultimate price. For the military leader, improving decision-making skills is not just a matter of professionalism, it can be a matter of life and death.

## DECISION-MAKING DEFINED

Decision-making is the process of developing a commitment to some course of action. Decision-making involves making a choice among several action alternatives. It involves more than simply the final choice among alternatives — we want to know how the decision was made in order to assess the validity of that choice. As well, the commitment mentioned in the definition above usually involves some commitment of resources — time, money, personnel.

In addition to conceiving of decision-making as the commitment of resources, decision-making also can be described as a process of problem-solving. From this perspective there are two broad classes of problems — well-structured problems and ill-structured problems. For a well-structured problem, the existing state is clear, the desired state is clear, and how to get from one state to another is fairly obvious. Because decision-making takes time and is prone to error, organizations and individuals attempt to program the decision-making process for well-structured problems. A program is simply a standardized way of solving a problem. As such, programs short-circuit the decision-making process by enabling the decision-maker to go directly from problem identification to solution. Programs usually go under labels such as rules, routines, standard operating procedures (SOPs), or even "rules of thumb."

The extreme example of an ill-structured problem is one in which the existing and desired states are unclear and the method of getting to the desired state (even if clarified), is unknown. Ill-structured problems are generally unique, that is, they are unusual and have not been encountered before. In addition, they tend to be complex and involve a high degree of uncertainty. As a result, they frequently arouse controversy and conflict among people who are interested in the decision.

The above description of an ill-structured problem is particularly apt when it comes to the modern battlespace. Military leaders at all levels today face highly complex, dynamic, and novel problem situations that they are called upon to resolve, but for which the known and practised solutions of doctrine will not suffice. These situations cover a wide range and variety, extending well beyond conventional combat. They are fundamentally social problems, comprising numerous individuals interacting in countless ways according to various motivations. Involving the interplay of human will, intellect, and creativity, these situations are essentially unknowable: no amount of information collection or analysis will reveal objective truth or provide the ability to predict events with certitude. Despite the most careful observation, these situations maintain the ability to surprise. They change unpredictably over time. Opponents adopt and quickly adapt methods designed to negate our advantages over them.

## DECISION-MAKING MODELS

Modern decision-making theories can be grouped into three categories: normative theories, so named because they describe how decision-making should be done; naturalistic theories, so named because they purport to describe and explain how individuals and organizations actually make decisions; and the systems thinking-systems analysis approach, designed to deal with highly complex human activity systems.

## **THE RATIONAL DECISION-MAKING MODEL**

The predominant normative theory involves the Rational Decision-Making Model. Formally, the model assumes perfect rationality. That is, it is a decision strategy that is perfectly informed, perfectly logical, and oriented toward economic gain. This model is based on the premise that human decision-making can be modelled in terms of formal processes predicted by normative theories of probability and logic. Of course, while perfect rationality is useful for theoretical purposes, the model rarely exists in the real world. The Nobel Prize–winning economist Herbert Simon recognized this reality and proposed that people must use the concept of "bounded rationality," rather than "perfect rationality."[1] That is, while they try to act rationally, decision-makers are limited in their capacity to acquire and process information. In addition, time constraints and political-social considerations act as bounds to rationality.

The (Bounded) Rational Decision Making Model can be depicted graphically as in Figure 1. The two particular concepts of framing and cognitive bias both illustrate the operation of bounded rationality. Framing refers to the (sometimes subtle) aspects of the presentation of information about a problem that is assumed by decision makers. A frame could include assumptions about the boundaries of a problem, the possible outcomes of a decision or the reference points used to decide if a decision is successful.

Cognitive biases are tendencies to acquire and process information in a particular way that is prone to error. These biases constitute assumptions and shortcuts that can improve decision-making efficiency but frequently lead to serious errors in judgment. A cognitive bias that contributes to incomplete information search is the tendency for people to be overconfident in their decision-making skills. This difficulty is exacerbated by

## Figure 1: The (Bounded) Rational Decision-Making Model

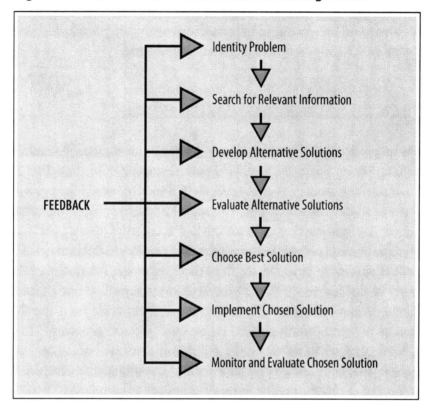

"confirmation bias," the tendency to seek out only information that conforms to one's own definition of, or solution to, a problem.

While the bounds of rationality often force us to make decisions with incomplete or imperfect information, too much information can also damage the quality of a decision. Information overload is the reception of more information than is necessary to make effective decisions.

The perfectly rational decision-maker can evaluate alternative solutions against a single criterion — i.e., economic gain. The decision-maker who is bounded by reality might have to factor in other criteria such as political acceptability and other members of the organization. The bottom line here is that the decision-maker working under bounded rationality frequently "satisfices" rather than "maximizes." *Satisficing* means that the decision-maker establishes an adequate level of acceptability for a solution and then screens solutions until s/he finds one that exceeds this level. When this occurs, evaluation of alternatives ceases and the solution is chosen for implementation.

## GROUP DECISION-MAKING IN THE RATIONAL MODEL

There are two basic brainstorming techniques in this model:

*Nominal Group Technique*

This technique is concerned with both generation and evaluation of ideas. Individuals in the group separately write down their ideas about a particular problem or decision. The group leader then lists each idea one at a time from each group member. These are then evaluated by the group as a whole.

*Delphi Technique*

The heart of Delphi is a series of questionnaires sent to respondents (experts in the field). Minimally, there are two waves of questionnaires. The first questionnaire is usually general in nature and permits free responses to the problem. The coordinator collates the responses and develops a second questionnaire that shares these responses and asks for suggested improvements. A final questionnaire might then be sent, asking respondents to rank or rate each improvement.

## THE CANADIAN FORCES' DECISION-MAKING MODEL

The CF's primary decision-making tools are the estimate and the operational-planning process (OPP). They are based on the Rational Decision-Making Model. The OPP consists of five stages: Initiation,

Orientation, Courses of Action Development, Plan Development, and Plan Review. According to Brigadier-General Stu Beare, a Canadian Army officer with extensive operational experience:

> Failure to grasp the true essence of an enemy, or to accurately visualize what he is both capable of and willing to do can lead, and has led, to unexpected or indeed, disastrous results. The OPP ignores the factors that determine an enemy's will to fight, and fails to consider the effects of will on enemy actions. The planning process assumes much with respect to a commander and staff's ability to predict enemy actions, principally by ignoring civil and political factors as well as the moral equation.[2]

To address these types of concerns in today's complex operational environment, considerable attention is now being paid to naturalistic decision-making theories.

## NATURALISTIC DECISION-MAKING THEORIES

Within the last 15 years, there has been recognition that the nature of the real world battlespace limits a human decision-maker's ability to implement truly analytical processes. This has led to the development of another approach that comprises what are called naturalistic or intuitive theories of decision-making. This approach is based on the premise that people use informal procedures or heuristics to make decisions within the restrictions of available time, limited information, and limited cognitive-processing capacity.

Three basic principles underlie naturalistic (intuitive) theories. The first is that decisions are made by sequential, holistic evaluations of potential courses of action against some criteria of acceptability rather than by feature-by-feature comparison of multiple alternatives along multiple dimensions. The second principle is that the decision-maker relies primarily on recognition-based processes to generate options and compare them to previous experiences (including actual on-the-job and training experiences). Thus, there is no exhaustive generation and comparison of alternatives. Instead, the decision-maker identifies potential courses of action by first assessing the situation, then recognizing past situations that are similar. Based on past experience, the decision-maker can recall previously taken courses of action and determine their acceptability to the current situation. The third principle is that decision-makers adopt a "satisficing" criterion, stopping the search when an acceptable course of action is identified rather than searching for an optimal solution. Real-world situations often demand very rapid responses and decision-makers may have to accept a solution that merely works, without considering whether a better solution exists.

An influential example of such a naturalistic theory is Gary Klein's Recognition-Primed Decision-Making Model. Klein conducted a significant program of research on decision-making in the military environment. He believes that the dynamic, uncertain, complex, and vague situations military leaders often find themselves in are not well suited to applying the rational models of decision-making. Klein suggests that decision-makers arrive at their decisions by matching the scenario they face to one previously experienced. If the solution arrived at is plausible, it is adopted, but it is not necessarily optimized as normative theory would require.

The Recognition-Primed Decision-Making Model (RPD) described by Klein has the advantage that most of us can see ourselves using it. The RPD Model describes how decision-makers can recognize a plausible course of action (COA) as the first one to consider. A leader's knowledge, training, and experience generally help in correctly assessing a situation and developing and mentally "war-gaming" a plausible COA, rather than taking time to deliberate and methodically contrast it with alternatives using a common set of abstract evaluation dimensions.

John F. Schmitt and Gary Klein developed the Recognition-Planning Model (RPM) from research on the RPD Model to codify the informal and intuitive planning strategies used by the U.S. Army and the USMC.[3] The RPM is illustrated in Figure 2.

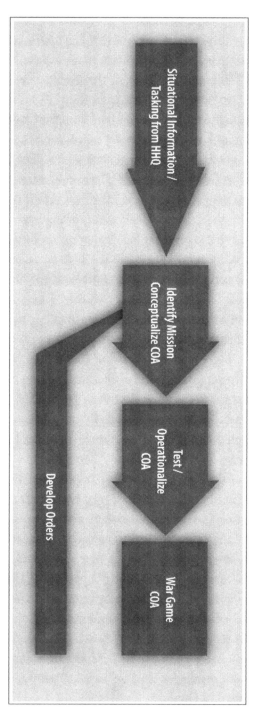

Figure 2: The Recognition Planning Model

## NARRATIVE-BASED MODELS

The decision-maker tries to construct a causal model for the situation in which s/he finds her/himself. A story is constructed to explain the situation and is then used to make a forecast about the future. The decision-maker chooses the option that best fits the model s/he has constructed. Clearly, in creating the story to explain events, decision-makers will be limited in their ability to do this by their view of the world. Personal experience, organizational setting, and broad cultural factors can influence an individual's ability to build the causal model that s/he uses to make the decision.

## SYSTEMS THEORY/SYSTEMS THINKING AND "WICKED" PROBLEM-SOLVING

After the Second World War, a powerful strand of systems thinking developed (Operational Research/Systems Engineering) now known as hard systems thinking. This approach is concerned broadly with engineering a system to achieve its objective. This approach had much influence on organizational theory as exemplified by Herbert Simon in his book, *The New Science of Management Decision*.[4] Overall, Simon sought a science of executive decision-making.

An alternative and richer perspective, developed in the 1980s and 1990s as an approach to intervention in human affairs, was called "soft systems methodology" (SSM). The usual distinction made between the two, hard and soft systems thinking, is that hard systems thinking tackles well-defined problems whereas the soft approach is more suitable for ill-defined, messy, or so-called "wicked" problems. Military problems, particularly at the operational and strategic levels, are decidedly "wicked." Another important distinction is that the soft approach also assumed that the process of inquiry into problematic situations can also be systematic. Thus, "soft systems methodology" is itself a system.

The intention of SSM is to make decision-making more transparent and explicit. Information about the problem and assessments of this information are captured in "rich pictures," which are used to express what is relevant and significant to the organization's decision-making. The "rich picture" leads to the "root definition" — a succinct encapsulation of the problem from a particular perspective. Root definitions are used as the basis to build conceptual models of fundamental activities that focus analysis on potential actions to improve the situation. By taking account of multiple perspectives and the messy nature of the real world, SSM seeks to find solutions that are both systematically desirable and culturally feasible. SSM involves managing debate between people so that learning may be facilitated, ideas evaluated, and plans for action developed.

Systemic Operational Design (SOD) is an alternative operational (campaign) design methodology derived from soft systems thinking in general and SSM in particular. SOD views operational design as a sub-component and key enabler of operational art that functions cooperatively with planning and execution through a continuous cycle of design, plan, act, and learn. SOD is a commander-centric, discursive approach to operational design that facilitates operational planning and execution by developing and articulating a hypothetical systems framework and logic within which planning can proceed. This implies a unique view of design and its relation to planning, a view that is consistent with the complex (wicked) nature of today's security environment.

SOD is a spiral and associative process that is comprised of seven steps of related discourse that build upon and inform one another. The seven steps lead to the articulation of a holistic operational design that enables detailed planning and execution.

The first step in SOD is "Systems Framing." The aim of Systems Framing is to rationalize the strategic directive and identify what has changed — what is the cause for intervention? The second step is "Rival as Rationale." The aim of this step is to identify those elements of the system — actors, strategies, and artifacts — that oppose the desired system trend expressed in the strategic guidance. The third step, "Command as Rationale," is to examine the tension that exists between the current command structure and that required by the emerging design based on the system, the rival, and the logic of the strategic directive. The aim of step four, "Logistics as Rationale," is to examine the tension that exists between the existing logistics system and that required by the emerging design. The fifth step, "Operational Framing," marks the transition from strategic logic to operational form and from problem-setting to problem-solving. Operational Framing identifies those broad conditions that, if achieved, would enable the operational form and logic to move the system in the direction desired by the strategic sponsor. The aim of sixth step, "Operational Effects," is to examine the identified conditions within the established systems logic that, if achieved, may transform the system in a positive manner in relation to the strategic directive. The Operational Effects discourse seeks to identify forms of manoeuvre that will generate effects in support of the broad conditions identified during the Operation Framing discourse. Finally, the seventh step, the "Forms of Function" discourse, is a translation of operational logic in the forms of conditions and effects into task and purpose in the form of a directed course of action (COA).

Systemic Operational Design is not a panacea for the challenges of contemporary operational design challenges. However, it does offer a viable alternative to the Operational Planning Process methodology, based on theoretical underpinnings that take the complex indeterminate nature of the contemporary operating environment and emerging science into account.

## SUMMARY

In modern war and conflict conducted at the tactical, operational, and strategic levels, there is a requirement for a mix of decision-making models and techniques. At the tactical level, problems tend to be structured and linear and, therefore, Rational Decision-Making approaches are predominant. Current thinking suggests that the Recognition Primed Model within this category is the most effective. Unstructured or "wicked" problems predominate at the operational and strategic levels. These problems are characterized by non-linearity and unpredictability. Under these circumstances, some versions of systems thinking decision models are required.

## NOTES

1. Herbert Simon, *The Sciences of the Artificial* (Cambridge, MA: MIT Press, 1996), 132.
2. Brigadier-General Stu Beare, Presentation at Canadian Forces College, Toronto, 17 May 2006.
3. Gary Klein and John Schmitt, "The Recognition-Primed Decision Model," *Military Review*, July 2004, 34.
4. Herbert Simon, *The New Science of Management Decision* (New York: Harper, 1960).


## SELECTED READINGS

Bryant, David, Robert Webb, and Carol McCann. "Synthesizing Two Approaches to Decision-Making in Command and Control." *Canadian Military Journal*, Summer 2003.

Checkland, Peter. *Systems Thinking, Systems Practice*. New York: John Wiley and Sons, 1999.

Johns, Gary and Alan Saks. *Organizational Behaviour*. Toronto: Prentice Hall, 2005.

Klein, Gary, et. al., eds. *Decision Making in Action: Models and Methods*. Norwood, NJ: Ablex Publishing, 1993.

Klein, Gary. "Strategies of Decision-Making." *Military Review*, May, 1989.

Klein, Gary and John F. Schmitt. "The Recognition-Primed Decision Model." *Military Review*, July 2004.

Knighton, Wing Commander R.J. "The Psychology of Risk and its Role in Military Decision-Making." *Defence Studies*, Vol.4, No. 3 (Autumn 2004).

McKown, Major Lyndon. "Improving Leadership Through Better Decision Making: Fostering Critical Thinking." A Research Paper Presented to the Research Department, Air Command and Staff College, March 1997.

Passmore, Major J.E. "Decision Making in the Military." *British Army Review*, No. 126 (Winter 2000–2001).

Schmitt, John, F. "A Systemic Concept for Operational Design." Paper given at the National Defense University, Washington, DC, 23 November 1997.

Simon, Herbert. *The Sciences of the Artificial*. Cambridge, MA: MIT Press, 1996.

# 16
# DISCIPLINE
## by Robert Edwards

Discipline is at the heart of the military profession and few professions are as dependent on discipline as the military. Whether during peace or war, it can spell the difference between military success and failure. Since the chief purpose of military discipline is to harness the capacity of the individual to the needs of the group, the probability of success on any particular mission will have a lot to do with the cohesion achieved by members of the group. The marshalling of individual wills and talents into a single entity enables military forces to face daunting challenges and great adversity. It enables military forces to achieve the unattainable through a sense of cohesion, unity of purpose, and concerted effort that would not be present in the sum of a collection of individual efforts. It also promotes effectiveness and efficiency in the military. Discipline seeks to elicit from individuals their best and most altruistic qualities. Its foundations are respect for leadership, appropriate training and expertise, and a military justice system where equity and fairness are clear to all. This chapter will highlight current doctrine and subject matter expertise on discipline.

## **WHAT IS DISCIPLINE?**

The word *discipline* as defined in the *Oxford Concise Dictionary* refers to "a sense of training of the mind and character as well as conforming to a system of rules for conduct."[1] In general, the word *discipline* would appear to have a distinct meaning when associated with the military, as opposed to its application to society at large. In the larger societal context, discipline has come to mean the enforcement of laws, standards, and mores in a corrective and, at times, punitive, way. However, it should be understood that the more important use of discipline in the military entails the application of control in order to harness the collective efforts of the group. The basic nature of discipline in its military application is more positive than negative, seeking to actively channel the individuals into a collective effort, thereby enabling the use of military force to be applied in a controlled and focused manner. Nevertheless, military force normally involves some level of aggression and violence. Hence, the means to control such military violence so that the right amount of force can be applied in exactly the right circumstances inevitably requires strong military discipline.

Military discipline involves psychological and ethical training toward voluntary and swift compliance with a system of rules of behaviour. In the defining document for Canada's military profession, *Duty with Honour: The Profession of Arms in Canada*, discipline is a fundamental Canadian military belief that is fully incorporated into the military ethos of the Canadian Forces (CF). The value of discipline, along with other fundamental Canadian military beliefs, forms a unique set of military values that make up the military ethos. This unique set of values, or military ethos, provide the general guidance required to conduct military operations in a style and manner that will earn soldiers, sailors, and air members of our military a highly regarded military quality — honour. Honour itself flows from practising the military ethos, and discipline still remains at the heart of executing military duty with honour.[2]

Yet military discipline still remains intrusive. The control of personnel is much more pervasive than in civilian society. Even in today's modern age, military discipline requires significant personal, physical, and mental rigour. Military discipline is intrusive in the sense of control required on one's physical body, actions, and attitude. It is necessary for military members to train for the right attitude or state of mind that will prepare them for the demands of combat. Accordingly, great emphasis is placed on mental alertness and physical control at all times, along with adherence to orders that will enable combat to be conducted in a way that reflects the military ethos. Of course, all this will require great strength of character so that military members conduct themselves in an exemplary manner twenty-four hours a day.

Discipline seeks to draw out the best of individuals, ideally relying on their sense of cooperation and teamwork to support the group and react effectively while under fire or in other crisis situations. This type of intrusive and rigorous discipline requires significant training; hence discipline is imposed in no uncertain terms on recruits. Recruits not only receive rigorous physical training, which enhances their physical strength, but they also are subjected to precise mental and physical control of their actions. However, the goal of effective military discipline is to gradually bring individuals to a point where, of their own volition, they willingly control their own conduct and actions according to military regulations and the values incorporated in the military ethos.

The traditional purposes of military discipline are to control the armed forces to ensure that it does not abuse its power, to ensure that members carry out their assigned orders efficiently and effectively — particularly in the face of danger, and to assimilate the recruit to the institutional values of the military. Military discipline is defined as "the military exercise of legal and coercive powers to control the behaviour of its members."[3] It should be noted that control of subordinates' conduct must conform to legal and other professional military norms. Furthermore, essential to military conduct is the increased reliance of self-discipline as opposed to imposed discipline in the development of military training. In other words, the overall aim of the rigorous military discipline imposed on recruits and junior members of the military is to develop their own sense of self-discipline (individual acceptance of the requirement to obey) so that they can be relied upon to act in accordance with general instructions without constant supervision.

## IMPORTANCE OF DISCIPLINE

The chief purpose of military discipline is to harness the capacity of the individual to the needs of the larger group. The sense of cohesion that comes from combining the individual wills of the group members gives unity of purpose to the group. It also allows large segments of armed forces to be strictly controlled in the conduct of operations in a manner that reflects the military ethos. Discipline is particularly important to military forces, which are routinely required to apply force or the threat of force in the course of their duties. Since the military holds the potential of significant destructive power on behalf of the state, it is reasonable to expect that they will be closely controlled. Moreover, all military members are subject to unlimited liability, and can be ordered into conditions that could result in their deaths or serious injuries. This readiness to fight, whether willing or otherwise, requires that the members of the military often suppress their own interests including, ultimately, the preservation of their own lives. During such times of mental and physical stress, it is essential that orders be clearly understood and carried out with the utmost dispatch. Vacillation or disagreement can result in chaos and disorder. Therefore, military discipline is key. The habit of obedience, as well as the presence of self-discipline that results from sound training, is not only essential in the execution of orders, but it can also spell the difference between courage and cowardice. When faced with danger, people naturally experience fear, which can lead to panic and clouded reasoning.

For example, the ship's company of a warship is carefully trained to follow strict military discipline in the event of a "sailor overboard" at sea. Individuals do not act unilaterally, instead they follow a carefully developed procedure which incorporates the alert, sailors proceeding to pre-designated stations at all life-saving equipment, followed by a rigorous adherence to recovery procedures which enables warships to recover members from the sea in minutes. Such rapid action in the open sea at all times of the year can only be achieved with a combination of collective and self-discipline from the entire ship's company and could not be achieved if members were left to their own devices.

## **DISCIPLINE AND MILITARY LAW**

Since the government controls the military, it should be no surprise that laws help regulate it. The military in a democracy is unique in that it holds the most destructive power of the state. This leads to a large number of rules, regulations, and laws designed not only to control the armed forces, but also to assist in ensuring that the values of broader society are maintained within the social fabric of the military. In the case of Canada, military members are subject to the Code of Service Discipline as set out in the National Defence Act, which is approved by Parliament. The legislation incorporates the basic principles of service life, including being subject to unlimited liability, obedience to authority and the obligation to obey lawful commands, subordination to those in authority, enforcement of discipline, and the obligation to maintain the welfare of one's subordinates. The broader scope of military law and the unique requirements of discipline were acknowledged by the Supreme Court of Canada in McKay v. R:

> Without a code of service discipline the armed forces could not discharge the functions for which they were created. In all likelihood those who join the armed forces do so in time of war from motives of patriotism and in time of peace against the eventualities of war. To function efficiently as a force there must be prompt obedience to all lawful orders of superiors, concern, support for, and concerted action with, their comrades and a reverence for and a pride in the traditions of the service. All members embark upon rigorous training to fit themselves physically and mentally for the fulfillment of the role they have chosen and paramount in that there must be rigid adherence to discipline.[4]

The code of service discipline establishes the standards of conduct expected of members of the armed forces. The conduct is enforced in part through a system of service tribunals, which is the military substitute for the system of civilian courts. These include military courts martial and summary trials. A court martial is a formal military court presided over by a legally qualified judge. The powers of punishment open to a court martial exceed those available to an officer presiding at a summary trial. Summary trials are presided over by superior commanders, commanding officers of bases, units or elements or delegated officers, and are the predominant form of disciplinary proceedings, since the responsibility to maintain discipline falls most directly on the commanding officer. In order to maintain good order and discipline, it is important to treat subordinates fairly. Hence, it is important to respect the values of the Canadian Constitution and also procedural fairness as reflected in the historic legal doctrine that applies to fair decision-making. The principle of procedural fairness is also reflected in a number of provisions in the Canadian Charter of Right and Freedoms. Military members are not required to give up the rights and obligations of citizenship; however, they do assume additional legal and professional obligations, liabilities, disabilities, and rights under military law, including, of course, being subject to military discipline.[5]

The laws that govern military activity are not solely domestic. International law also regulates affairs between states and acts as a standard of conduct and morality between nations. Therefore, international laws such as the Charter of the United Nations, the Hague Conventions governing military operations, the Geneva Conventions and Additional Protocols I and II to the Geneva Conventions reflect the standard of conduct required by military forces. The military, its officers and non-commissioned officers being agents of the state, are bound by Canada to follow the provisions of international law when conducting military operations. A CF Code of Conduct,[6] which acts as a guideline for members of the Canadian Forces in the Law of Armed Conflict, has been developed for all military operations. This code ensures that the conduct of members complies with the intent of all the international laws that Canada has agreed to respect.[7]

Since the habit of obedience requires a compliance with all but unlawful orders, no breach of orders can be overlooked. Failure to comply with even minor orders and regulations involves a lack of authority. If obedience cannot be ensured by willing compliance, then it must be enforced by corrective action. In some cases, the disciplinary problem can be addressed through administrative action. Minor breaches are usually dealt with by summary trial and result in the imposing of minor punishment. It is by correcting the minor breaches that compliance with all lawful orders is ensured and discipline is maintained. In becoming disciplined members of the armed forces, the individuals conform to the institutional requirements of the military, thereby making it more likely that they will fight when required to do so.

Military discipline requires that it be clear from the perspective of subordinates that all legal orders must be followed, including those orders that expose them to danger. Every officer and non-commissioned member (NCM) must obey the lawful commands and orders of a superior officer. Usually, there will be no doubt about such orders. However, even when the subordinate is unsure as to the lawfulness of an order, it should be obeyed unless it is manifestly unlawful. For example, an order to shoot unarmed children walking in a street in a village is a manifestly unlawful order and should not be obeyed. In a case where there are conflicting orders with a previous lawful command or order, an officer or non-commissioned member shall orally point out the conflict to the superior officer who gave the most recent command or order.[8] An order or command is any form of communication, written, oral, or by signal, which conveys instructions from a superior to a subordinate.[9]

## **HOW TO INSTILL DISCIPLINE**

The essential requirement for and the intrusive nature of discipline in the military is reflected in the wide-reaching responsibility for all members of the armed forces to ensure that discipline is maintained. No member of the military can be a bystander or remain neutral as to whether other members respect and obey authority. The acquiescence to insubordinate behaviour is harmful to the creation and maintenance of a habit of obedience. The obligation on all members of the military to enforce military law (and thereby obedience to orders and instructions) is reflected in the requirement for all officers and NCMs to become acquainted with, observe, and enforce the code of service discipline, regulations, rules, orders, and instructions that pertain to the performance of their duties.

The existence and maintenance of military discipline continues to be a very important ingredient in even the most modern and highly technical military units. In a unit where the discipline is high, the objectives of the group will be more readily attained. An example of the importance of unit discipline took place during the Korean War at Kowang-San (Hill 355) on 2 May 1953 when the 3rd Battalion of The Royal Canadian Regiment withstood an overwhelming and devastating Chinese attack. During this battle, Lieutenant Ed Hollyer and his platoon, amongst others in "C" Company, held their positions to the death, even to the point of calling in artillery fire on their own position. Their determined and disciplined defence provided the commanding officer with the time to organize a counterattack that retook the position, inflicted heavy casualties on the enemy, and forced the Chinese to withdraw to their side of no man's land. General West, commander of the Commonwealth Division, hastened to express his pride and admiration for all ranks of Third Battalion in their ordeal. Even though the unit had only just arrived in Korea and this battle was their baptism of fire, veterans of the battle credit the discipline of all ranks in the unit in directly contributing to this victory despite the loss of 2 officers and 23 NCMs killed, 27 wounded, and 8 others missing in action.[10]

The officer or NCM who has a working understanding of human behaviour is better equipped to obtain good performance and instill discipline. The leader recognizes that people are influenced by their attitudes, perceptions, and individual differences. An awareness of group behaviour will assist the leader to understand the leader–follower relationship and, therefore, to enhance his or her ability to influence and discipline subordinates. Discipline may be exercised through training, encouragement, authoritative direction and guidance, supervision, coercive feedback, and, punishment, if necessary.

Discipline does not simply consist of the imposition of punitive sanctions. The maintenance of discipline in Canada's modern armed forces relies on a blend of both collective discipline (imposed discipline) and self-discipline (individual acceptance of the requirement to obey or conform). Within each of the Army, Navy, and Air Force, the nature of military service may require a different blend of collective and self-discipline. The maintenance of discipline, and, in particular, the development of self-discipline, requires that a positive role be played by military leaders.

The basis of discipline training in the military is imposed discipline that new recruits receive in their first few months of service. The requirement for trained obedience demands both mental and physical effort. It consists in cultivating the necessary mental alertness so that the intent of the order can be grasped, and the requisite mental and physical reaction obtained to ensure that the order is carried out with dispatch.

In difficulty and danger, people tend to react in accordance with pre-established behaviour and, therefore, the execution of orders will become habitual even during periods of high stress, such as during combat conditions. Hence, recruits are taught to strictly obey commands, to work as members of a team, to dress smartly, to adhere to military standards, to be alert, to show initiative and, ultimately, to perform effectively on their own, guided by regulations, their new sense of military ethos, and self-discipline. Gradually, as more experience is attained, these recruits will come to accept the patterns of military discipline on themselves, so that they willingly conduct themselves according to all the values of a military ethos. Such self-discipline is generally considered to be based upon "pride in service, a willing obedience that is given to superior character and training, skill, education and knowledge."[11]

Leaders often will encounter a measure of resistance to the rigours of military discipline. Due to the constant demands of the work, it sometimes appears difficult to maintain a high standard of discipline over the course of military operations. Initially, the leader must continue to demand appropriate military discipline and counsel individuals or the group where appropriate. When resistance continues, one must resort to penalties as prescribed by the regulations. Repeated offences by subordinates also represent a breakdown in leadership to a certain extent. Wherever possible, it is best to concentrate on the preventative and to ensure full understanding of the need for discipline among subordinates, rather than taking remedial action after the fact. As indicated earlier, there is no substitute for sound discipline in a military unit. It could easily make the difference between cowardice and courage, as well as success or failure.

Discipline must be enforced fairly. A lack of fairness on the part of a superior can seriously undermine cohesion, morale, and discipline of subordinates, thereby impacting negatively on unit effectiveness.[12] A good leader sees that all orders apply equally to all personnel — there are no favourites. When there is an offence, steps must be taken to get all the facts. This includes listening to the subordinate's story before deciding on the nature of discipline to be applied. The leader should point out faults when they occur, but when this approach fails the leader should take appropriate action, depending on the seriousness of the incident. In the case of formal charges, the various military legal procedures must be followed. Subordinates should always understand the reasons behind orders wherever possible. Moreover, the leader should make the effort to find out why an order was not followed. In this way, the leader is able to give constructive advice and direction, which helps to build good discipline.[13]

There will be times in military operations when this is not possible, due to an urgent need for subordinates to execute orders without delay. Nevertheless, subordinates should understand the intent of orders wherever possible, including in military operations when feasible. A concerted breach of discipline by a number of subordinates is a serious matter and must be addressed immediately. Such breaches represent a serious breakdown in discipline, and also reflect on leadership within the unit.

All breaches of discipline will require an investigation. Depending on the circumstances of the case, simple breaches may only require an informal unit investigation. For more complicated cases, disciplinary investigations could involve a summary investigation or a board of inquiry. In other cases, the most appropriate investigating body could be the National Investigation Service or the military police. In all cases, the purpose of the investigation is to reconstruct events, gather evidence, ascertain the elements of an alleged offence, and identify those responsible. A timely and objective investigation may also provide the best record upon which to justify a decision not to proceed with charges where they are not warranted.[14]

Traditionally, discipline has (erroneously) implied punishment — a negative approach — but in military training, the emphasis is on the positive, constructive aspects of discipline. Nonetheless, the progression from imposed discipline to total self-discipline sometimes requires the administration of penalties. According to the logic of punishment that is administrated fairly, offenders ideally should see the justice of their being punished, learn from experience, and go on to better performance in the future because of a new understanding about the need for service discipline.

Unfortunately, this is not always the case. The member under punishment may not see the incident in the same light at all. For example, stewing about fancied injustices or severe punishments can create negative results in members' performance. Therefore, it is important "to make the punishment fit the crime" and, for that matter, the individual, as well. Every effort must be made to instill discipline in a positive way. This is done by maintaining discipline through personal example, skill, integrity, and professional knowledge. Such military members will exemplify professional military service as "duty with honour" in accordance with the Canadian military ethos.[15]

## **ROLE OF LEADERSHIP IN DISCIPLINE**

The role of leadership in military discipline is pivotal. Both officers and NCMs have specific obligations to instill discipline in their subordinates. The ability to command respect and to effectively control subordinates is based, in a large part, on the leadership ability of the officers and NCMs. The authority to maintain discipline at the unit level is concentrated in the hands of commanding officers. The decision to centre disciplinary power in the hands of the commanding officer reflects a balance between the level of responsibility and professional status of that officer on the one hand, and the degree of personal identification and control over subordinates on the other hand.[16] Among the factors affecting the ability to fight are primary group allegiances, unit esprit, and leadership. The enhancement of fighting effectiveness through primary group allegiance, unit esprit, and leadership is most effective at the lower levels of the military structure, where personal contact and allegiances are strongest. Hence, there is a need for strong military discipline in units.

However, discipline plays a vital role at all levels within the military. Too frequently discipline is left to the concern mainly of the lower levels, a matter primarily attended to by NCMs and needed only at unit level. But discipline is important for the proper functioning of the chain of command throughout the military. Undisciplined staff officers or commanders who hold themselves above the rigours of military discipline can do more harm to the collective effort of the military than can any soldier.[17] Officers and NCMs must be able to command respect of their subordinates not solely through their rank or appointment, but through their personal example, professional knowledge and expertise, skill and integrity.

The disciplined leader exhibits self-discipline, thus setting the example and professional tone for subordinates. The disciplined leader has planned and organized her or his task and mission, and knows what needs to be done and how to achieve the mission. Yet, the leader is open to suggestions and input from subordinates when and where appropriate. The leader should always make his or her expectations clear and in accordance with professional standards. Orders should always be enforced firmly but fairly. NCMs play an extremely important role in passing on values, instilling discipline, and bolstering morale. They act as the eyes, ears, and backbones of the disciplinary process.[18]

The special responsibility of officers and senior NCMs to maintain institutional standards means they will be held to a higher accountability for the same breach of discipline than lower-ranking members of the military. That is not to say that every transgression by an officer or senior NCM warrants formal disciplinary action. The greater number of decisions and increased requirement to exercise discretion, which comes with higher rank, often enhances the chances for mistakes. The corrective action taken must reflect the nature of the error. However, there is an increased sense of accountability to maintain sound discipline amongst subordinates, which all officers and NCMs must bear in the military.[19]

## DISCIPLINE IN THE NEW SECURITY ENVIRONMENT

The nature of CF operations continues to evolve within the new security environment. Military missions present new challenges, often involving complex situations in failing and failed states. "While the majority of missions abroad will involve peace support and humanitarian assistance," concludes a recent DND study, "current trends in the nature of conflict reinforce the need for a robust combat capability."[20] In addition, the pace of technology continues to grow, with much greater emphasis on sensing and precision-guided capabilities in an increasingly lethal operational environment. Furthermore, CF operations are encompassing a multitude of dimensions, which require coordination through a range of governmental and even non-governmental authorities. Many of these operations are taking place in well-populated areas where the risk of collateral damage is significant. Yet, there is an ever-increasing demand for accountability and scrutiny in every aspect of operations, which only serves to increase pressures on actions taken by CF members in the field.

At the same time, military members are better informed, better educated, enjoy a higher standard of living than ever before, and feel strongly about their rights and freedoms as a Canadian citizen. Merely "using" them without sincere interest in their welfare will meet with resistance. In addition, our troops are normally more distributed over the battlefield due to the extended capabilities of our weapons, hence are subject to more independent decision-making. But, there is still a need for discipline throughout the armed forces. Robust combat capability must be exercised through discipline. Orders must be followed. Our armed forces hold a significant inventory of lethal weapons on behalf of the state, and must exercise full control and use of these weapons according to government guidelines. The fact that operations have never been more complex and are no longer fought in isolated battlefields only exacerbates the need for discipline, particularly since troops are widely disbursed across the battlefield, and often operate in or near well-populated areas.

## SUMMARY

The key to maintaining discipline in the new security environment remains in strong leadership, appropriate training and expertise, and a military justice system where fairness and equity are clear to all.[21] In addition, the basis for military discipline must properly reflect the fundamental military values and unique beliefs codified in the Canadian military ethos.[22] Soldiers today are more suspicious and critical of motives of their superiors if it is not clear to them how their actions relate to the mission. In this sense, the commander's intent should be clear to all military forces. Wherever possible, leaders at all levels must assist in explaining the intent of orders to their troops, along with enforcing the discipline required to execute the mission cohesively in a concerted effort. In military operations, there always will be times where firm discipline, and direct and unambiguous orders will need to be given, followed, and executed. However, every effort must be made to motivate and lead subordinates to willingly accept discipline and practise self-discipline for the purpose of the mission. In this way, discipline allows for compliance with the interests and goals of the military organization while instilling shared values and common standards. Above all, leaders encourage the concept of self-discipline, which reduces the need for externally imposed discipline and supports independent action and initiative.[23]

## NOTES

1. Katherine Barber, ed., *The Canadian Oxford Dictionary* (Don Mills, ON: The Oxford University Press, 1998), 398.
2. Canada, *Duty with Honour: The Profession of Arms in Canada* (Kingston: DND, 2003), 32.
3. Canada, *Leadership in the Canadian Forces: Conceptual Foundations* (Kingston: DND, 2005), 130.
4. Canada, *Military Justice at Summary Trial* (Ottawa: DND, V2.1, 15 February 2006), 1–10.

5. *Ibid.*, 1–6.
6. Canada, *Code Of Conduct for Canadian Forces Personnel* (Ottawa: DND, B-GG-005-027/AF 023, 15 March 2007), 3–1.
7. *Ibid.*, 1–3.
8. Canada, *Queen's Regulations and Orders Article 19.015 and 19.02*, http://www.national.mil.ca/admfincs (accessed 16 March 2007).
9. North Atlantic Treaty Organization (NATO), International Standardized Agreements (Int Stand A), http://www.nato.int/docu/stanad/aap-6-2007.pdf (15 March 2007).
10. G.R. Stevens, *The Royal Canadian Regiment, Vol. 2* (London: London Printing Company, 1967), 266.
11. Canada, *Military Justice at Summary Trial*, 1–10.
12. *Ibid.*, 1–5.
13. Canada, *The Officer Handbook*, 62.
14. Canada, *Military Justice at Summary Trial*, 5–1.
15. Canada, *Duty with Honour*, 77.
16. Canada, *Military Justice at Summary Trial*, 1–13.
17. Canada, *Dishonoured Legacy: The Lessons of the Somalia Affair* (Ottawa: Public Works and Government Services, 1997), 431.
18. Canada, *The Officer Handbook*, 64.
19. Canada, *Military Justice at Summary Trial*, 1–14.
20. Canada, *Future Force: Concepts for Future Army Capabilities* (Kingston: Directorate of Land Strategic Concepts, 2003), 22.
21. Canada, *Military Justice at Summary Trial*, 1–10.
22. Canada, *Duty with Honour*, 54.
23. Canada, *Leadership in the Canadian Forces: Leading People* (Kingston: DND, 2007), 18.

## SELECTED READINGS

Canada. *Code of Conduct for Canadian Forces Personnel*. Ottawa: DND, 2007.

Canada. *Duty with Honour: The Profession of Arms in Canada.* Kingston: DND, 2003.

Canada. *Leadership in the Canadian Forces: Conceptual Foundations.* Kingston: DND, 2005.

Canada. *Leadership in the Canadian Forces: Doctrine.* Kingston: DND, 2005.

Canada. *Military Justice at Summary Trials.* Ottawa: DND, 2006.

Canada. *Officer Handbook.* Winnipeg: DND, 1985.

Canada. *Queen's Regulations and Orders.* Ottawa: DND, 2007.

Canada. *Dishonoured Legacy: The Lessons of the Somalia Affair.* Ottawa: PWGCS, 1997.

Horn, Colonel B. "A Timeless Strength: The Army's Senior NCO Corps." *Canadian Military Journal,* Vol. 3, No. 2 (Summer 2002).

# 17
# DIVERSITY
by Justin Wright

The issues of diversity and multiculturalism have become increasingly prevalent over the past few decades, especially in the context of globalization and increased engagement on the international stage. Diversity has gained particularly important status in Canada. For example, the creation of legislative acts such as The Employment Equity Act (1986) (EE Act) and The Multiculturalism Act (1988) have positioned Canada as the first country to adopt an official policy of multiculturalism. Canada's status as an officially multicultural nation is a reflection of the value that its government places on diversity, inclusiveness, and respect for the dignity of all people. It should come as no surprise, then, that as organizations necessarily operate within the context of their host nation,[1] these organizations are obligated and committed to the promotion and enactment of Canada's espoused values.[2] In this way, diversity impacts all forms of organizations in Canada, including the Canadian Forces (CF), which officially came under the EE Act in 2002. Thus, it is a leader's responsibility to embrace diversity and its associated values, to ensure those values remain an integral part of the Canadian workplace, and to recognize the various ways in which diversity can strengthen and enhance the nation's ability to remain competitive and successful at the international level.

## DIVERSITY DEFINED

The Directorate of Human Rights and Diversity (DHRD)[3] states that: "Diversity is any collective mixture characterized by differences and similarities or all the ways in which we differ. Diversity includes variations within a group such as race, ethnicity, religion, age or gender, encompassing differences in natural abilities, personalities and physical characteristics."[4]

In other words, the notion of "diversity" captures the various differences in the way that each of us is situated in the social world and in relation to one another, the sum of which makes up each of our unique biographies. In practical terms, every group has some degree of diversity, in that each individual within a given group is in some way unique from the other group members.

The applications of the principles of diversity are, importantly, distinct from Canada's Employment Equity legislation. Though related, Employment Equity is a legislative tool designed to bring fairness, or equity, to the workplace in regards to four specifically designated groups: Aboriginal peoples, women, members of visible minorities, and persons with disabilities.[5] In contrast, diversity is concerned with: "… adopting practices to manage diverse work teams while respecting and allowing for differences. Diversity is not restricted to members of the four designated groups; it is inclusive of everyone in the workforce. Diversity fosters a work environment that values flexibility and innovation in accommodating the varying needs of all individuals."[6]

Importantly, diversity is conceptualized as not only a way to be respectful of the differences of all people within our society, but is also viewed as a way of developing a more flexible, innovative, and adaptive working environment. Fostering innovation and flexibility in the workplace through diversity has become an important element for many Canadian organizations in a global context, including the proposed direction of the CF within a new security environment.[7]

In another example, conceptualizing diversity as an advantage rather than a burden has gained momentum in the field of pedagogical and educational studies. For example, author Jui-shan Chang asserts that incorporating cultural diversity into the core structure of her various university classes allows her students to greatly enhance their educational experience by asking them to approach the material from different cultural perspectives. Emphasizing the strengths of cultural diversity in these classroom settings also encourages her students to take pride in their own cultural identities and to respect the cultural differences of their peers.[8]

## **THE NEED FOR DIVERSITY**

As indicated above, diversity is important to Canadian organizations because they are a reflection of Canada and its espoused values and beliefs. In other words, diversity is of vital importance because, "unlike other liberal democracies, Canada's official multiculturalism and related policies enshrine diversity and associated operational practices both in its legal statutes and in its Constitution."[9] The CF, in particular, not only has a responsibility to uphold our nation's enshrined values and beliefs, but also "the legitimacy of the profession of arms requires that it embody the same values and beliefs as the society it defends."[10] As CF members come from and return to civilian society, the value placed on diversity in that society must be reflected in the institution that purports to defend those same values and the people who hold them. Moreover, the obligation of Canadian organizations to embody the values of diversity is mandated by law. Diversity, inclusiveness, and respect for the dignity of all people are reflected in such Canadian policies and legislation as: The Official Languages Act (1969), The Canadian Human Rights Act (1977), The Canadian Charter of Rights and Freedoms (1982), The Employment Equity Act (1986), The Multiculturalism Act (1988) and The Canadian Forces Employment Equity Regulations (2002).

Another reason why diversity is a key issue for the Canadian workplace is the shifting trends in Canada's demographics. First, Canada is now characterized by an aging population. As the "baby-boomer" generation approaches retirement age, staff shortages are projected to occur across the board, which will lead to lower levels of unemployment and increased challenges for organizational recruiters.[11] Compounding this dilemma is Canada's current economic growth and its correlated creation of new jobs, which makes professional recruitment still more competitive and difficult. For example, Royal Military College professor

Christian Leuprecht sums up this dilemma in a military context with the assertion that, "It is easier for the CF to recruit during macro-economic downturns and in geographic areas with high chronic unemployment levels than it is in times of high economic growth and in areas where unemployment levels are low."[12] The implication here is that the Canadian professional recruitment strategy has to expand and adapt to the current demographic trends in Canada — low unemployment, economic growth, an aging population, declining birth rates, and higher proportions of visible minorities in the population. In particular, declining birth rates will have an important impact on recruitment:

> In 1950, 92 percent of Canada's population growth was a function of natural increase. At the height of the baby boom, in 1959, the fertility rate averaged 3.9 children (live births) per woman ... Fertility currently registers at 1.6 children per woman and is projected to drop to 1.2 during the next two decades. To maintain its current population size, Canada will have to take in an estimated 260,000 immigrants per year with the result that, by 2050, over half the country's population will have been born abroad.[13]

In short, if the country's potential recruitment pool is being replenished by immigration rather than by procreation, then the face of the Canadian workforce will inevitably have to become more diverse in order to sustain itself. In light of this reality, it is clear that we must be prepared to shift toward an organizational environment that can not only manage a diverse workforce, but can also thrive on it.

## **IMPACT OF DIVERSITY**

Given the impact that diversity has and will continue to have in Canada, it follows that it must be given more attention if Canadian organizations are to remain competitive and successful. In the broadest sense, this can be understood in terms of cohesion; it is vital to foster a climate of trust and respect to ensure the cohesiveness of a team or workgroup, and in a group made up of diverse individuals, this includes a responsibility to demonstrate respect for all of those individuals regardless of their differences (i.e., gender, ethnicity, sexual orientation, religion, etc.).[14] However, it is also crucial for leaders to look beyond the view that diversity is something that must be overcome or "dealt with" and to begin to recognize that a more diverse work force can enhance efficiency and organizational success:

> Embracing diversity is about valuing individual differences and utilizing them for the benefit and efficiency of the team without compromising the organizational culture and mission. Individuals working for an organization fostering a strong sense of equal opportunity in an atmosphere free of discrimination are more productive and team oriented. They better understand their individual and group tasks and remain focused on the organizational mission. This results in strong unit cohesion, team effort and esprit de corps.[15]

One of the best examples of the unique ways in which diversity has directly enhanced organizational success or, in a military context, operational effectiveness, is the use of American Indian Code Talkers during the Second World War. These communication units were able to help the Allied Forces transmit virtually unbreakable coded messages through their unwritten native languages.[16] What this example illustrates is not that we should be looking for ways to exploit diverse groups, but by remaining open to the strengths of a diverse workforce, organizations will engender an environment in which diverse members can bring unique skills and solutions forward to help enhance efficiency and contribute to success.

Recent research in cross-cultural psychology has suggested that, "... the creative tensions associated with diversity may encourage mutual inspiration and facilitate learning. Diversity ensures richness of input that may facilitate creative and innovative work outcomes."[17] It is arguable that, in the context of today's multinational global market, the importance of a creative and innovative work force has never been more pronounced. In a military context, counterinsurgency focused operations and, in particular, the reality of the "Three Block War"[18] has led to the recognition of the importance of a unit's ability to remain flexible and innovative.[19] By being an inclusive society, Canadians can draw on the wider and more diverse cross-section of their population's human resources, and organizations such as the CF will be able to enhance their ability to remain flexible and to adapt in the new global security environment.[20] Moreover, one of the main issues in current counter-insurgency operations (e.g. Afghanistan) is the development of a positive relationship with the local population in-theatre. This campaign to "win the hearts and minds" of another culture may potentially be more successful if the CF were to focus more rigorously on insights derived from the perspective of a diverse force.

## MANAGING DIVERSITY: HOW TO CREATE A DIVERSE ENVIRONMENT

As discussed above, it is the mandate of our leaders to foster a climate of trust and respect among their staff, and to ensure that the dignity of all members of the organization is protected. Within the CF, this mandate is expressed as a key component of the Canadian Forces' military ethos.[21] Moreover, our leaders' responsibility to promote and uphold the value of diversity is a reflection of the obligation of all Canadian organizations to the values and beliefs of Canada.[22] The following discussion offers more detail on some of the major ways in which diversity impacts Canadians today.

*Ethnicity*

Although the term "ethnicity" is commonplace in the Canadian lexicon today, the concept of "ethnicity" is complex, and consists of a wide range of competing and contested iterations. Social anthropologist Richard Jenkins comments:

> The ancient Greek word *ethnos*, the root of "ethnicity," referred to people living and acting together in a manner that we might apply to a "people" or a "nation": a collectivity with a "way of life" — some manners and mores, practices and purposes — in common, whose members share something in terms of "culture." Thus the anthropologist Fredrik Barth (1969) defined ethnicity as "the social organization of culture difference." Ethnicity is not only a relatively abstract collective phenomenon, however: it also matters to individuals. To quote another anthropologist, Clifford Geertz, ethnicity is "personal identity collectively ratified and publicly expressed."[23]

This excerpt illustrates that the concept of ethnicity is both complex and broadly applied, which makes it problematic for organizational and governmental policy-makers in Canada. Understanding the nuances of ethnicity as a concept, let alone ethnic diversity as a practical way of life, is a very tall order. It is therefore not surprising that the conceptualization of ethnicity in Canadian policy tends to be narrowly focused. One of the most well-known ways of approaching the concept of ethnicity in Canadian policy is through the issue of visible minorities.

As outlined earlier, the demographics of the Canadian population are changing. One of the most rapidly growing components within the Canadian population is that of visible minorities. In a Canadian legislative context, the term "visible minority" is very specific, and refers to Black, Chinese, Japanese, Korean, Filipino, South Asian, Southeast Asian, West Asian, North African or Arab, Latin American, and persons of mixed origin.[24] However, it should also be noted that, in the academic literature on diversity, the ascription of the term "visible minority" or "ethnic minority" is not based on a difference in race alone, but is also a reflection of a difference in culture, language, and/or associated values and beliefs.[25] Leuprecht reports that, "Between 1986 and 2001, the proportion of Canada's visible minority population [as defined by the EE Act] doubled, to 13.4 percent."[26] To reiterate, declining birth rates and growing populations of visible minorities mean that the potential pool for recruitment has, and will continue to become, more diverse. If organizations are to maintain their required numbers, they can no longer afford to look to the traditional segments of the population alone for recruitment. This means that organizations must adapt so that they can properly accommodate and respect the needs of a workforce that will increasingly comprise visible minorities. Legally speaking, adapting to diversity means that leaders must ensure that their organizations adhere to the mandated regulations set out in the legislative acts cited above. Informally, leaders should strive to adapt by actively encouraging all employees to share their unique perspectives and experiences in order to stimulate a more flexible and innovative team and therefore contribute to efficiency and remaining competitive. This goes beyond an organization's attempt to achieve a representative, diverse workforce and demands that organizations need to include diversity as part of their strategy, policy, practices, and behaviours.[27]

*Gender*

The term "gender integration" has become deeply entrenched in Canadian culture. Gender equality today is considered a core Canadian value, one that has been hard-earned (see Chapter 22 on Gender). Author Bey Benhamadi reflects:

> The history of Canadian women has been a struggle for equality between the sexes. Several historic facts reveal the hard battle they have had. For example, there was the right to vote in federal elections, gained in 1918, which then spread through provincial assemblies, and the overturning of the ruling which prevented them from sitting in the Senate a few years later, when only 4 percent of women worked outside the home.[28]

Canada's stance on the rights of women, enshrined in the legislative acts and policies cited above, places us at the front of the international community. Yet, despite this progress, continues Benhamadi:

> ... there are still areas needing improvement, as can be seen from certain socioeconomic indicators: 70 percent of women are still concentrated in traditionally female sectors of activity and still remain the minority in the sciences and engineering. In addition, even in the case of university graduates working full time, in 1998 women only received the equivalent of 70 percent of the remuneration of their male colleagues. In brief, women more often than men occupy posts requiring fewer qualifications.[29]

In other words, the movement for gender equality in Canadian culture has gained considerable ground, but has yet to achieve its goal. For Canadian organizations to truly incorporate diversity into their strategy, policy, practices, and behaviours, leaders need to begin to recognize the various ways in which difference may be a source of strength, and may lead to a more cohesive, dynamic, and competitive workforce. Doing so will require more than merely accepting a legislative mandate; a cultural shift is required in which how we as a nation think about concepts like gender needs to fundamentally evolve.

At a practical level, this involves breaking down more traditional views concerning the roles and abilities of different groups. For example, the role of women in the CF has changed dramatically due to a Canadian Human Rights Commission (CHRC) Tribunal, which directed that women were to be fully integrated into all CF roles except service on submarines.[30] Although this was a huge step forward for women's rights in the Canadian military, the restriction on service on submarines was still indicative of a need to re-conceptualize the roles and abilities that our culture attributes to women. Subsequently, that restriction on submarine service was lifted in 2001.[31] By continuing to challenge our understanding of diversity, Canada will progress steadily toward a more inclusive culture, and Canadian organizations will benefit from a more flexible and innovative workforce.

*Sexual Orientation*

Sexual orientation has become an extremely controversial topic worldwide. The movement for acceptance of alternate sexual orientation sparks some of the most heated debate and, for some, challenges the core of fundamental beliefs. One of the more popularized examples of the controversy surrounding homosexuality in the context of military service is encapsulated in the U.S. military's "Don't Ask, Don't Tell" policy,[32] in which homosexuals are not banned from service outright, but disclosure of their sexual orientation is nonetheless discouraged.

In Canada, the acceptance of homosexuality is mandated within the legislative acts and policies cited earlier. Sexual orientation is no longer a justifiable reason for discrimination in relation to active employment in the Canadian workplace. In fact, "The only justifiable reasons for discrimination are failure to meet basic eligibility standards and the inability to meet fair selection criteria that are clearly linked to the capacity to adapt to and perform one's duties ..."[33] In other words, it is accepted that sexual orientation has no bearing on a person's ability to perform his or her job, and to treat them otherwise is discriminatory. In practical terms, avoiding discrimination based on sexual orientation should be viewed by leaders as a way of increasing their organization's efficiency and competitive edge, by strengthening team cohesion through inclusiveness.

*Religion*

On the world stage today, one of the most important determinants in international and inter-cultural relations is religion.[34] For most, religion is simply the way in which their personal and cultural beliefs and values are framed, expressed, transmitted, and lived. For others, understood in the West as fundamentalists or extremists, religion dictates not only their behaviour, but the ways in which their way of life must supersede all others. In Canada, religious freedom is protected by the Charter and by legislation. Reflecting Canadian policy, The Directorate of Military Gender Integration and Employment Equity (DMGIEE)[35] stated that, "In accordance with the Canadian Charter of Rights and Freedoms and the Canadian Human Rights Act, the policy of the Canadian Forces is to accommodate the fundamental religious requirements of its members." This includes demonstrating respect for "… major religious and spiritual requirements and tenets, including celebrations and observances, as well as dress, dietary, medical and health requirements."[36] As was the case with sexual orientation, religion has no bearing on a person's ability to perform his or her job. More importantly, by fostering a climate of respect and inclusiveness, leaders will find that religious diversity can also lead to important implications for flexibility, creativity and innovation, and can therefore be conceptualized as a way to maintain a competitive and successful organization.

## **DIVERSITY IS A LEADERSHIP ISSUE**

The discussion above should illustrate that diversity is an important issue today, and that it impacts organizations from the macro down to the micro level. In particular, it has been emphasized that it has become necessary for organizations worldwide to begin to evolve and adapt to diversity and multiculturalism, and that this is particularly the case for the Canadian workplace given the status that diversity has achieved in Canadian legislation and national policy. In reality, the idea is not that far-fetched; showing respect for others and treating diverse groups as equals is actually a very old idea, one that can be found at the heart of most religious and philosophical ideologies throughout history and across cultures.[37] However, an organization also has an obligation to maintain the health and safety of its members, and to achieve success within an acceptable range of resource expenditure. To that end, leaders must be able to balance their duty to accommodate diversity with the realities of workplace health and safety regulations and the overall goals of their respective organizations. In some cases, maintaining this balance means the denial of accommodation. The Canadian Human Rights Commission states:

> Accommodation can only be denied if a rule, standard or practice is based on a bona fide occupational requirement (BFOR), or on a bona fide justification (BFJ). This means that an employer or service provider can only deny accommodation if it does something in good faith for a purpose connected to the job or service being offered, and where changing that practice to accommodate someone would cause undue hardship to the employer or service provider, considering health, safety and cost. The Supreme Court of Canada has decided the steps an employer or service provider must take in order to show a BFOR or BFJ (*Meiorin* and *Grismer* test).[38]

In other words, it is understood that although respecting the differences of all people is one of the key components of Canada's espoused values, an organization is first and foremost responsible to ensure its members' well-being and safety. For example, a standard operational requirement in the CF is training in the use of gas masks. In a case where an individual's religious dress may inhibit their ability to effectively use a gas mask, the leader's primary responsibility is to the individual's safety rather than his or her religion. The duty of a leader to accommodate is also superseded if the cost to do so would be considered "undue": "The cost of a proposed accommodation would be considered "undue" if it is so high that it affects the very survival of the organization or business, or it threatens to change its essential nature. The mere fact that some cost, financial or otherwise, will be incurred is insufficient to establish undue hardship."[39]

Above all, an organization must remain committed to efficiency, remaining competitive and, ultimately, to achieving success. Although it is vital to recognize that in many cases the accommodation of diversity holds the potential to enhance an organization's ability to meet its goals, leaders must be able to discern when accommodation creates undue hardship, thereby jeopardizing efficiency and competitive edge. In a military context, the ability to balance the duty to accommodate is that much more poignant, as military institutions must do so in relation to the realities of war. In a generic, organizational context, accommodating diversity despite health risks or disproportionate cost may lead to a loss of revenue or injury. Making the same mistake in a military setting can lead to casualties, collateral damage, and dire geopolitical consequences.

## **SUMMARY**

The issue of diversity has become a crucial element for the future of the Canadian workplace, particularly in terms of professional recruitment. Also, it is a value that has gained important status in Canada through its enshrinement in Canadian legislation. Lastly, it is important for leaders to recognize that diversity is not just something that has to be tolerated or "dealt with." Rather, leaders need to conceptualize diversity as an element of Canadian culture that has the potential to lead to a more adaptive and innovative work force, and therefore, if approached properly, can be turned into a way of enhancing organizational efficiency and honing the competitive edge. By actively striving to make diversity a part of its strategy, policy, practices, and behaviours, the Canadian Forces can strengthen its operational effectiveness, both at home and in other countries around the world.

## **NOTES**

1. Charles Cotton, Rodney Crook, and Franklin Pinch, "Canada's Professional Military: The Limits of Civilization," *Armed Forces & Society*, Vol. 4, No. 3 (1978): 365–390.

2. Canada. *Leadership in the Canadian Forces: Conceptual Foundations* (Kingston: DND, 2005).
3. DHRD is the organization responsible for Canadian Forces policy concerning Employment Equity and diversity.
4. Canada, Department of Justice, Employment Equity Act and CF EE Regulations can be found at http://lois.justice.gc.ca/en/showtdm/cr/SOR-2002-421//.
5. *Ibid.*
6. *Ibid.*
7. Canada. *Leadership in the Canadian Forces: Conceptual Foundations* (Kingston: DND, 2005).
8. Jui-shan Chang, "A Transcultural Wisdom Bank in the Classroom: Making Cultural Diversity a Key Resource in Teaching and Learning," *Journal of Studies in International Education*, Vol. 10, No. 4 (2006): 369–377.
9. Christian Leuprecht, "Demographics and Diversity Issues in Canadian Military Participation," in *Challenge and Change in the Military: Gender and Diversity Issues,* eds. Franklin C. Pinch, Allister T. MacIntyre, Phyllis Browne, and Alan C. Okros (Kingston: Canadian Defence Academy Press, 2004), 122.
10. Canada. *Duty with Honour: The Profession of Arms in Canada* (Kingston: DND, 2003), 28.
11. Christian Leuprecht, 124.
12. *Ibid.*
13. *Ibid.*, 133.
14. Canada. *Leadership in the Canadian Forces: Conceptual Foundations* (Kingston: DND, 2005).
15. Canada, Department of Justice, Employment Equity Act and CF EE Regulations can be found at http://lois.justice.gc.ca/en/showtdm/cr/SOR-2002-421//.
16. David Segal and Mady Segal, "America's Military Population," *The Diversity Factor*, Vol. 13, No. 2 (2001): 18–23.
17. Karen Van der Zee et al., "The Influence of Social Identity and Personality on Outcomes of Cultural Diversity in Teams," *Journal of Cross-Cultural Psychology*, Vol. 35, No. 3 (2004): 283–303.

18. "Three block war" refers to the entire spectrum of tactical challenges ranging from humanitarian assistance to peacekeeping to traditional warfighting that could be encountered in the span of a few hours and within the space of three contiguous city blocks. Simply put, military personnel must be able to transition from humanitarian operations, peace support, or stability tasks to high intensity mid-level combat, potentially all in the same day all in the same area of operations. See General Charles C. Krulak, "The Strategic Corporal: Leadership in the Three Block War," *Marine Corps Magazine*, January 1999.
19. Lieutenant-Colonel Dave Kilcullen, "Complex Warfighting," *Future Land Operating Concept* (Australian Army, Unclassified Draft Developing Concept, 07 April 2004), 7.
20. Judd Sills, "Experts Weigh in About What Makes Diversity Initiatives Effective," *The Diversity Factor*, Vol. 13, No. 4 (2005): 23–29.
21. Canada. *Leadership in the Canadian Forces: Conceptual Foundations* (Kingston: DND, 2005).
22. Canada. *Duty with Honour: The Profession of Arms in Canada* (Kingston: DND, 2003).
23. Richard Jenkins, "Ethnicity," in *The Blackwell Encyclopedia of Sociology*, ed. George Ritzer (Blackwell Publishing, 2007). Also available at http://www.blackwellreference.com (accessed on 19 March 2007).
24. The Employment Equity Act, 1986.
25. Christopher Dandeker, and David Mason, "Diversifying the Uniform? The Participation of Minority Ethnic Personnel in the British Armed Services," *Armed Forces & Society*, Vol. 29, No. 4 (Summer 2003): 481–507.
26. Christian Leuprecht, 129.
27. Judd Sills, "Experts Weigh in about What Makes Diversity Initiatives Effective," *The Diversity Factor*, Vol. 13, No. 4 (2005): 23–29.
28. Bey Benhamadi, "Governance and Diversity within the Public Service in Canada: Towards a Viable and Sustainable Representation of Designated Groups (Employment Equity)," *International Review of Administrative Sciences*, Vol. 69 (2003): 510.

29. *Ibid.*
30. Karen D. Davis and Brian McKee, "Women in the Military: Facing the Warrior Framework," in *Challenge and Change in the Military: Gender and Diversity Issues,* eds. Franklin C. Pinch et al. (Kingston: Canadian Defence Academy Press, 2004), 53.
31. *Ibid.*
32. Aaron Belkin, "Don't Ask, Don't Tell: Is the Gay Ban Based on Military Necessity?" *Parameters*, Summer 2003, 108–119.
33. Franklin C. Pinch, "Diversity: Conditions for an Adaptive, Inclusive Military," in *Challenge and Change in the Military: Gender and Diversity Issues* (Kingston: Canadian Defence Academy Press, 2004), 179.
34. Keith Roberts, *Religion in Sociological Perspective,* 4th ed. (Toronto: Wadsworth, 2004).
35. The DMGIEE is now the Directorate of Human Rights and Diversity (DHRD).
36. "Religion in Canada," Catalogue No.: D2-147/2003, http://www.forces.gc.ca/hr/religions/engraph/religions00_e.asp (accessed on 28 February 2007).
37. Keith Roberts.
38. "Duty to Accommodate," *Canadian Human Rights Commission*, http://www.chrc-ccdp.ca/preventing_discrimination/duty_obligation-en.asp (accessed 06 February 2007).
39. *Ibid.*

## SELECTED READINGS

Belkin, Aaron. "Don't Ask, Don't Tell: Is the Gay Ban Based on Military Necessity?" *Parameters*, Summer 2003, 108–119.

Benhamadi, Bey. "Governance and Diversity Within the Public Service in Canada: Towards a Viable and Sustainable Representation of Designated Groups (Employment Equity)." *International Review of Administrative Sciences*, Vol. 69 (2003): 505–519.

*Canadian Human Rights Commission*, http://www.chrc-ccdp.ca/, last updated 12 March 2007.

Dandeker, Christopher and David Mason, "Diversifying the Uniform? The Participation of Minority Ethnic Personnel in British Armed Services," *Armed Forces & Society*, Vol. 29, No. 4 (Summer 2003): 481–507.

*Directorate of Human Rights and Diversity*, http://hr3.ottawa-hull.mil.ca/dmgiee/, last modified 29 November 2005.

Pinch, Franklin C., Allister T. MacIntyre, Phyllis Browne, and Alan C. Okros, eds. *Challenge and Change in the Military: Gender and Diversity Issues*. Kingston: Canadian Defence Academy Press, 2006.

Segal, David R., and Mady Wechsler Segal. "America's Military Population." *Population Bulletin*, Vol. 59, No. 4 (December 2004): 1–44.

Thomas Jr., R. Roosevelt. *Building a House for Diversity: How a Fable about a Giraffe and an Elephant Offers New Strategies for Today's Workforce*. (New York: Thomas & Associates Inc., 1999).

Van der Zee, Karen, Nelleke Atsma, and Felix Brodbeck. "The Influence of Social Identity and Personality on Outcomes of Cultural Diversity in Teams." *Journal of Cross-Cultural Psychology*, Vol. 35, No. 3 (2004): 283–303.

# 18
# ETHICS
## by Daniel Lagacé-Roy

As the world becomes more ambiguous and uncertain, individuals are challenged with a range of ethical situations and faced with difficult choices to make. This complex world tends to blur the meaning of ethics. Ethics becomes "intangible" or "vague" and is therefore regulated by "grey zones." In most instances, the difficulty in embracing the "right" conduct resides in the choice between two conflicting values or the lesser of two evils. It is therefore important to provide a comprehensive way of clarifying what someone "ought to do" when faced with difficult choices, and to shed some light on what exactly ethics means.

## WHAT IS ETHICS?

The *Oxford Dictionary of Philosophy* defines ethics as "the study of the concepts involved in practical reasoning: good, right, duty, obligation, virtue, freedom, rationality, choice."[1] To be more precise, ethics (from the Greek *ethos*), as a sub-field of philosophy, aims at clarifying the nature of right and wrong and how someone ought to live. In more practical terms:

> … the study of ethics can offer two things. First of all, it helps one appreciate the choices that others make, and evaluate the justification they give for those choices. But secondly, it involves a reflective sharpening of one's own

> moral awareness — a conscious examination of values and choices, of how these have shaped one's life so far, and (more importantly) of how they can be used to shape the future.[2]

This quotation points out that ethics involves the promotion of certain values (e.g., good life) and therefore implies making choices. A "value" is a belief that is centrally important and, for this reason, is critical in guiding decisions and actions.[3] More importantly, it is also the result of a certain evaluation and implies a standard of worth. Values can be classified as having an intrinsic worth, meaning "good" for their own sake (e.g., self-respect), or an instrumental worth, meaning "good" only because of their consequences (e.g., money). Not everyone holds the same values; however, individuals learn through parents, teachers, role models, as well as societal norms, the difference between an action that is considered right and one that is considered wrong. As for choices, they arise when individuals have to choose between complex situations. The choice between right and wrong seems less difficult than choosing between two wrongs or between two rights. These difficult situations (e.g., choosing between two wrongs or two rights), called "ethical dilemmas," are at the core of studying ethics.

## WHAT ETHICS IS NOT?

Ethics is different than the term "morals" (from the Latin *mores*). However, these two terms are often used interchangeably. Although this is acceptable, the distinction between ethics and morals is manifest in their particular aim: *morals* applies to rules of conduct, customs, or beliefs by which people live and *ethics* is the study of the meaning of those rules, customs, or beliefs. For instance:

> When we speak of moral problems, then, we generally refer to specific problems, such as: Is lying in this situation right? or: Is stealing at this moment wrong?

> Ethical problems are more general and theoretical. Thus What makes any act, such as lying or stealing, right or wrong? and What makes any entity good? are ethical problems.[4]

Ethics is not defined by "religion" and therefore does not abide by religious convictions. Undeniably, religion has played a crucial role in the development of ethical thought and has been closely associated with ethics. However, religion examines ethics through moral theology and defines the concepts of right and wrong according to a religious authority. Ethics, in today's secular world, views religious beliefs as possible "angles" that provide answers on how an ethical dilemma could be addressed or answered.

Ethics is not a "system" that is "inapplicable to the real world."[5] In other words: "It would be a mistake to regard ethics as a purely 'academic' study, having no intimate connections with the daily lives of men [and women]." [6] Ethics does not happen in a vacuum. It impacts everyone in all aspects of work and personal life. Moreover, it involves the concern of others and how individuals relate to each other. Taking "others" into consideration is emphasized by the ethicist Peter Singer, who said, "Suppose I begin to think ethically, to the extent of recognizing that my own interests cannot count for more … than the interests of others. In place of my own interests, I now have to take into account the interests of all those affected by my decision."[7]

Ethics is not "relative and subjective." Relativism and subjectivism base their arguments on the premises that the ethical nature of an act depends on the decision made by a particular society (i.e., relative to the society or a culture) or by a singular person (i.e., it is an expression of someone's own opinion). These two perspectives are defendable: societies' values and beliefs differ from one another, and every ethical decision takes into account personal judgment and opinion. However, relativism and subjectivism should not be the only criteria on which ethical decisions are based. These two views highlight the fact that if ethics is defined as only relative to a society or a personal point of view, the need to question the meaning of what is "right" and what is "wrong" is no longer relevant. A particular society or a specific person therefore answers that question.

## WHAT IS AN ETHICAL DILEMMA?

First, the meaning of the word *dilemma* is a situation in which a choice has to be made between two equally valid alternatives. Those alternatives, either desirable or undesirable, are the disputable arguments for a satisfactory solution. Second, some dilemmas are more challenging than others and present an ethical aspect. The ethical nature of a dilemma is the result of an evaluation that carries either:

    a) uncertainty;
    b) conflicting values or;
    c) that causes harm regardless of the action chosen.

Uncertainty represents the most common type of ethical dilemma. It refers to a problematic situation where "the right thing to do" is not clear. There is no simple choice between right and wrong. Equally valid reasons can support two or more possible solutions to resolve the dilemma. For example, in a much-supervised environment, reporting that a colleague is being harassed by a supervisor might be very difficult to do because it could cause retaliation from the superior in question.

The competing values dilemma involves a situation in which values compete with each other. In *The Last Days of Socrates*,[8] Plato (427–348/7 BC) recounts one famous example of this type of conflict. A young man called Euthyphro contemplates the possibility of prosecuting his father for having murdered a slave.[9] The dialogue between Socrates and Euthyphro reveals that the young man wants to be an honest citizen by reporting a murder; however, by doing so he is disloyal to his father. This situation reflects an account of conflict between two values, one of honesty and the other of loyalty.

The harm dilemma is a situation in which any possible solution will cause harm or injury to others. This type of dilemma is often described as a "lose-lose situation." For example, in military operations, the possibility of harming civilians while trying to protect your troops is a situation that is sometimes inevitable.

In certain circumstances, dilemmas are deemed "personal" because the course of action (right or wrong) is clear, but personal values such as self-justice, friendship, or self-interest contribute to the difficulty of acting.[10] While a personal dilemma does not constitute an ethical dilemma, this type of situation is difficult, nonetheless. For example, reporting a fellow colleague who has falsified a claim is the right thing to do: however, on a personal level, it remains a difficult situation to act upon because it might jeopardize a friendship and produce tension in the work environment.

## **ETHICAL FOUNDATIONS**

The subject of ethics requires the clarification of general concepts (e.g., difference between ethics and morals), along with basic knowledge of some philosophical theories (e.g., utilitarianism). More important, a critical reflection on ethical issues entails the awareness of theories that have laid down the foundations on which ethical arguments are based. These theories (or schools of thought) are usually divided into two fields: normative and non-normative ethics. However, when the subject of ethics is discussed, the normative branch is usually the one that people refer to as their philosophical background. This branch of ethics is highlighted below, along with a brief explanation of non-normative ethics.

Normative ethics is a reasoned evaluation of human conduct through explanation and justification. This branch of ethics determines the "rightness" and "wrongness" of an action based on:

    a) a person's self-disposition;
    b) the non-consequences;
    c) the consequences.

The philosopher Aristotle (384–322 BC) associates the idea of a person's self-disposition with the development of virtues. This great thinker, the tutor of Alexander the Great and author of *Nicomachean Ethics*, argues that virtuous persons perform ethical actions. However, being virtuous does not come naturally: one needs to be educated and trained. For Aristotle, a virtue (e.g., honesty, fairness) depends on someone's self-inclination to "be" honest or fair, for example. In other words, such philosophy indicates that a virtuous person will know and do what is right.

In more modern debates on ethics, German philosopher Immanuel Kant (1724–1804) recognized that the ethical status of an act did not rely on consequences (i.e., non-consequences). According to Kant, an act has an ethical worth in accordance with absolute obligations (e.g., keep your promises, always tell the truth). These obligations, called categorical imperatives, are universal duties that must be fulfilled because they compel someone to act, no matter the circumstances or personal desires.

Such strict adherence to obligations has raised criticisms of Kant's theory as a rule-based approach to ethics. One criticism came from one utilitarian theorist John Stuart Mill (1806–1873), who stated that the rightness of an act relied on the consequences or the results of that act. The statement, "the end justifies the means," captures some aspects the utilitarian approach. However, it is important to note that the consequences of an act have to promote the greatest happiness for the greatest number as opposed to simply promoting one's own self-gratification.

The non-normative branch of ethics consists of either:

a) a factual investigation of ethical behaviour (e.g., how do people in fact behave) which is used in social sciences such as anthropology and sociology; or

b) an analysis of the terminology used in ethical discourse (e.g., what is the meaning of the word "right") which is mostly used in the field of philosophy.[11]

What is important to remember is that normative and non-normative ethics are not completely foreign from one another. While they are separated in theory, they can be associated in practice when debating and analyzing an ethical dilemma. For example, someone might come to the decision that an action is right because of the consequences (i.e., normative ethics). Yet, the process by which this evaluation was made might have been influenced by factors, such as peer pressure or work environment. The recognition and investigation of those factors belong to the non-normative branch of ethics.

## ETHICAL DECISION-MAKING

Ethical dilemmas can sometimes be difficult to resolve, while some are less challenging than others (e.g., having to choose between life and death); some may appear "okay" because they do not hurt anyone (e.g., reporting a fellow student who is cheating); or some may indicate the clear action, but are difficult to carry out (e.g., reporting abuse). By comparing these three examples, one may argue that because a life is at stake and because someone is being abused, the first and the third situations are greater dilemmas than the second, reporting someone who is cheating. It may be very comforting to rationalize ethical dilemmas in such a fashion. However, most ethical dilemmas are not usually that easy to explain. Therefore, it is sometimes useful to follow a framework that provides the tools for distinguishing all aspects of a dilemma.

The framework presented below is a value-based approach to decision-making[12] and is designed to examine all of the factors that influence an ethical dilemma. It allows anyone to walk through a sequence of questions that guide the internal process of analyzing an ethical dilemma and to move to an appropriate course of action. This sequence of questions includes:

a) the assessment of a situation;
b) the ethical considerations; and
c) options and risks.

## THE ASSESSMENT OF A SITUATION

This is a general summary of the scenario in which facts, ethical concerns, and issues are taken into account. The facts are a description of events or circumstances of a situation as a person experiences it. What is the situation all about? For example, as a member of a promotion board, you notice a discrepancy in one of the files submitted.

The ethical concerns are issues that question the ethical nature of a situation and are perceived as problematic. What makes this situation an ethical issue? For example, you are concerned that the individual was not properly assessed.

The personal factors refer to personal values and moral responsibilities of the impact of someone's decision (e.g., to act or not to act) on others and on himself/herself. Is there a sense of personal involvement or self-identification with this situation? For example, the individual supposedly wrongly assessed was a previous employee.

The environmental factors refer to the work environment. It includes perceptions of what is acceptable and unacceptable and what is considered "your business" and "none of your business." What are the roles assigned and expected by everyone in that work environment? For example, as a new member of this promotion board, you might be expected to just "listen" and "learn the ropes" of the trade.

It is important to remember that facts, ethical concerns, personal, and environmental factors are not always obvious when experiencing an ethical dilemma and that perceptions can also play a central part in evaluating if a situation is deemed to be ethical or unethical.

## THE ETHICAL CONSIDERATIONS

These involve identifying ethical values (e.g., loyalty, duty) that come into play in an ethical situation and determining the type of ethical dilemma. In most situations in life, we do not pay constant attention to the influence of values: they are usually taken for granted. However, when a problematic situation arises (i.e., a dilemma), they tend to surface and may lead to conflicts (e.g., loyalty competing with integrity). They are good indicators of how a situation is perceived, and they assist in recognizing the type of dilemma that a situation entails. The gathered information helps to choose between three types of dilemmas: uncertainty, competing values, and harm.

As mentioned earlier, some situations are not necessarily ethical dilemmas. They are personal dilemmas that emphasize personal values (e.g., self-justice, self-interest). It is sometimes difficult to see the difference between ethical and personal dilemmas. It is therefore essential to pay attention to the assessment of a situation in order to identify the "exact" facts, values, and ethical concerns at "stake" in a particular situation.

## **OPTIONS AND RISKS**

In ethical situations, options are considered the "best solutions" for courses of action and range from acting upon a situation to not acting. Options are often guided or influenced by regulations, rules, care for others, personal sense of what is "right" and "wrong," outcomes, self-interest, etc. The best course of action takes into consideration the risks associated with its application. These risks could be at the personal level (e.g., poor evaluation), at the managerial level (e.g., credibility tarnished), or at the institutional level (e.g., loyalty and trust in superiors questioned). The best course of action (i.e., the option to act upon) will be based on these options and risks.

## **THE IMPORTANCE OF TEACHING ETHICS**

In today's world, ethics in various professions such as medicine, law, and the military has become paramount. The central focus of teaching the subject of ethics is to find ways to engage and prepare their members in attaining ethical behaviours. In most professions, studying ethics is a difficult endeavour because it is not easy to identify the appropriate teaching method for their members. For example, the military profession cannot take this issue lightly, since the teaching of ethics is further challenged by the context of today's military operations. Present Canadian Forces (CF) operations are carried out in an environment where ethics is hard to define. This ambiguous, complex, and chaotic environment in which extremes are present (e.g. crimes against humanity, global terrorism, non-state actors) tends to blur (even re-define) the meaning of ethics. Ethics becomes "intangible" or "vague" and is therefore regulated by "grey zones" and interpreted as if a gap exists between ethical conduct and the disruptive threats posed by a "contemporary enemy."[13]

Therefore, the task of teaching ethics is to help members of any profession to think through their own ethical values and standards; to learn ethical reasoning and critical thinking; to know their obligation, duty, and responsibly as members of a profession; to develop their knowledge of concepts and principles; and to develop skills in decision-making. In that context, teaching ethics requires an educational tool that sensitizes members to all aspects of an ethical dilemma. One educational tool that can be used when teaching ethics is the case study method.

A case study describes a situation or event in a chronological sequence and positions the reader as the decision-maker. This type of method is more formal than just the passing down or telling of stories. Although such stories are useful, they only provide the perspective of a single person (the "storyteller") delivering a tale without, in many instances, the involvement of the listeners. Using this approach, a storyteller relates a point of view and describes the manner in which the ethical dilemma was interpreted and resolved. As a result, the listeners gain no insights into "what they should do" if they were in a similar situation. In addition, there are no guarantees that the storyteller acted ethically.

Conversely, the benefits of the case study method is that members are more engaged in the discussion, which allows them to better understand the critical elements that constitute an ethical dilemma and how it should be approached. Another important benefit is that a case study brings about the crucial role of ethics in everything the members do and makes them realize how the overarching impact of ethics will be expressed in their leadership. This important exercise also provides an opportunity for members to broaden their view beyond the scope of the automatic "right answer." They realize that a choice between "right" and "wrong" is only one facet of a course of action. Other facets, such as choosing between "two wrongs" or "two rights," are more difficult and require more complex moral reasoning. The process of moral reasoning involves personal values, a sense of right and wrong and an understanding of "what ought to be" when faced with a difficult situation.

## **ETHICS AND LEADERSHIP**

As referred to earlier, the hallmark of ethics is to clarify concepts like "right" and "wrong" and how "someone ought to live." This initial understanding of the meaning of ethics leads to its practicality and to the subject of leadership as a comprehensive example of how ethics influence the way leaders lead. In that context, ethics refers to the understanding of ethical behaviour and climate (i.e., environment) applied to leadership. According to university professor Dr. Joanne B. Ciulla, ethics is at the heart of leadership.[14] In the same vein, it is arguable that ethics is not only essential for leaders but, in fact, defines leadership.

Leaders in any organizations are required to embrace strong ethical values (e.g., honesty, integrity) and to be exemplars of ethical behaviour (e.g. can be trusted) by establishing and emphasizing the ethical climate (e.g., confidentiality) that is necessary to influence ethical behaviour. This ethical climate can be defined in general terms as norms, standards, expectations, and practices accepted and reinforced by leaders. In order to become an exemplar of ethical behaviours, a leader has to become the embodiment of good practices, such as responsibility and accountability.[15]

The importance of ethics is captured in the practices of responsibility and accountability. Leaders are responsible for ensuring that ethical conduct is practiced and that unethical behaviors, "no matter how subtle, no matter how private,"[16] are not. Leaders will be held accountable for not exercising their leadership in fostering an ethical climate. In order words, leaders in any organizations are challenged with the duty of establishing the necessary conditions (e.g., opportunities to discuss ethical concerns) for an environment that demands ethical conduct.

## **SUMMARY**

Ethics is at the core of everything individuals do in their daily activities, ranging from driving to work, dealing with colleagues, resolving issues, to talking to friends and family members. This impact on daily decisions affects the way individuals interact with each other. In essence, ethics is the central focus of how someone ought to behave, to act, to respond: essentially how someone ought to live.

This chapter highlighted many aspects of ethics that are summarized here in five points. First, ethics requires a broader understanding of the concepts that underline its meaning. Ethics is often difficult to define because of its association with other terms such "morals" or its close relationship with religion, as an example. The distinction between "what ethics is" from "what ethics is not" shows that the aim of ethics is to clarify the nature of "right" and "wrong."

Second, clarifying what makes an act "right" or "wrong" often happens in the context of ethical dilemmas. As defined in this chapter, an ethical dilemma is a situation in which a person has to choose between conflicting values (e.g., honesty versus loyalty) or between two actions, both of which are undesirable options (e.g., lose-lose situation). Ethical dilemmas happen all the time.

Third, ethics, as a systematic study, has a long history. Philosophers and great thinkers have spent their lives pondering the question of how "someone ought to live" and how to determine if an act is "right" or "wrong." For example, one ethical system of thought argues that an act is considered "right" according to the decisions made based on the acceptable or desired consequences. Therefore, someone will evaluate his/her decision according to the likely outcomes.

Fourth, it is often useful to consider a framework (e.g., ethical decision-making model) when analyzing or discussing an ethical dilemma. A framework is a tool that helps to identify the important factors (e.g., personal factors, values) that are at play in an ethical dilemma. Frameworks are often used with case studies in the teaching of ethics in the classroom.

Fifth and finally, ethics and leadership are intertwined. Leaders have to demonstrate strong ethical values (e.g. honesty) that can be emulated by their followers. They have the duty to create an ethical climate that allows members to voice their concerns about ethical issues and to explicitly encourage ethical behaviour. In that context, members in any organization will do what is right and will go beyond their own self-interests for the good of their institution.

## NOTES

1. Simon Blackburn, *Oxford Dictionary of Philosophy* (Oxford, UK: Oxford University Press, 1996), 126.
2. Mel Thompson, *Teach Yourself: Ethics* (London: Hodder Education, 2006), 10.
3. Canada, *Leadership in the Canadian Forces: Conceptual Foundations* (Kingston: DND, 2005), 133.
4. Manuel Velasquez, *Philosophy : A Text With Readings* (Belmont, CA: Walsworth Publishing Company, 1994), 432.
5. Peter Singer, *Practical Ethics,* 2nd ed. (Cambridge, UK: Cambridge University Press, 1993), 2.
6. Richard H. Popkin and Avrum Stroll, *Philosophy Made Simple* (New York: Broadway Books, 1993), 7.
7. Op.cit., 13.
8. Plato was a student of Socrates. Harold Tarrant, ed., *The Last Days of Socrates,* trans. Hugh Treddenick (London: Penguin Classics, 2003).
9. Socrates (469–399 BC) is considered the one who began the discourse on ethics. He discovered that this discourse was closely related to human conflict. "Euthyphro" in Plato, *The Last Days of Socrates*, 9–12.

10. Kidder and Bloom make reference to this type of "dilemma" as "moral temptation." They write: "The term 'moral temptation' indicates that this should not be referred to as a dilemma, but is a situation in which the decision-maker is tempted to allow his or her own interests or needs to outweigh those of the organisation or of others involved in the situation." Cited in Th. A. van Baarda, and D.E.M. Verweij, eds., *Military Ethics: The Dutch Approach — A Practical Guide* (Leiden, The Netherlands: Marinus Nijhoff Publishers, 2006), 132.
11. Manuel Velasquez, *Philosophy: A Text with Readings*, Op.cit., 434–436.
12. The information and methodology presented here stem mainly from the CF perspective on ethical decision-making as outlined by the Canadian Forces Defence Ethics Program (DEP), *Introduction to Defence Ethics*, 2nd ed. (2005).
13. Kenneth W. Watkin, "Humanitarian Law and 21st Century Conflict: Three Block Wars, Terrorism and Complex Security Situations," in *Testing the Boundaries of International Law*, eds. Susan C. Breau and Agnieszka Jachec-Neale (London: The British Institute of International & Comparative Law, 2006), 2.
14. Joanne B. Ciulla, "Leadership Ethics: Mapping the Territory," in *Ethics: The Heart of Leadership*, ed. Joanna B. Cuilla (Westport, CT: Quorun Books, 1998) 3–25.
15. R. Richardson, D. Verweij, and D. Winslow, "Moral Fitness for Peace Operations," *Journal of Political and Military Sociology*, Vol. 32, No. 1 (Summer 2004): 109.
16. K.G. Penney, "A Matter of Trust: Ethics and Self-Regulation Among Canadian Generals," in *Generalship and the Art of the Admiral*, eds. B. Horn and S. Harris (St. Catharines, ON: Vanwell Publishing Limited, 2001), 156.

## SELECTED READINGS

Canada. *Ethics in the Canadian Forces: Making Tough Choices* (Instructor Manual). Kingston: DND, 2006.

Canada. *Ethics in the Canadian Forces: Making Tough Choices* (Workbook). Kingston: DND, 2006.

Ciulla, Joanna B., ed. *Ethics: The Heart of Leadership.* Westport, CT: Quorum Books, 1998.

Cook, Martin L. *The Moral Warrior.* New York: State University of New York Press, 2004.

Gabriel, Richard A. *The Warrior's Way: A Treatise on Military Ethics.* Kingston: Canadian Defence Academy Press, 2007.

Gensler, Harry J., Earl W. Spurgin, and James C. Swindal. *Ethics: Contemporary Readings.* London: Routledge, 2004.

Lang, Anthony F. Jr., Albert C. Pierce, and Joel H. Rosenthal, eds. *Ethics and the Future of Conflict.* Upper Saddle River, NJ: Pearson Prentice Hall, 2004.

Orend, Brian. *The Morality of War.* Peterborough, ON: Broadview Press, 2006.

Rabinow, Paul, ed. *Michel Foucault: Ethics, Subjectivity and Truth*, Volume 1. New York: The New Press, 1997.

Singer, Peter, ed. *Ethics.* Oxford, UK: Oxford University Press, 1994.

———. *Practical Ethics,* 2nd ed. Cambridge, UK: Cambridge University Press, 1993.

Van Baarda, Th. A. and D.E.M. Verweij, eds. *Military Ethics: The Dutch Approach — A Practical Guide.* Leiden, The Netherlands: Martinus Nijhoff Publishers, 2006.

# 19
# FATIGUE
by Bernd Horn

The impact of fatigue on military operations is indisputable.[1] In short, it can cause catastrophic failure and/or death. In its clutches, military personnel become less effective, have lower morale, and are more prone to debilitating combat stress reactions. No one is immune. Not surprisingly, commanders and leaders must pay special attention to fatigue to ensure it does not impact on combat effectiveness. As such, they require an understanding of fatigue, its causes, effects, and means of mitigating its potential impact.

## WHAT IS FATIGUE?

Fatigue is defined as a change in an individual's psychophysiological state due to sustained performance. This change has both subjective and objective manifestations, which include an increased resistance against further effort, an increased propensity toward less analytic information processing and changes in mood.[2] In essence, "fatigue is a term used to describe a constellation of adverse, unwanted effects that can be traced to the continued exercise of an activity."[3] In the simplest of terms, fatigue is the deterioration of performance over time.

## CAUSES OF FATIGUE

Fatigue is the result of one, or any combination of, the following:

* intense emotional and mental strain;
* strenuous and/or prolonged physical exertion;
* inadequate food and water intake and/or food lacking nutrition;
* unfavourable environmental conditions; and
* sleep loss.

## **SIGNS OF FATIGUE**

Commanders and leaders must be able to recognize signs of fatigue if they are to minimize its effects. Extreme fatigue leads to physical deterioration. As such, major physical signs of fatigue and sleep loss are:

* vacant stare with sunken, bloodshot eyes;
* eye strain, sore or heavy eyes, dim and blurred vision;
* droning and humming in the ears;
* paleness of skin;
* slurred speech;
* headaches;
* slowed responsiveness;
* lowered body temperature and heart rate;
* lack of energy or vitality, drowsiness;
* unstable posture — swaying, chin dropping;
* intermittent loss of muscular strength, stiffness, cramps;
* faintness and dizziness; and
* difficulty making fine movements.

Other signs of fatigue include:

* poor work output — performance may falter or even stop — increase in accidents;

* neglect of routine tasks — even though the tasks may be of critical importance;
* decrease in overall interest — lower morale, lack of motivation, little interest in tasks;
* decreased willingness to initiate action — even interaction with team members;
* decreasing performance standards in others may be accepted;
* degraded communications due to lapses in both short-term memory and attention to detail;
* weakened ability to make judgments and grasp even simple instructions due to mental confusion, forgetfulness, and lack of initiative;
* loss of sense of humour, and uncharacteristic moodiness, irritability, or argumentativeness;
* decreased attention to personal hygiene and self-care;
* greater acceptance of risk;
* frequent daydreaming; and
* a sense of pessimism or fatalism.

## THE PARADOX OF FATIGUE

An important point to note is that the debilitating effects of sleep loss may not be apparent to those who are sleep-deprived. This lack of insight and judgment is one of the dangerous aspects of sleep loss. Commanders and leaders are especially prone to this reality since they are normally fixated on operational matters to the exclusion of most other things. It is for this reason that individuals at all levels must be continually monitored.

## THE EFFECTS OF CHRONIC FATIGUE ON PERFORMANCE

Fatigue has a dramatic effect on operational capability. Some analysts believe that the high proportion of "friendly fire" casualties among coalition forces in the closing stages of the 100-hour Operation Desert Storm was due largely to fatigue. Sleep loss may not seriously degrade gross motor physical performance (e.g. marching, digging) for 18–40 hours. After 40 hours, even physical performance will rapidly degrade. Mental abilities and fine motor coordination skills, however, usually deteriorate to unacceptable levels after only 18 hours of sustained work. Research has shown that after 17 hours of sustained wakefulness, cognitive psychomotor performance decreases to a level equivalent to the performance impairment observed at the blood alcohol concentration of 0.05 per cent (the equivalent of a moderate state of intoxication).[4] After 36 hours of continuous wakefulness, most tasks involving perceptual, mental or fine motor skills degrade sharply. After 72 hours of sustained operations, almost all individuals will exhibit significant and persistent symptoms of fatigue that preclude effective combat performance.

In summary, fatigue causes:

* Communication difficulties. Tired individuals may not be able formulate coherent messages and they may omit important information transmitted to them in orders or reports;
* Inability to concentrate. Individuals have difficulty maintaining attention and are easily confused, therefore preventing them from following complex directions or performing numerical calculations;
* Mood changes. These normally include increased irritability and can entail depression and apathy;
* Increasing omissions and carelessness. Individuals begin to skip tasks, miss events, and make mistakes;
* Decreased vigilance. Individuals become less alert and may fail to detect threats;
* Slowed comprehension. Individuals take longer to understand any form of information;
* Encoding/decoding information. Individuals take longer to transform data or to process information;
* Hallucinations;

- * Muddled thinking. Reasoning becomes slow and confused;
- * Faulty short-term memory;
- * Slowed perception. Individuals are slow to understand things seen or heard; and
- * Slowed responses. Individuals may be slow in translating direction into action.

## **COMBATING FATIGUE**

The amount of sleep an individual requires varies as a result of a number of variables such as age, workload, and level of fitness.[5] In addition, unusual physical, mental, and emotional demands will increase the short-term need for sleep. Normally the youngest and oldest appear most prone to fatigue on operations.[6] Commanders should be aware that individuals differ in the following ways:

Amount of sleep required to maintain adequate performance:

- * depth of sleep and degree of sleep inertia;
- * patterns of sleep (e.g. "early bird" versus the "night owl");
- * ability to fall asleep; and
- * ability and motivation to remain awake when fatigued.

As a guideline, general combat effectiveness can be maintained on four hours of sleep for every 24-hour period. Most individuals can maintain an acceptable level of physical performance for 9–14 days on such a schedule. However, the risk for mission-critical errors is increased, but still remains at acceptable levels. In addition, an extended rest (i.e. three days) would be necessary at the end of the period. Nonetheless, every opportunity to have short sleeps should be utilized. The longer the "nap," the greater the benefits in improving alertness, mood, and performance. Even short naps of 10–30 minutes (although not ideal) are still better than nothing.

In general, fatigue prevention can be achieved by:

- imposing sensible work demands and schedules;
- maintaining an appropriate diet;
- fostering morale; and
- whenever possible, reduce sleep debt.

## **LEADERSHIP AND FATIGUE**

Commanders and leaders often pride themselves at being present at the helm or command post for extended periods of times. In many ways, they see themselves as immune to fatigue and, for some, sleep-denial is a manner of showing toughness and self-discipline. However, research has shown that mental tasks requiring complex reasoning, detailed planning, prolonged concentration, and quick or difficult decision-making are generally more tiring than even strenuous physical exercise. As such, commanders and leaders actually often require more rest than their subordinates.

The reality that commanders and leaders actually need more sleep, however, is problematic. Their desire to lead by example and the idea of going to sleep when others are still awake and working can be seen as not sharing in the hardship or working as hard as their subordinates. Nonetheless, this attitude is counterproductive. In fact, it endangers the mission. "So while it was difficult to rest while the soldiers were still fighting out there, I knew I had to," explained General Fred Franks during Operation Desert Storm. He added, "The troops did not need a tired, fuzzy-thinking commander the next day when we made our final move." Franks asserted, "The best I could do was get some rest."[7]

In essence, commanders and leaders owe it to their subordinates to ensure they are as well rested as possible so that their thinking, analytical skills and decision-making are as clear as possible. The practice of sleep denial is normally counterproductive. As sleep debt increases, so too do the physical, psychological, and behavioural consequences of sleep loss. No one is immune. Although each person has differing requirements for the amount of sleep needed, in the end, everyone needs rest. As a result, commanders and leaders must ensure that they are capable of passing responsibility to others and, short of crisis, trusting those appointed to run the show while they are resting.

In the end, commanders and leaders must consider sleep needs and fatigue as critical components of their operational planning. As such, a sleep/rest schedule of key decision-makers should always be in place to ensure the chain of command does provide for rested individuals who can think clearly and make timely, sound, and reasoned decisions as required, as well as recognize signs of degradation in others and take over as required.

## **SUMMARY**

Fatigue is the result of a number of factors such as intense emotional strain, mental exertion, physical exertion, poor and/or inadequate diet, strenuous and unfavourable environmental conditions, and sleep loss. Fatigue can degrade operational effectiveness within hours and its impact is often insidious as fatigued individuals normally do not realize or acknowledge that their performance, or that of those around them, is deteriorating, because their judgment becomes impaired. For military commanders and leaders, an awareness of fatigue and its effects must be of paramount importance. First, they must ensure fatigue does not impair their personnel and, as such, they must consciously plan, in accordance with the situation and circumstances, for as much rest and sleep as possible. Similarly, due to the mental nature of their primary tasks, commanders and leaders are often more prone to fatigue, and since the consequences of their actions, specifically decisions, are more serious, they must ensure that they also get adequate rest and sleep so that they can make sound, clear, and timely decisions.

## NOTES

1. This chapter is based primarily on Operational Effectiveness Guide 98-1: "Fatigue, Sleep Loss, and Operational Performance," April 1998.
2. Dimitri van der Linden, Michael Frese, and Theo F. Meijman, "Mental Fatigue and the Control of Cognitive Processes: Effects on Perseveration and Planning," *Acta Psychologica*, Vol. 113 (2003): 46.
3. Samantha K. Brooks, Don G. Byrne, and Stephanie E. Hodson, "Non-Combat Occupational Stress and Fatigue: A Review of Factors and Measurement Issues for the Australian Defence Force," *Australian Defence Force Journal*, No. 145 (November/December 2000): 43.
4. Brooks et al., 43–44.
5. Level of fitness acts as a moderator of both sleep need and adaptability to disrupted or deprived sleep.
6. With increasing age, the duration of sleep need decreases, however, shortfalls in sleep will be more disruptive to general functioning. Also, there is a tendency to perform best in morning hours and increased difficulty adjusting to changes in sleep and work schedules.
7. Quoted in Major P.J. Murphy and Captain C.J. Mombourquette, "Fatigue, Sleep Loss, and Operational Performance," Personnel Research Team, Operational Effectiveness Guide 98-1 (April 1998): 9.

## SELECTED READINGS

Brooks, Samantha K., Don G. Byrne, and Stephanie E. Hodson. "Non-Combat Occupational Stress and Fatigue: A Review of Factors and Measurement Issues for the Australian Defence Force," *Australian Defence Force Journal*, No. 145 (November/December 2000): 35–45.

Caldwell, J.A., Lynn Caldwell, David Brown, and Jennifer Smith. "The Effects of 37 Hours of Continuous Wakefulness on the Physiological

Arousal, Cognitive Performance, Self-Reported Mood, and Simulator Flight Performance of F-117A Pilots." *Military Psychology*, Vol. 16, No. 3 (2004): 163–181.

Curry, Justin. "Sleep Management and Soldier Readiness: A Guide for Leaders and Soldiers," *Infantry*, May–June 2005, 26.

Linden, Dimitri van der, Michael Frese, and Theo F. Meijman. "Mental Fatigue and the Control of Cognitive Processes: Effects on Perseveration and Planning," *Acta Psychologica*, Vol. 113 (2003): 45–65.

Meijman, Theo F. "Mental Fatigue and the Efficiency of Information Processing in Relation to Work Times." *International Journal of Industrial Ergonomics* 20 (1997): 31–38.

Murphy, Major P.J. and Captain C.J. Mombourquette. "Fatigue, Sleep Loss, and Operational Performance." Personnel Research Team, Operational Effectiveness Guide 98-1, April 1998.

Webster, Donna M., Linda Richter, and Arie W. Kruglanski. "On Leaping to Conclusions When Feeling Tired: Mental Fatigue Effects on Impressional Primacy." *Journal of Experimental Social Psychology*, Vol. 32 (1996): 181–195.

# 20
# FEAR
by Bernd Horn

Fear is a concept with which everyone is familiar. Arguably, it needs no explanation since almost everyone has felt its debilitating grip. In essence, fear knows no boundaries. As Colonel S.L.A. Marshall, the well-known American combat historian explained, "Fear is general among men."[1] Research has conclusively confirmed that everybody experiences it. Significantly, "fear," adjudged scholar Elmar Dinter, "is the most significant common denominator for all soldiers."[2] As a result, it is incumbent on military leaders to fully understand what it is, its manifestations, as well as the actions necessary to counterbalance the negative influence of fear.

## WHAT IS FEAR?

In the simplest of terms, fear is an emotion — "a state characterized by physiological arousal, changes in facial expression, gestures, posture, and subjective feeling."[3] However, once evoked, fear is an extremely powerful emotion that causes a number of bodily changes to occur such as rapid heartbeat and breathing, dryness of the throat and mouth, perspiration, trembling, and a sinking feeling in the stomach. It can also have more obvious, if not embarrassing, manifestations such as uncontrolled urination or bowel movements.

This bodily reaction is caused by the activation of the sympathetic division of the autonomic nervous system as it prepares the body for emergency action — the fight or flight reflex. In short, it prepares the body for energy output. It does this by way of a number of bodily changes (which need not occur all at once):

1. blood pressure and heart rate increase;
2. respiration becomes more rapid;
3. pupils dilate;
4. perspiration increases while secretion of saliva and mucous decreases;
5. blood-sugar level increases to provide more energy;
6. the blood clots more quickly in case of wounds;
7. blood is diverted from the stomach and intestines to the brain and skeletal muscles; and
8. the hairs on the skin become erect causing goose pimples.[4]

These changes all have a specific purpose. As already stated, the sympathetic system activates the body for emergency action by arousing a number of bodily systems and inhibiting others. For example, sugar is released (by the liver) into the bloodstream for quick energy; the heart beats faster to supply blood to the muscles; the respiration rate increases to supply needed oxygen; digestion is temporarily inhibited (thus, diverting blood from the internal organs to the muscles); pupils dilate to allow in more light; perspiration increases to cool the agitated body; and the blood flow to the skin is restricted to reduce bleeding.[5]

Of great importance is the fact that these bodily changes that occur actually assist the affected individual. "The man who recognizes fear can often make it work in his favor," concluded war reporter Mack Morriss, "because fear is energy. Like anger, fear shifts the body into high gear."[6] John Dollard, a social anthropologist, explained in his seminal research into the subject that fear "is a danger signal produced in a man's body by his awareness of signs of danger in the world around him."[7]

## TYPES OF FEAR

Overall, there are two general types of fear. The first is acute fear that is generally provoked by tangible stimuli or situations (e.g., a loud bang or a snake suddenly slithering by) that normally subside quite quickly when the frightening stimuli is removed or avoided. The second type of fear is chronic fear. This is generally more complex and may or may not be tied to tangible sources of provocation, for example, an individual who persistently feels uneasy and anxious for unidentified reasons, such as the fear of being alone.[8]

Regardless of the type of fear — it need not be immediate or the result of personal experience. Fear is a learned reaction. "Men and animals," reported John T. Wood, "experience fear in the face of present, anticipated, or imagined danger or pain."[9] Similarly, Jeffrey Alan Gray, professor of psychology at the Institute of Psychiatry in London, asserted, "Fear ... is due to the anticipation of pain."[10]

## MANIFESTATIONS OF FEAR

The most common symptoms of fear are: pounding of the heart and rapid pulse, tenseness of muscles, sinking feelings, dryness of mouth and throat, trembling, and sweating.[11] Studies have also established that fear in younger and unmarried soldiers is marginally less than in older, married ones and that junior officers and non-commissioned officers show a little less fear than the other ranks.[12] Not surprisingly, overwhelmingly, most people appear to be more susceptible to fear when they are alone.[13]

## WHAT CAUSES FEAR?

Everyone is prone to fear, although individuals tend to react differently to varying stimuli. The first major cause for fear, in a military context, is the fear of the unknown and the unexpected. "What a man has not seen," stated the ancient Greek general Onasander, "he always expects will be greater than it really is." Retired combat veteran and military theorist, Major-General Robert Scales Jr., opined, "... soldiers fear most the enemy they cannot see."[14] Not surprisingly, anecdotal evidence indicates that fear increases in foggy conditions, or when it is dark, or with the loss of orientation following an unexpected enemy attack from the rear.[15]

A second major cause of fear originates from a feeling of hopelessness. This is often due to a belief of, or actual, inability in the face of danger to influence the probable outcome of events. Simply put, it is caused by a feeling of being threatened without the power to do anything about it. "Fear," asserted Professor S.J. Rachman from the Institute of Psychiatry at the University of London, "seems to feed on a sense of uncontrollability: it arises and persists when the person finds himself in a threatening situation over which he feels he has little or no control." Research demonstrated that, "Being in danger when one cannot fight back or take any other effective action, being idle or being insecure of the future, were the elements that tended to aggravate fear in combat."[16]

Noise is yet another common stressor and major cause for trepidation. An airborne officer reported that, in Tunisia in 1942, he witnessed a group of American ammunition carriers shocked into inactivity "simply by the tremendous noise of real fighting. Instead of getting the ammunition forward to a machine gun, these men were huddled together, hugging the ground, shaking — pitifully unaware that their route was protected by a hill."[17]

A fourth cause, associated with noise, but adding a significant additional factor that creates fear, is immobility due to shelling or fire. Samuel Stouffer, in his monumental study of the American soldier in the Second World War, reported that many veterans testified that the "severest fear-producing situation they encountered in combat was just such immobilization under artillery or mortar fire."[18]

Yet another cause of fear is deprivation. Obviously, all soldiers need sleep, food, and drink regardless of their level of physical fitness. Practical experience in conflicts since the Second World War has demonstrated that the physical and psychological factors that lowered morale and sapped the courage of participants were fatigue, hunger, and thirst.[19] Paradoxically, there is a symbiotic relationship between fatigue and fear. The more fatigued a person is — the more susceptible to fear they become. And, the greater their fear, the greater is the drain on their energy. "Tired men fright more easily," observed Colonel S.L.A. Marshall in his decades of battlefield studies, and he concluded, "... frightened men swiftly tire."[20]

Extreme fatigue ultimately makes it impossible for some men to continue to function. Psychologist F.C. Bartlett insisted, "There is perhaps no general condition which is more likely to produce a large crop of nervous and mental disorders than a state of prolonged and great fatigue." This is the result of four factors: (1) Physiological arousal caused by the stress of existing in what is commonly understood as a continual fight-or-flight arousal condition; (2) cumulative loss of sleep; (3) the reduction in caloric intake; and (4) the toll of the elements such as rain, cold, heat, and dark of night.[21]

A sixth cause is the fear of killing others. Western culture that inculcates in individuals, from an early age, the value of life and the abhorrence of killing others is deeply rooted in the psyche of soldiers. Another root cause of fear, particularly for military personnel, is the threat of being killed or wounded. Although this fear is self-explanatory and completely understandable, it does not appear to be the predominant cause of fear expressed by combatants. Israeli military psychologist Ben Shalit was surprised to find the low emphasis on fear of bodily harm and death, and the great emphasis on "letting others down."[22] Research and anecdotal evidence demonstrate that this finding is virtually universal in all militaries.

## THE IMPACT OF FEAR

Fear can have a debilitating effect if not managed correctly. British Colonel Ian Palmer, a Professor of Defence Psychiatry, reported, "Unfortunately, once given into, fear colours our cognitions leading us to expect the worst." He explained, "This contagion and catastrophic thinking is a real threat to the effectiveness of military operations." Palmer noted that fear may often lead to "freeze" and "flight" (i.e. panic) reactions.[23]

It is these effects, namely the consequences of fear, which can be so devastating. First, fear affects performance. Studies have shown that physical manifestations include "weight loss, tremors, abuse of alcohol and tobacco, insomnia, nightmares, cardiac irregularities, loss of confidence, and general nervous breakdown."[24] Moreover, after decades of battlefield studies, Colonel Marshall determined that "in the measure that the man is shocked nervously, and that fear comes uppermost, he becomes physically weak." He added that the "body is drained of muscular power and of mental coordination."[25]

Professor Dollard discovered that fear also led to over-caution. Of those he questioned, 59 percent stated that there were occasions when they were too cautious and had their efficiency reduced by fear. Moreover, fear can lead to panic. He also found that 75 percent of the veterans he interviewed believed that "fear can be contagious [and] that it can be transmitted from one soldier to another."[26]

Direct performance aside, fear can also cause severe emotional stress and psychiatric breakdown. Professor Anthony Kellett's examination of Second World War studies led him to believe that "More than anything else, fear itself is the critical ingredient in psychiatric breakdown in combat ... causes a strain so great that it causes men to break down."[27] Stouffer's seminal work reported that 83 percent of those questioned asserted that they had the experience of seeing "a man's nerves 'crack up' at the front."[28] This is significant as one study revealed that 70 percent of 1,700 American veterans surveyed in Italy in 1944, reported that they became nervous or depressed, or their morale suffered, at the sight of another man's psychiatric breakdown.[29]

Finally, fear can also impact adversely on decision-making. Research has shown "that during stressful combat-like training, every aspect of cognitive function assessed was severely degraded, compared to the subjects' own baseline, pre-stress performance." Moreover, the magnitudes of the deficits were greater than those typically produced by alcohol intoxication or treatment with sedating drugs. The study team concluded that "on the battlefield, the severe decrements we measured ... would significantly impair the ability of warfighters to perform their duties." Specifically, the team determined that extended periods of pressure and fear lead to over-reaction, an increase in wrong decisions and inconsistency.[30] Similarly, Professor Dinter noted that fear and exhaustion will also reduce the willingness to make decisions at all.[31]

Significantly, fear has a cumulative effect. Dollard's research indicated that fear increases in proportion to the duration of the engagement and the number of frightening incidents endured by an individual.[32] Scottish historian Hew Strachan concluded, "The battle-hardened veteran was a mythical figure." He discovered that "sustained exposure to danger did not harden a soldier but eroded his limited resources."[33] Marshall explained, "Sustained fear is as degenerative as prolonged fatigue and exhausts the body energy no less."[34] Similarly, Lieutenant-Colonel David Grossman determined from his research that, "In sustained combat this process of emotional bankruptcy is seen in 98 percent of all soldiers who survive physically."[35] Another contemporary report concluded, "All soldiers have a breaking point beyond which their effective performance in combat diminishes."[36] Quite simply, even the most psychologically strong person will eventually succumb. Simply put, no one ever becomes accustomed to fear — it is just a matter of trying to control it.

One study conducted during the Second World War by Lieutenant-Colonel J.W. Appel and Captain G.W. Beebe observed, "Each moment of combat imposes a strain so great that men will break down in direct relation to the intensity and duration of their exposure … the average point at which this occurred appears to have been in the region of 200–240 aggregate combat days." The British estimated that a rifleman would last for about 400 combat days — the longer period being attributable to the fact that they tended to relieve troops in the line for a four-day rest after approximately 12 days.[37] Another study confirmed that at 200–240 days of combat, the average soldier became "so overly cautious and jittery that he was ineffective and demoralizing to the newer men."[38]

## CONTROLLING FEAR

The key factor in respect to fear is not whether someone experiences fear, but rather how they control it. Quite simply, there are strategies for controlling fear and limiting the potential insidious effects it can have on individuals. The first step is simply acknowledging that fear is a normal occurrence. The existence of fear must not be repressed by individuals, nor should those who articulate their fear be ridiculed. Rather, the topic should be discussed. Research has indicated that eight out of ten combat veterans felt that it is better to admit fear and discuss it openly before battle. The belief that "the man who knows he will be afraid and tries to get ready for it makes a better soldier," was shared by 58 percent of those surveyed.[39] "If it [fear] is allowed to back up in a man, unspoken and unaired in any way," explained war correspondent Mack Morriss, "it can form a clot and create an obstacle to normal action."[40]

Another vital method for controlling fear is training and education.[41] Flavius Renatus asserted in 378 AD, "The courage of the soldier is heightened by the knowledge of his profession." Knowledge is the key as it provides confidence — not only in oneself, but in one's comrades, equipment, and tactics. This is achieved through realistic training, as well as a complete understanding of the realm of conflict. This reduces the fear of the unexpected and the unknown. It is for this reason that the British parachute school has adopted the motto, "Knowledge Dispels Fear." Its importance is such that General Slim asserted, "Training was central to the discipline soldiers needed to control their fear, and that of their subordinates in battle; to allow them to think clearly and shoot straight in a crisis, and to inspire them to maximum physical and mental endeavour."[42] Moreover, realistic training (e.g. battle simulation, full combat loads, non-templated enemy action, intense tempo, stress, physical exertion, and fatigue) creates reasonable expectations of how far an individual or unit can go and how long they can fight. It is also valuable to the extent it inculcates in soldiers the realization that they can survive on the battlefield.

As such, it is critical to add the element of ambiguity and the unknown in all training activities. In addition, training should be conducted at night, in poor light, and unknown surroundings. Moreover, it should include situations where things go wrong. This will assist with inoculating individuals to the fear of the unknown and accustom them to dealing with adversity. The beneficial effect of realistic training is undisputed. Research and studies have shown that "the general level of anxiety in combat would tend to be reduced insofar as the men derived from training a high degree of self-confidence about their ability to take care of themselves ... troops who expressed a high degree of self-confidence before combat were more likely to perform with relatively little fear during battle."[43]

The issue of confidence is an important one. Confidence is perhaps the greatest source of emotional strength that a soldier can draw upon. "With it," insisted behavioural expert Bernard Bass, "he willingly faces the enemy and withstands deprivations, minor setbacks, and extreme stresses, knowing that he and his unit are capable of succeeding."[44] Numerous studies have shown that well-led and cohesive units tend to have fewer stress casualties than units lower in these qualities.[45] Self-confidence can be achieved through training, education, and fitness, as well as through sound leadership, team cohesion, and dependable equipment.

The value of training is also derived from its ability to create an element of developing instinctive reactions. Drill, for instance, is utilized to teach the instinctive reaction of a body of troops to commands. "What is learnt in training," insisted commando commander Lord Lovat, "is done instinctively in action — almost without thinking down to the last man."[46]

In the same manner, discipline and response to leadership are crucial variables shaping attitudes among combat veterans. The role of discipline is one of providing a psychological defence that helps the soldier to control fear and ignore danger through technical performance. "It is a function of discipline," extolled Field Marshall Bernard Montgomery, "to fortify the mind so that it becomes reconciled to unpleasant sights and accepts them as normal every-day occurrences ... Discipline strengthens the mind so that it becomes impervious to the corroding influence of fear ... It instills the habit of self control."[47]

Another vital method for controlling fear is the maintenance of routine and habit. The adherences to simple daily routines, such as the ritual of shaving, provide a sense of normalcy, that is, in essence, reassurance to individuals. This is vital in maintaining an equilibrium that allows individuals to perform consistently. Lord Lovat declared, "Habit is ten times nature."[48]

There are a number of other strategies that can also be used to manage fear in a positive manner. Humour is the most important form of self-discipline and acts to release tension. Second World War veteran Howard Ruppel observed, "When circumstances become unbearable, the experienced soldier with some sense of humor and the ability to laugh at one's self has a better chance to retain his sanity than the serious minded fellow."[49]

For others, religion and faith provide a foil for fear. Another veteran, Max Kocour, revealed that faith among combat men was usually a general belief in God and was not centred around any particular religion or denomination. "We developed faith," he divulged, "regardless of religions, which had been created by man, we felt we were on the right side of faith, under the protection/care of a truly fine Supreme Being."

Others rely on more artificial tools for controlling their angst and fear. Alcohol and drugs are a time-honoured way of dealing with pain, fear, and stress — often more widespread than generally acknowledged. Although drugs and alcohol have often been used to help, the success is always of marginal value. It alleviates anxiety only temporarily, but, more importantly, it reduces the ability to act in a rational and coordinated manner. In addition, there are often long-term consequences of use.

A more effective tool for the management of fear, and one with less harmful side effects, is the timely and accurate passage of information. In the chaos of battle, information is almost a means of power. Individuals are hungry for anything that may shed light on events that are about to impact on their futures. Quite simply, knowledge dissipates the unknown and dampens groundless rumours. "If a soldier knows what is happening and what is expected of him," explained a veteran British officer, "he is far less frightened than the soldier who is just walking towards unknown dangers."[50] Theodore Roosevelt insisted, "Fear can be checked, whipped and driven from the field when men are kept informed."[51]

The passage of information is predicated on effective communications that are equally vital to staving off the effects of fear. It is critical to keep personnel informed as much as possible about virtually everything. It is not only the content of the message that is important but also the process itself. Regular communications ensures that everyone knows that they are not alone — that they are still part of a team. It is for this reason that communications should always be maintained at all cost. Initially, during the Second World War, the Allies believed that German and Japanese night attacks were amateurish and disorganized because of the excessive amount of yelling that was used. However, they later discovered that this was deliberate — not only as a means of control but also as a tool to manage fear.

Also essential for managing fear is simple activity. Once again, Dollard found that veteran soldiers quickly learn that to be busy means to be less afraid: "When fear is strong, keep your mind on the job at hand."[52] Major-General T.S. Hart, former director of medical services in the UK agreed. "There is no doubt," he asserted, "that inactivity at a time of tension breeds fear and that the best antidote … is purposeful action."[53]

Yet, another powerful tool for controlling fear is strong group cohesion/primary group relationships. As already noted, the greatest fear felt by most combat soldiers is the fear of letting down their comrades. This is a powerful impetus not to allow fear to create panic. S.L.A. Marshall asserted, "I hold it to be one of the simplest truths of war that the thing which enables an infantry soldier to keep going with his weapons is the near presence or the presumed presence of a comrade."[54]

This sense of obligation, coupled with a sense of responsibility for ensuring the well-being of others, also generates a feeling of responsibility for upholding the reputation of the unit. This sense of responsibility also helps to alleviate fear. Creating demanding expectations of combat behaviour in members and then linking a soldier's self-esteem to the reputation of the unit and the welfare of his fellow soldiers is a powerful control mechanism. Many believe that a man behaves as a hero or coward according to the expectations of others of how he is to behave. "The overwhelming majority of men," reported Dollard, "felt that they fought better after observing other men behaving calmly in a dangerous situation."[55] Marshall insisted, "No matter how lowly his rank, any man who controls himself automatically contributes to the control of others." He added, "Fear is contagious but courage is not less so."[56] This was also born out in research. Dollard revealed that 94 percent of those veterans surveyed stated they fought better after observing other men behaving calmly in a dangerous situation.[57]

Leadership is also a critical element in controlling fear. Dollard noted that 89 percent of those surveyed emphasized the importance of getting frequent instructions from leaders when in a tight spot.[58] Furthermore, evidence clearly indicates that leaderless groups normally become inactive.[59] Not surprisingly, Samuel Stouffer found that "cool and aggressive leadership was especially important" in pressing troops forward in dangerous and fearful situations such as storming across a beach raked by fire.[60]

This finding is based on the fact that "role modelling" has an extremely important influence on a person's reaction to threatening situations. With regard to the evocation of courageous behaviour, American enlisted men in the Second World War told interviewers that leadership from in front was very important.[61] Most research has reinforced the intuitive deduction that "men like to follow an experienced man ... [who] knows how to accomplish objectives with a minimum of risk. He sets an example of coolness and efficiency which impels similar behaviour in others." In this regard, the presence of strong thoughtful leadership creates "a force which helps resist fear."[62]

But, this effect is only present if there is trust in the leadership. Soldiers must believe that leaders mean what they say. Body language, tone, and eye contact all betray insincerity. Actions must match words. In the end, it comes down to setting the example. Leaders must never ask, or expect, troops to do that which the leaders are unwilling to do themselves. Stouffer's study showed that what the officers did, rather than what they said, was important.

## SUMMARY

In the end, fear is a real and powerful force. But, the essence of the issue is not whether one experiences fear, but rather how it is dealt with. It can be controlled and utilized to benefit the effectiveness of individuals and units in times of danger. Conversely, the failure to recognize the reality of fear and its effects can have serious repercussions that manifest themselves at the most disadvantageous moments. As such, it is important to ensure that the necessary steps are taken to ease anxiety and fears. For example, it is important to discuss the issue to ensure that the perceptions and expectations of leaders and subordinates alike are realistic. Leaders must also imbue confidence in individuals, teams and their equipment, and must develop strategies to allow all to feel a sense of control over their destinies, regardless of activity or operation. Moreover, leaders must develop contingency planning and undertake additional training and education so that individuals are better able to cope with the unknown or unexpected. Although the very complex and unpredictable human nature of war mitigates against assurances that an individual or a group will not allow fear to take control of their actions, the overwhelming and compelling human desire not to let down comrades, combined with solid leadership, training, and strong unit cohesion, provide the necessary counterbalance to fear.

## NOTES

1. John Keegan, *The Face of Battle* (London: Penguin, 1976), 71.
2. Elmar Dinter, *Hero or Coward* (London: Frank Cass, 1985), 12.
3. Dennis Coon, *Introduction to Psychology*, 8th ed. (New York: Brooks/Cole Publishing Company, 1998), 429.
4. Rita Atkinson, Richard Atkinson, Edward Smith, Daryl Bem, and Susan Nolen-Hoeksema, eds., *Hilgard's Introduction to Psychology*, 12th ed. (New York: Harcourt Brace College Publishers, 1996), 379–380.
5. David M. Myers, *Psychology*, 4th ed. (Holland, MI: Worth Publishers, 1995), 433. See also Coon, 431.
6. John C. McManus, *The Deadly Brotherhood: The American Combat Soldier in World War II* (Novato, CA: Presidio Press, 1998), 251.
7. John Dollard, *Fear in Battle* (Westport, CT: Greenwood Press, Publishers, 1944), 56. Dollard's research was based on his study of 300 American volunteers who fought in the Spanish Civil War.
8. S.J. Rachman, *Fear and Courage* (San Francisco: W.H. Freeman

and Company, 1978), 6.
9. John T. Wood, *What Are You Afraid Of?* (Englewood Cliffs, NJ: Prentice Hall, 1976), 22.
10. Jeffrey Alan Gray, *The Psychology of Fear and Stress*, 2nd ed. (Cambridge, UK: Cambridge University Press, 1987), 19.
11. Dollard, 2.
12. *Ibid.*, 24.
13. Rachman, 84.
14. Major-General (retd) Robert H. Scales Jr., *Yellow Smoke: The Future of Land Warfare for America's Military* (New York: Rowman & Littlefield Publishers Inc., 2003), 168.
15. Dinter, 18 and 98; and Wood, 28–29.
16. Rachman, 50–52.
17. Don Wharton, "Bringing the War to the Training Camps," *The Reader's Digest*, Vol. 42, No. 254 (June 1943): 37.
18. Samuel A. Stouffer, *The American Soldier: Combat and Its Aftermath*. Vol. II (Princeton, NJ: Princeton University Press, 1949), 83; and Rachman, 82.
19. Major P.B. Deb, "The Anatomy of Courage," *Army Quarterly*, Vol. 127, No. 4 (October 1997): 405.
20. S.L.A. Marshall, *The Soldier's Load and the Mobility of a Nation* (Quantico, VA: The Marine Corps Association, 1950), 46.
21. Lieutenant-Colonel David Grossman, *On Killing* (New York: Little, Brown and Company, 1996), 69.
22. *Ibid.*, 52.
23. Colonel Ian Palmer, "The Emotion That Dare Not Speak Its Name," *The British Army Review* 132 (Summer 2003): 33.
24. Allan D. English, "A Predisposition to Cowardice? Aviation Psychology and the Genesis of 'Lack of Moral Fibre,'" *War and Society*, Vol. 13, No. 1 (May 1995): 17.
25. Marshall, *The Soldier's Load*, 41.
26. Dollard, 28; and Rachman, 76.
27. Anthony Kellett, *Combat Motivation — Operational Research and Analysis Establishment (ORAE) Report No. R77* (Ottawa: DND, 1980), 268.

28. Stouffer, Vol. 2, 124–25, 134, 208–9. See also Rachman, 61, 76–78.
29. *Ibid.*, Stouffer and Rachman.
30. H.R. Lieberman, G.P. Bathalon, C.M. Falco, J.H. Georgelis, C.A. Morgan III, P. Niro, and W.J. Tharion, "The Fog of War: Documenting Cognitive Decrements Associated with the Stress of Combat," abstract in *Proceedings of the 23rd Army Science Conference,* December 2002.
31. Dinter, 82.
32. Dollard, 22.
33. Quoted in Brigadier-General Denis Whitaker and Shelagh Whitaker, *Rhineland: The Battle to End the War* (Toronto: Stoddart, 2000), 351.
34. Marshall, *The Soldier's Load*, iii.
35. Grossman, 84.
36. Jeremy Manton, Carlene Wilson, and Helen Braithwaite, "Human Factors in Field Training for Battle: Realistically Reproducing Chaos," in *The Human Face of Warfare: Killing, Fear and Chaos in Battle,* eds. Michael Evans and Alan Ryan (St. Leonard's, Australia: Allen & Unwin, 2000), 188.
37. Richard Holmes, *Acts of War: The Behaviour of Men in Battle* (New York: The Free Press, 1985), 215.
38. William Ian Miller, *The Mystery of Courage* (Cambridge, MA: Harvard University Press, 2000), 61.
39. Dollard, 2–3, 24.
40. McManus, 251. It must be noted that some psychologists believe that an unnecessary focus on fear or sharing your feelings about being afraid may actually enhance your fears.
41. Training is defined as "a predictable response to a predictable situation," as opposed to education which is "the reasoned response to an unpredictable situation — critical thinking in the face of the unknown." Professor Ronald Haycock, "Clio and Mars in Canada: The Need for Military Education," presentation to the Canadian Club, Kingston, Ontario, 11 November 1999.
42. Robert Lyman, *Slim, Master of War* (London: Constable, 2004), 78.

43. Rachman, 63–64.
44. B.M. Bass, *Leadership and Performance Beyond Expectations* (New York: Free Press, 1985), 69.
45. J.G. Hunt and J.D. Blair, *Leadership on the Future Battlefield* (New York: Brassey's, 1986), 215.
46. Will Fowler, *The Commandos at Dieppe: Rehearsal for D-Day* (London: HarperCollins, 2002), 55.
47. Field Marshall Bernard L. Montgomery, "Discipline from Morale in Battle: Analysis," in Canada, *The Officer: A Manual of Leadership for officers in the Canadian Forces* (Ottawa: DND, 1978), 66.
48. Quoted in Fowler, 55.
49. McManus, 247.
50. Lieutenant-Colonel Colin Mitchell, *Having Been a Soldier* (London: Mayflower Books, 1963), 41.
51. Quoted in Canada, *CDS Guidance to Commanding Officers* (Ottawa: DND, 1999), 230.
52. Dollard, 3.
53. Kellett, 281. Dollard's study found that 71 percent felt fear most acutely just before going into action (from not knowing what to expect).
54. Miller, 214.
55. See Dollard, 28; and Rachman, 76.
56. Miller, 209.
57. Dollard, 28.
58. *Ibid.*, 44.
59. Dinter, 92.
60. Stouffer, 68.
61. Kellett, 299.
62. Dollard, 44.

## SELECTED READINGS

Dinter, Elmar. *Hero or Coward.* London: Frank Cass, 1985.

Dollard, John. *Fear in Battle.* Westport, CT: Greenwood Press, Publishers, 1944.

Evan, Michael and Alan Ryan, eds. *The Human Face of Warfare: Killing, Fear and Chaos in Battle.* St. Leonard's, Australia: Allen & Unwin, 2000.

Gray, Jeffrey Alan. *The Psychology of Fear and Stress,* 2nd ed. Cambridge, UK: Cambridge University Press, 1987.

Holmes, Richard. *Acts of War: The Behaviour of Men in Battle.* New York: The Free Press, 1985.

Horn, Colonel Bernd. "Revisiting the Worm: An Examination of Fear and Courage," *Canadian Military Journal,* Vol. 5, No. 2 (Summer 2004): 5–16.

McCoy, Colonel B.P. *The Passion of Command: The Moral Imperative of Leadership.* Quantico, VA: Marine Corps Association, 2006.

Palmer, Colonel Ian. "The Emotion That Dare Not Speak Its Name," *The British Army Review* 132 (Summer 2003).

Rachman, S.J. *Fear and Courage.* San Francisco: W.H. Freeman and Company, 1978.

Wood, John T. *What Are You Afraid Of?* Englewood Cliffs, NJ: Prentice Hall, 1976.

# 21

# FOLLOWERSHIP

by Brent Beardsley

An often-stated truism of leadership is that before one can become an effective leader, one must first learn to be an effective follower and retain that characteristic throughout their career. Every leader, regardless of their rank, position or appointment, must be an effective follower. "Followership" is a relatively new term for this well-established and widely accepted concept. The Canadian Forces (CF) adopted the term from the U.S. Air Force for use in its new leadership doctrine as contained in *Leadership in the Canadian Forces: Leading People*.[1]

## WHAT IS FOLLOWERSHIP?

Effective followership is defined in CF doctrine as "fully committed service to the achievement of mission success by being receptive, implementing change, and helping to build a culture consistent with the CF military ethos."[2] Effective followers, as military professionals, must first, foremost, and always be effective and valued members of a unique profession and as such, must believe and consistently live up to the ethos of their respective profession (See Chapter 28 on The Military Professional in this handbook).

Effective followers:

1. get the job done professionally, effectively, legally, and ethically;

2. are valued, active, effective, and essential members of their team, thinking and acting in support of the larger team;
3. are professionally committed to service to the organization before self;
4. are receptive implementers of change; and
5. are living the military ethos.[3]

## **MEILINGER'S 10 RULES OF GOOD FOLLOWERSHIP**

Colonel Philip Meilinger of the U.S. Air Force authored a seminal paper on followership in which he provided a list of 10 rules for effective followers. Each of these rules speaks to one or more of the core Canadian Military Values of Duty, Loyalty, Integrity, and/or Courage:

1. *Don't blame your boss for an unpopular decision or policy: Your job is to support, not undermine.*

    This rule speaks to Loyalty. Effective followers are loyal in their personal allegiance to their country, to the rule of law, and to their superiors, peers, and subordinates alike. Loyalty is essential to build trust in all directions. Unless a decision or policy is manifestly unlawful or clearly unethical, effective followers must be loyal regardless of the unpopularity of the decision or policy. They must treat the direction or policy as if it was their own.

2. *Fight with your boss if necessary: But do it in private, avoid embarrassing situations, and never reveal to others what was discussed.*

    This rule speaks to Courage with the interpersonal relationship that an effective follower must build with their respective superior(s). An effective leader expects effective followers to have the moral courage to objectively and logically debate, discuss, and even to challenge their direction before it is formally issued or even after it is issued, but only in private and always in a confidential manner. The effective follower provides critical input before the decision is made, accepts the direction issued, and only brings criticism to the leader in a respectful and private manner.

3. *Make the decision, then run it past the boss — use your initiative.*

   This rule speaks to Duty in the effective use of initiative. Effective followers clearly understand and accept the requirement for using their initiative to solve unexpected problems or to take advantage of unanticipated opportunities in a dynamic organization. They also understand the limits of the use of their initiative. This rule contributes to trust. Trust by the leader that the effective follower understands their intent and trust by the follower that they will be supported for using their initiative.

4. *Accept responsibility whenever it is offered.*

   This rule speaks to Duty in seeking and accepting responsibility in a dedicated and disciplined manner in order to effectively contribute to mission success.

5. *Tell the truth and don't quibble. Your boss will be giving advice up the chain of command based on what you said.*

   This rule speaks to Integrity. Tell the truth, the whole truth and nothing but the truth. If the full truth cannot be told, explain why. The entire basis of trust between leaders and followers is based on integrity between the two in all of their interpersonal relationships.

6. *Do your homework. Give your boss all the information needed to make a decision; anticipate possible questions.*

   This rule speaks to Duty in the requirement for effective followers to ensure their leaders are fully briefed and prepared with the timely, accurate, and relevant information they require to fully inform their leaders in making timely, accurate, and relevant decisions.

7. *When making a recommendation, remember who will probably have to implement it. This means you must know your own limitations and weaknesses as well as your strengths.*

   This rule speaks to the Duty of self-assessment. As a general rule, the follower who makes recommendations should be the one who takes ownership of that recommendation and fully and successfully implements the recommendation. However, should the follower not possess the expertise or experience to effectively implement the recommendation, this weakness should be fully stated to the leader. The ultimate responsibility of who will implement remains with the leader, but in order to determine who will implement, the leader must be fully informed as to the strengths and weaknesses of followers.

8. *Keep your boss informed of what's going on in the unit; people will be reluctant to tell him or her about their problems and successes. You should do it for them, and assume someone else will tell the boss about yours.*

   This rule speaks to Duty and Loyalty. No leader wants to be surprised by a personal crisis of one of their followers and it is the duty of each and every follower to ensure the leader is fully informed of the problems, strengths, weaknesses, and potential crises in their organization. Use this duty as an opportunity to praise those who deserve praise and to inform the leader of those who may require special attention or support.

9. *If you see a problem, fix it. Don't worry about who would have gotten the blame or who now gets the praise.*

    This rule speaks to Integrity, Loyalty, and Duty. Effective followers think first and foremost of the organization and then of themselves. They meet and defeat problems or challenges to the organization before they become crises and without regard as to who was responsible or who should get the blame or credit. The effective and efficient functioning of the organization in achieving mission success is the only concern of the effective follower.

10. *Put in more than an honest day's work, but don't ever forget the needs of your family. If they are unhappy, you will be too, and your job performance will suffer accordingly.*

    This rule speaks to Duty. Effective followers understand that they must take care of themselves and of their families, along with other obligations in addition to their professional obligations. Effective followers must live balanced lives that satisfy the personal, family, and professional duties and obligations. Ultimately, the follower must determine that balance ensure that all responsibilities are prioritized and met.[4]

## SUMMARY

Effective followers need to reverse role as effective leaders must also reverse role. The effective leader should consistently be asking themselves what type of effective follower they want in their organization and then to assume that role with their leaders. The effective follower should consistently be asking themselves, if they were the leader, what type of followers would they want in their organization and then to assume that role. Effective followership is a relatively new term for a truism that has been known and accepted throughout history in professional militaries. It is based on the key idea that everyone who serves, regardless of their rank or appointment, is a follower and must master and maintain effective followership in the performance of their duties. The rules of followership are clearly based on the core values of the ethos of the profession. Effective followers must clearly understand, believe, and, in all that they say and do, live the core values of their profession. In doing so, professionals will be effective followers in their organization.

## NOTES

1. Canada, *Leadership in the Canadian Forces: Leading People.* (Kingston: DND, 2007), 9.
2. *Ibid.*, 9.
3. *Ibid.*, 9.
4. Philip Meilinger, "The Ten Rules of Good Followership," *AU 24 Concepts for Air Force Leadership* (Montgomery, AB: Air University, 2001).

## SELECTED READINGS

Canada, *Duty with Honour: The Profession of Arms in Canada.* Kingston: DND, 2007.

Canada. *Leadership in the Canadian Forces: Leading People.* Kingston: DND, 2007.

Kelley, R.E. *The Power of Followership: How to Create Leaders People Want to Follow, and Followers Who Lead Themselves.* New York: Doubleday-Currency, 1992.

Meilinger, P.S. Col. "The Ten Rules of Good Followership," AU-24 *Concepts for Air Force Leadership.* Maxwell: Air University Press, 2001.

# 22
# GENDER
by Karen D. Davis

Gender is understood and conceptualized in many ways, all of which have an impact on how individuals understand relationships among people, how organizations develop policies, and how leaders make decisions that impact women and men in various ways. Most often gender is understood as a substitute for male and female as categories of "sex." However, the meanings and implications of gender reach beyond two essential categories of male and female. In addition, the ways in which gender is understood will vary across time and context. It is difficult to think of a context in which gender has been more salient than it has been in military organizations such as the Canadian Forces. During the past three decades, human rights legislation, and demographic and social change have guided the sanctioned inclusion of women, gays, and lesbians in the Canadian military, and the organization has responded through change to policies, practices, and leadership. These changes, while welcome and embraced by many, do create challenges to effective leadership in organizations. The discussion below highlights various understandings of gender and the implications for leaders in organizations.

## GENDER DEFINED

Beginning in the latter part of the 20th century, scholars of gender studies have struggled with dichotomous notions of male/female, man/woman, masculinity/femininity, and what those dichotomies mean to individuals, organizations, and societies. Judith Butler, for example, has attempted to address several questions. Is gender an essential attribute that a person is said to be? If gender is socially and culturally constructed, what is the manner or mechanism of the construction? Can gender be constructed differently or does its "constructedness" imply social determinism? Are individuals agents in generating and transforming their own gender?[1]

Within the academic literature, gender has been commonly understood as a socially constructed concept for more than 25 years;[2] however, contemporary understandings of gender are rooted within the status of women relative to men in society. Understandings of gender can also change over time, and organizations respond to these changes in various ways. For example, until the late 1980s the Canadian Forces had an internal Directorate of Women; in 1990 an external Advisory Board on *Women* (emphasis added) in the Canadian Forces[3] was established, and in 1993 that same board became the Minister's Advisory Board on *Gender Integration* (emphasis added). Until approximately 1990, the current and historical participation of women in Canada was understood largely in terms of the gaps and differences between women and men. The shift in focus from women to gender is representative of the increasing awareness of the importance of the relations between women and men on the experiences and status of both women and men. The shift in language, however, leaves many still unaware of the significance of gender in developing awareness of society, its organizations, and its people.

Conceptions of gender based upon categories of sex identify the biological differences between women and men.[4] From this perspective, there are two unique and mutually exclusive categories of gender informed by assumptions of biological determinism: male and female. Such understandings of gender assume that "the male/female dichotomy is natural; being masculine or feminine is natural and not a matter of choice; all individuals can (and must) be classified as masculine or feminine"; and these differences are relatively immutable — exceptions are frequently considered deviant in some way.[5] Although less prevalent than in the past, this understanding continues to be reinforced by culture, myth, law, education, and religion.[6] Indeed, military policies restricting the participation of women, based upon the status of *woman* or *female*, reinforce gender as a biologically determined and dichotomous category. It is important for leaders to clarify their understandings of gender, and thus the combined effect of their leadership and gender on social and institutional relationships.[7]

In her historical analysis of gender, Joan W. Scott traces contemporary usage of the term *gender* to the feminist movement and its rejection of biological determinism implicit in the use of terms such as *sex* or *sexual difference*.[8] Alternatively, the use of the term gender was introduced to "insist on the fundamentally social quality of distinctions based on sex."[9] In its simplest recent usage, gender has become a synonym for women. The term gender sounds somewhat more neutral and objective than "women" and thus dissociates itself from what has been broadly understood as the strident politics of feminism.[10] Regardless of attempts to avoid feminist labels and consequences, the use of the term gender has persistently created perceptions that activities conceived from a gender perspective are really about women, if not about advocacy for women. In most organizations if the title of a leadership activity, for example, has the word gender in it, it is quite likely that most of the attendees will be women rather than men, as it will be broadly perceived that it reflects perspectives of women and is, therefore, of interest only to women. Unfortunately, this type of response also underlines the notion that perspectives reflecting gender or women, however it is understood, are of little value relative to mainstream leadership activities. Interestingly, after completing his extensive study on war and gender, political scientist Joshua Goldstein asserted "Men *should* pay more attention to gender. We learn about ourselves by doing so."[11]

Gender refers not only to women and men, but also to the relationships between them, thus introducing "a relational notion that women and men [are] defined and understood in terms of one another, and no understanding of either [can] be achieved in an entirely separate manner."[12] The relational aspect of gender is also of considerable importance to leaders. Most experienced leaders will recognize the dilemma of implementing a policy or accommodation for one individual or group of employees, only to create resentment among those who do not share the same characteristics and thus are not in a position to benefit from the policy. Leaders also understand the importance of shared experience in developing cohesive and productive teams and as such approach leadership from what can be referred to as *gender-neutral*, *gender-free*, or a *gender-blind* perspective. In many situations, this approach will be effective as well as necessary in demonstrating that all members of the team are equally important to the team as well as sharing the load in an equitable manner. However, this approach breaks down when assumptions are made about the extent to which the experiences and perspectives of team members are shared beyond their immediate role on the team. That is, regardless of gender, the experiences that men and women bring to the organization will differ depending upon an endless array of factors, including gender role expectations in their formative years, race, class, religion, ethnicity, sexual orientation, abilities, interests, etc. In addition, there are undeniable biological differences between men and women and the various needs that those differences create. The challenge then is to address gender through a *gender-inclusive* approach that abandons socially constructed assumptions about women and men.[13]

## **LEADERSHIP AND GENDER**

Understanding gender and gender relationships is an important leadership competency with real implications for leaders and the performance of their teams. Consciously or unconsciously, leadership competency in this domain acts as a powerful conduit of the attitudes and values that prevail in the work environment. An approach that is not inclusive or not perceived to be fair will create divisions among team members that will often be counterproductive, if not disruptive, to team dynamics.

Perhaps the greatest challenge is the accommodation of women, and mothers in particular, into traditionally all-male organizations. While policies such as parental leave can go a long way toward alleviating the perceived inequalities, the fact remains that women are biologically different from men, and, therefore, some women may require maternity-related absence from the work force. However, maternity represents just one dimension of becoming and being an active parent for the better part of two decades. In addition, this does not impact all women, and parenting is a significant aspect of the lives of many men. It has also been argued that biology is less determined than is often believed; that is, biology provides diverse potential that is limited to a great extent by culture and social determination.[14] In any case, leadership and organizational response to the various demands and opportunities of parenting that place undue focus on the absence of women from the workplace for maternity are responding to underlying assumptions of biological determination, without recognizing the ways in which beliefs about separate gender roles are informing their response.

In fact, the role of parents, regardless of whether the parent is male or female, is shaped by social expectations that are reinforced within organizations. Pointing out that gender is manifest in personal identities and social interactions among women and men, sociologist Judith Lorber attributes institutions with the "establishment of patterns and expectations for individuals …."[15] That is, although we observe gendered relationships among women and men, the claim is that institutions, not individuals, create gendered processes. Also, Lorber does not believe that gender is synonymous with men's domination of women, and instead asserts there are cross-cutting racial and class statuses within gender status "that belie the universal pattern of men's domination and women's subordination implied by the concept of patriarchy,"[16] and early feminist conceptions of gender. In other words, it is not safe to assume that the experiences of all women or all men will be the same simply because of their gender. The implications for leaders in organizations are particularly significant from this perspective as it underlines the importance of the influence of policies, practices, and leadership within organizations on the ways in which women and men will understand gender and relate to one another both within and across gendered categories. Within an organization such as the military, for example, junior non-commissioned women and men will often share more of the same experiences and perspectives on military service than junior non-commissioned men will share with male senior officers. When women and men take on senior leadership roles in organizations, their authority, responsibilities, and accountabilities continue to shape their experience as they develop from a leader in the organization to a leader representing the organization, its values, and its culture. Essentially, the way in which leaders understand and address issues, including gender-related issues, is a unique aspect of diversity and organizational culture.

## **THE GENDER OF THE LEADER**

Organizational culture can also have a significant impact on what followers expect from leaders, including what is considered to be effective leadership or effective leadership style. In their review of gender and leadership, psychologists Linda Zugec and Karen Korabik concluded that leaders who are both task-oriented and person-oriented, possess a wide repertoire of behaviours, and are more behaviourally flexible,[17] function more effectively than leaders who are focused on either people or tasks, are less flexible, and rely on a particular leadership style. Essentially, an effective leader relies on a repertoire of abilities and behaviours that represent characteristics that have been labelled as both masculine and feminine in nature.[18] However, Zugec and Korabik also identify a range of issues that have had a negative impact on women in leadership roles in military organizations: organizational culture; token status in male-dominated settings; occupational segregation; and gender stereotyping, which can lead to negative perceptions of the performance of female leaders. For example, in a recent study of military leadership characteristics at the United States Air Force Academy, it was found that male cadets possessed strong masculine stereotypes of successful leaders and perceived a lack of similarity between women and leaders.[19]

Even though women's leadership effectiveness is often perceived differently than that of their male counterparts, there is overwhelming evidence to indicate that there is no significant difference in the leadership abilities and potential of women and men. The fact that women are often perceived differently reinforces understandings of gender as socially and culturally constructed; that is, regardless of ability, commonly held expectations of appropriate roles and behaviours for women and men will influence gender relations. Feminine leadership traits in women[20] and men may be perceived negatively when masculine styles are given more value. Women may also be perceived negatively when they exhibit masculine leadership characteristics and behaviours[21] that challenge social expectations of women and femininity. Expectations that fall within the male/masculine versus female/feminine dichotomy, without recognizing differences across gender, can create adversarial relationships among and between women and men, as well as limit the effectiveness of leaders, followers, and the team overall.

## **IMPLICATIONS FOR LEADERS**

It is important that leaders recognize the basic influences that shape the behaviours of men and women throughout their social and professional lives. Cristina Trinidad and Anthony H. Normore, specialists in leadership and education, summarize those influences as: socialization; culture of origin; and organizational culture.[22] These three factors can result in differences in leadership style as well as differences in perceptions of appropriate leadership style across gender and ethnicity. Effective leaders establish a climate of respect for women and men, regardless of gender differences, thus allowing all members of the team to contribute to their full potential and develop an understanding of what is expected and effective in terms of leadership in the organization.

There is ample evidence suggesting that flexibility in leadership style and approach is one of the hallmarks of an effective leader, and more importantly that "effective leadership is not the domain of either gender and both can learn from the other."[23] Mixed gender organizations, including the military, have a particular opportunity to develop this flexibility within their leadership cadre. While there are obvious benefits, this is a challenging goal. In a study of the social construction of gender in organizations, sociologists Mats Alvesson and Yvonne Due Billing note that gender is most usefully understood as "a number of dynamic, ambiguous and varying phenomena …"[24] Given this potential complexity, it is tempting to seek understanding through categorizing within as few categories as possible; however, this complexity also serves as a caution to oversimplifying and labelling individual behaviours and organizational processes related to leadership.

Changing the expectations within a culture that has relied on homogenous interpretations of leadership styles and expectations can be difficult and may require non-traditional strategies. In the education field, for example, where most teachers are women and most administrators and leaders are men, mentoring and networking have proven particularly important to the success of women leaders.[25] While the participation of women in previously male-dominated organizations is increasing, once in leadership and operational roles, women frequently find themselves isolated from other women, including women who have successfully advanced into senior leadership roles.[26] All leaders can learn from the experience of successful women as well as ensure that their male and female followers have an opportunity to learn from the experiences of women and men who are effective leaders.

## **SUMMARY**

Exercising leadership, regardless of the characteristics of followers, means acting upon assumptions regarding expected and appropriate gender roles and behaviours. By continuously challenging those assumptions and increasing gender awareness, leaders can ensure that the full potential of all members of the team, regardless of gender, is leveraged in achieving overall mission success. If a particular activity, requirement, reward, or issue is gender-neutral, expectations for all members of the team should also be gender-neutral. If different individuals or groups will be impacted in different ways, a gender-inclusive strategy that considers a range of needs may be required. Leadership is leadership, regardless of gender. However, the challenge is in gaining an effective understanding of the ways in which gender is at work on your team, as well as how you, the leader, are influencing team relationships and performance.

## NOTES

1. Judith Butler, *Gender Trouble: Feminism and the Subversion of Identity* (NewYork and London: Routledge, 1990), 7.
2. Mark Hussey, Preface to *Masculinities: Interdisciplinary Readings* (Upper Saddle River, NJ: Prentice Hall, 2003), iii.
3. The Minister's (Minister of National Defence) Advisory Board was established in response to 1989 Canadian Human Rights Tribunal ruling that directed external monitoring of gender integration in the Canadian Forces.
4. Status of Women Canada, *Gender-Based Analysis: A Guide for Policy-Making* (Ottawa: 1998).
5. Mark Hussey, ix.
6. *Ibid.*
7. This assertion expands upon Joan W. Scott's observation, in "Gender: A Useful Category of Historical Analysis," *American Historical Review,* Vol. 91, No. 5 (1986): 1069, that thinking about gender is often not done precisely or systematically and thus the need to clarify and specify how one thinks "about the effect of gender in social and institutional relationships …"
8. Joan W. Scott, "Gender: A Useful Category of Historical Analysis," 1054.
9. *Ibid.*
10. *Ibid.*, 1056.
11. Joshua Goldstein, *War and Gender: How Gender Shapes the War System and Vice Versa* (New York: Cambridge University Press, 2001), xiv.
12. Joan W. Scott, "Gender: A Useful Category of Historical Analysis," 1054.
13. See, for example, Franklin C. Pinch, "Diversity: Conditions for an Adaptive, Inclusive Military," in *Challenge and Change in the Military: Gender and Diversity Issues,* eds. Franklin C. Pinch, Allister T. MacIntyre, Phyllis Browne, and Alan C. Okros (Kingston: Canadian Defence Academy Press, 2004), 171–194.
14. Joshua Goldstein, *War and Gender.*

15. Judith Lorber, "Paradoxes of Gender," *Masculinities: Interdisciplinary Readings* (Upper Saddle River, NJ: Prentice Hall, 2003), 3.
16. *Ibid.*, 4; Joshua Goldstein, in *War and Gender*, defines patriarchy as "... social organization based men's control of power," 2.
17. Linda Zugec and Karen Korabik, *Multiple Intelligences, Gender, and Leadership Effectiveness in Male-Dominated Versus Gender-Balanced Military Units: A Review of the Literature* (Toronto: Defence Research and Development Canada, 2004), 16.
18. *Ibid.* Zugec's and Korabik's analysis categorizes the integration of masculine and feminine characteristics as androgyny.
19. Lisa A. Boyce and Ann M. Herd, "The Relationship between Gender Role Stereotypes and Requisite Military Leadership Characteristics," *Sex Roles*, Vol. 49, No. 7/8 (October 2003): 373.
20. See for example, Angela R. Febbraro, *Women, Leadership and Gender Integration in the Canadian Combat Arms: A Qualitative Study* (Toronto: Defence Research and Development Canada, 2003).
21. *Ibid.*
22. Cristina Trinidad and Anthony H. Normore, "Leadership and Gender: A Dangerous Liaison?" *Leadership and Organization Development Journal*, Vol. 26, No. 7 (2005): 577.
23. Steven H. Appelbaum, Lynda Audet, and Joanne C. Miller, "Gender and Leadership? Leadership and Gender? A Journey through the Landscape of Theories," *Leadership and Organization Development Journal*, Vol. 24, No. 1 (2003): 49.
24. Mats Alvesson and Yvonne Due Billing, "Beyond Body-Counting: A Discussion of the Social Construction of Gender at Work," in *Gender, Identity and the Culture of Organizations*, eds. Iris Aaltio and Albert J. Mills (London and New York: Routledge, 2002), 74.
25. Cristina Trinidad and Anthony H. Normore, "Leadership and Gender: a Dangerous Liaison?," 582.
26. See for example, Karen D. Davis, introduction to *Women and Leadership in the Canadian Forces: Perspectives and Experience* (Kingston: Canadian Defence Academy Press, 2007), 2.

## SELECTED READINGS

Aaltio, Iris and Albert J. Mills, eds., *Gender, Identity and the Culture of Organizations*. London and New York: Routledge, 2002.

Davis, Karen D., ed., *Women and Leadership in the Canadian Forces: Perspectives and Experience*. Kingston: Canadian Defence Academy Press, 2007.

Febbraro, Angela R. *Women, Leadership and Gender Integration in the Canadian Combat Arms: A Qualitative Study*. Toronto: Defence Research and Development Canada, 2003.

Goldstein, Joshua. *War and Gender: How Gender Shapes the War System and Vice Versa*. New York: Cambridge University Press, 2001.

Hussey, Mark. Preface to *Masculinities: Interdisciplinary Readings*. Upper Saddle River, NJ: Prentice Hall, 2003.

Pinch, Franklin C., Allister T. MacIntyre, Phyllis Browne, and Alan C. Okros, eds. *Challenge and Change in the Military: Gender and Diversity Issues*. Kingston: Canadian Forces Leadership Institute, 2004; second printing, Canadian Defence Academy Press, 2006.

Scott, Joan W. "Gender: A Useful Category of Historical Analysis," *American Historical Review* Vol. 91, No. 5 (1986): 1053–1075.

Status of Women Canada. *Gender-Based Analysis: A Guide for Policy-Making*. Ottawa:, 1998.

Zugec, Linda and Karen Korabik. *Multiple Intelligences, Gender, and Leadership Effectiveness in Male-Dominated Versus Gender-Balanced Military Units: A Review of the Literature*. Toronto: Defence Research and Development Canada, 2004.

# 23

# GRIEF

by Rhonda Gibson and Robert D. Sipes

C ombat involves death.

So when I teach, one of the things I believe we need to do is embrace the word "kill." You will read 100 military manuals, and you'll never see the word "kill." It's a dirty four-letter word. It's an obscene word. And yet it's what we do ...[1]

In a relatively brief period of time, Canadian Forces (CF) soldiers have transitioned from fifty years of peacekeeping and being largely passive witnesses to conflict, to active combat and engagement in the incredible acts of violence that war demands. Since 2002, Canada has lost more than 87 soldiers and a member of the Canadian Diplomatic Corps in Afghanistan, not to mention the scores of serious combat injuries. Moreover, Canada has suffered the second-highest number of combat causalities of all contributing countries.[2] Without debate, the tempo and duration of dangerous military operations have increased. Given this shift, the challenge for the CF is to become better prepared in a technical or psychological sense, to deal with death and grief. This, however, represents a significant undertaking since researchers have yet to confirm the long-term effects of military personnel serving in such operations.[3]

Sudden death as a traumatic event can shatter a person's belief that he or she and the environment is safe. "It takes only a fraction of a second to shatter that sense of well-being, a brief moment when we have an accident or an instant when a blood clot forms in a loved one's aorta. When such an event happens, the world we knew suddenly seems to vanish, and suddenly we see ourselves as the vulnerable creatures we always knew we were but somehow never fully acknowledged. Much as we have prepared ourselves for that moment, when it does arrive we are going to feel emotional pain."[4]

With certainty, death in combat will occur. It is therefore incumbent on leaders to best prepare themselves and their soldiers, through knowledge and skill acquisition, to process loss. This preparation will help to maintain soldier fitness and well-being and, ultimately, mission success. To do so, however, requires that leaders be versed in the theory of grief and grief reactions so that they can help provide soldiers with the tools required to successfully manage death in theatre. This chapter reviews the psychological process of grief resulting from death and loss in theatre, with a focus on the education phase of "grief inoculation,"[5] the goal being to better prepare leaders to understand and respond to their personal and subordinate needs.

## GRIEF DEFINED

The word *grief* is derived from the Latin word *gravare*, meaning to burden or to cause distress.[6] At face value, it is a concept with which most people are familiar or have experienced. Grief as a concept and a predictable human response, however, is without a consensus definition.[7] It is vague and ambiguous.[8] The *Webster New World Dictionary* defines grief as: "Intense emotional suffering caused by a loss, disaster, misfortune; acute sorrow; deep sadness."[9] In this sense, grief is a response or, more likely, an adaptational response to loss.[10] Traumatic grief refers to the experience of the sudden loss of a significant and close attachment. This type of grief is likely experienced in combat.

## HUMAN RESPONSE TO DEATH

Exposure to death can be both disturbing and frightening. Although the relationship between exposure to traumatic death and mental health reactions is neither well documented, nor understood, it will certainly have a profound effect on those who bear witness to it.[11]

Grief, in response to death, is one of the most intense and enduring emotions one can experience and it can manifest itself across several dimensions: cognitive, physiological, behavioural, social, emotional, and spiritual. Specifically, grief reactions can include:

* Cognitive effects: confusion, difficulty making decisions, impaired thinking, difficulty with problem-solving, memory interference, memory loss, and poor concentration;

* Physiological effects: nausea, muscle tremors, sweating, dizziness, fatigue, sleep difficulties, nightmares, restlessness, decreased libido, slower speech pattern, restlessness, extreme hunger or lack of appetite, trouble concentrating, crying, headaches, stomachaches, weakness, shortness of breath, or tightness in the chest, and dry mouth;

* Behavioural effects: excessive drinking or drug use, loss of motivation, change in appetite, restlessness, and poor self-care;

* Social effects: legal, financial, and work problems, as well as social isolation from family members or friends;

* Emotional effects: anxiety, fear, anger, irritability, guilt, despair, hopelessness, helplessness, depression, numbing, hostility, preoccupations, intrusive thoughts, and disbelief; and

* Spiritual effects: loss of faith, lack of belief in "a superior being," feelings of abandonment, and search for meaning.

In response to death, soldiers may feel overwhelmed and unable to cope with the inevitable feelings of powerlessness and helplessness that result. Soldiers may also think that they are alone in experiencing these feelings and they may become discouraged and begin to question if there is any meaning in the midst of their suffering. As such, they may search for meaning and purpose in what they do.

## **WHAT LEADERS NEED TO KNOW**

No amount of knowledge will completely prepare leaders for bereavement. Knowledge, however, will enable leaders to help guide their subordinates, and hopefully themselves, through the grieving process. At a minimum, leaders should understand:

* that grief is a normal process and responding to grief and death is an individual phenomenon;
* theories of grief and recovery from grief;
* that there are many factors that affect the grieving process;
* barriers to recovery;
* barriers to seeking assistance;
* what happens if grief is ignored; and
* survivor guilt.

*Grief is a Normal Process*

What is important for leaders to understand is that people will display grief and that it is a normal and multi-faceted response to loss. Further, there is no "right way" to grieve — it is an individual experience. Individuals will develop a method of grieving that fits them and their perception of their particular loss. Some will grieve quickly, others take a long time to finish, and some grieve openly, others silently. Researchers have long supported the notion that grief over the loss of a loved one would generally last one year but that the symptoms of grief would wax and wane over time. Recent models of grief, however, suggest that the second year following the loss can be just as emotionally painful and even harder to deal with emotionally, as the bereaved person feels less comfortable sharing their pain and loss. Initially, grief can seem to be overwhelming and people can feel they have no control over their emotions. With time, people find that they have developed an increased ability to choose when they access the memories and emotions. What is important for leaders to remember is that this process does become easier with time. There are, however, no quick fixes; no short cuts. An old African saying describes it as:

> There is no way out of the desert except through it. Knowledge of the grief process gives us a very generalized map of the terrain we have to cover. Each of us will take a different route. Each will choose his own landmarks. He will travel at his own unique speed and will navigate using the tools provided by his culture, experiences and faith. In the end, he will be forever changed by his journey.[12]

Grief begins whenever there is a loss or a perception of impending loss.[13] Because it is painful, people want to get over it, and often do so by denying the pain. Unfortunately, admitting to psychological pain can be perceived as a personal shortcoming or major weakness from society in general and likely more so in the CF. To resist this attitude, leaders must acknowledge and promote the understanding that showing pain or grief does not represent a weakness or personal shortcoming. The experience of grief is not to be feared or avoided. Grief is a healing process that ultimately brings us comfort for our pain.

### Theories of Grief

Several theories of grief have been proposed over the last century. When examining theories of grief, it is important to remember that grief is not the same for everyone, although there are certain commonalities. Theories are only guides; they cannot describe the unique reactions and experiences of an individual. Since there is no right way to grieve, there is no fixed process through which everyone passes following a death. As such, leaders cannot seek to impose a "cookie cutter" approach when reacting to the grieving soldier.

Several researchers have studied and attempted to explain the complicated process of grief. One of the most comprehensive models is Rando's theory.[14] In 1993, psychologist Therese Rando defined a six-stage model for grief that was based on her work as a traumatologist and thanaologist at the Institute for the Study and Treatment of Grief. Rando's six stages for Grief include:

1. Recognize. The individual experiences the loss and understands that it happened;
2. React. The individual reacts emotionally to the loss;
3. Recollection and re-experience. This involves reviewing memories of a lost relationship;
4. Relinquish. Individuals begin to put loss behind them and realize that the world has truly changed and there is "no turning back";
5. Re-adjust. This represents a return to daily routines with the loss feeling less acute; and
6. Reinvest. Individuals will re-enter the world building new relationships — they accept the loss has occurred and move on.

This model, like others, attempts to put into words what is not directly observable. Words, however, are just symbols. For this reason, those who are experiencing grief may find one set of words that better capture their internal reality. Words that work for one person may not work for another. Thus, respecting individual differences is essential in assisting individuals with the process of dealing with grief.

*Factors Affecting the Grief Process*

Research has demonstrated that to fully understand the multidimensionality of grief, the leader must consider:[15]

* <u>The Psychological Level</u>. What are the coping abilities, beliefs, feelings, and psychological characteristics of the person who has experienced a loss? For example, although there may be maladaptive responses to death or loss in theatre, soldiers can be quite resilient. Resiliency can be defined as the process and outcome of successfully adapting to adverse life events and especially highly stressful or traumatic events.[16] Resiliency is the interaction between one's beliefs, approaches, behaviours, and physiology that can facilitate a positive and fast recovery from adverse events. Factors that affect resiliency include: individual characteristics, social ties, and effective coping strategies. Resiliency does not prevent an individual from experiencing difficulty or distress; it only facilitates the adaptation process. Resiliency is a psychological tool that will aid in the management of fear and anxiety during a stressful time.

* <u>The Interpersonal Level</u>. What was the type of relationship with the deceased — family, friend, co-worker, allies; or no direct relationship but impacted because of beliefs (i.e. "children and civilians are innocent"), the style of responding to the situation, the type of support received (positive is supportive, negative is "get over it").

* <u>The Social-Cultural Level</u>. What norms, roles, and rituals are available to the bereaved? For example,

perceived social antecedents such as social support, degree of acceptance, security, cultural/ethnic/religious/philosophical background, support of family and friends, support from community, government and coalition forces can all affect the grief process in a positive or negative way. Other social factors include: the number and type or role of the dead person and the social reorganization that will be required (replacements and unit's reaction to the replacements — "the new guy syndrome"), the unit's participation in the rituals accompanied by a death in theatre, the presence of military rules, norms, values, styles, and past experiences that might inhibit the grief process,[17] and the total impact on the unit — each unit is more than the sum of its parts.

Other factors that can affect the grieving process include whether the loss was sudden, if it was violent, the nature of death, multiple deaths, unspeakable deaths, status of the victim, and the length and quality of the relationship with the deceased.

*Barriers to Recovery and Seeking Assistance*

Grief is part of the human experience and, if expressed, it can be tolerable. If suppressed, it will rise up to haunt us, surprise us, and shape our lives in ways we cannot control.[18] We often teach our soldiers that, if they follow the rules, they will reduce the risk of death. Combat, however, certainly brings home the reality that death is an ever-present possibility. This can create a loss of faith and belief in leadership and their very training and skills as a soldier.

There are several factors or barriers that can influence the grief response and impede successful completion of the mourning process. They include but are not limited to:

* a soldier's emotional health (i.e., previous mental health issues, psychiatric disorders, limited or absent coping strategies, unresolved past losses, previous experience, tendency toward self-destructive or suicidal behaviours, and overwhelming self-blame for events not under their control);

* a perceived or received lack of support systems (i.e., disconnection from normal support systems, an absent or unhelpful family, and, limited or no access to professionals); physical health (amount of energy depleted, amount of rest, sleep, exercise, nutrition, drugs, alcohol, cigarettes, and caffeine);

* demographic factors (age, maturity, gender); and

* situation-specific factors such as death during chaotic firefights including "friendly fire," and equipment failure.

Another potential barrier that may impede recovery is the societal perception that soldiers and leaders may have accepted that "it is not okay to openly express their grief, which has a profound effect on the morale in the Army."[19] Human beings, in general, want to avoid grief, but it is the pain that they truly want to avoid. As already mentioned, grief itself is the healing process that ultimately brings comfort for the pain. Although the feelings accompanying the grief process may seem overwhelming, soldiers are often very resistant to acknowledge their feelings or seek help. Barriers to seeking assistance include:

* denial — being unable or unwilling to accept that a loss has taken place;

* fear that the admission of needing help, even to oneself, is a sign of weakness. Soldiers expect

themselves to be as strong emotionally as they are physically, and do not want to appear weak, or as "sick, lame, or lazy";

* not knowing that help is available;

* fear of lack of confidentiality; and

* concern over the effect of seeking assistance on their evaluations and the resulting impact on one's career.

In addition, soldiers often believe that they can take care of their own and do not always welcome help from professionals or outsiders. Soldiers, as well as leaders, often believe that they must "suck it up and continue," no matter what the circumstances. Although situational demands may require a "suck it up" attitude at times, this does not mean that the grieving process can or should be ignored.

Leaders need to be aware of these factors that may affect the grief response in their soldiers. There is, *and it cannot be reinforced enough as a basic principle of leadership,* no substitute for knowing one's soldiers and taking an active interest in their welfare. Although death traditionally has been dealt with in a reactive manner, without doubt, being proactive and taking steps to build soldier resiliency is a much better approach.

*Grief Ignored*

Grief, if unresolved or ignored, can lead to the development of complicated grief[20] or "the intensification of grief to a level where the person is overwhelmed and resorts to maladaptive behaviors."[21] Complicated grief will affect unit morale, mission readiness, and the individual soldier's ability to perform in the field.

One of the key stages in the grieving process involves psychological disorganization and integration of the grieving in a reorganization process. In combat, psychological disorganization is rarely allowed for obvious operational reasons. Since this critical step (disorganization) is not often possible, the soldier may psychologically revert to previous stages of the grief process (e.g., shock, anger, denial, numbing) in an effort to gain completion of the process. This interruption of the process coupled with other combat experiences, can result in the soldier being "stuck" in the process (i.e., complicated grief).[22] Complicated grief is a risk factor for a variety of other mental health disorders including: Major Depressive Disorder, Post Traumatic Stress Disorder (PTSD), Traumatic Grief, Panic Disorder, and Generalized Anxiety Disorder. Other maladaptive behaviours such as alcohol abuse and substance disorders, smoking, poor nutrition, suicidal or homicidal thoughts, and physical health problems such as cardiovascular and immunological dysfunction can result from complicated, unresolved grief.[23]

Symptoms of complicated grief include:

* preoccupation with the deceased;

* intrusive thoughts;

* memories that are upsetting;

* avoiding reminders of death;

* feeling life is empty;

* longing for the person;

* seeing the person who has died;

* anger about the death;

* feeling it is unfair to live when the other person has died;

* disbelief about the death;

* feeling stunned or dazed; and

* difficulty trusting and caring about others.

To mitigate complicated grief, leaders at all levels must intervene when a death occurs. Leaders must validate soldiers' feelings and experiences and "make an appointment" to further handle the issues after the incident. It's akin to the defusing process in Critical Incident Stress Management (CISM) and can be done fairly effectively "on the fly." Moreover, it is paramount to monitor the different layers of grief. Grief can hit those individuals harder if they were directly responsible for giving the orders that lead to a death or multiple deaths.

*Survivor Guilt*

In addition to the risk of experiencing complicated grief, soldiers can also experience survivor guilt. Survivor guilt was first noted in those who survived the Holocaust.[24] Survivor guilt is defined as a deep feeling of guilt often experienced by those who have survived some catastrophe that took the lives of others. It derives in part from a feeling that they did not do enough to save the others who perished, and in part from feelings of being unworthy relative to those who died. Survivor guilt, like grief, is a normal response to a traumatic event. It is difficult for human beings to feel grateful for being alive while at the same time feeling intense sorrow for those who did not survive. Survival is an achievement and that although one may have had limited choices during the traumatic event, the individual has the choice of how they will respond emotionally after the event.[25] Survival can be used for personal growth.

*Resiliency*

Although there may be maladaptive responses to death or loss in theatre, soldiers can be quite resilient. Resiliency can be defined as the process and outcome of successfully adapting to adverse life events and especially highly stressful or traumatic events.

## RECOMMENDATIONS FOR LEADERS

In light of the current tempo and nature of CF missions, the death of soldiers is inevitable. The leaders' role is to effectively assist soldiers to prepare for potential loss, help them understand and process their grief reactions to ensure their ability to focus on mission success, and successful reintegration after the mission. Leaders can best help their soldiers through self-knowledge — knowledge of both their individual soldiers and the nature of the process of grief and bereavement.

Leaders represent an incredible resource to their subordinates in assisting with the grief process. Prior to deployment, leaders must ensure that:

* a comprehensive screening is completed with the soldier and his or her family, the goal being to be sure that the necessary stability exists that will permit the soldier to be deployed. This speaks to ensuring a certain degree of soldier resiliency;

* pre-operational preparation for deployment includes educating military families of support resources available (e.g., Family Resource Centres). This should also include how communications will be managed between the chain of command and the family, media, the unit, and rear party. In addition, leaders should ensure that family members are offered education on coping with the stress of

deployments and death notification procedures; and

* they provide training on the grief process, both normal and dysfunctional, and reactions to loss. Where possible, resiliency training should be used to prepare/mitigate the actual impact when death occurs. This could include information on Critical Incident Stress Management.

In addition leaders must:

* understand the grief process, specifically, signs/symptoms, and adverse reactions;

* acknowledge that grief is a normal process and that individuals will deal with grief in their own way (i.e., there is no right way to grieve). Leaders must understand that the length of time that each individual will grieve varies;

* understand that early intervention is required and that the dangers of unresolved grief, (i.e., complicated grief) at some point will likely not only affect individual soldiers, but the unit as a whole;

* be aware of factors that affect the grief process;

* have knowledge of the potential barriers to accessing services and ensure these barriers are mitigated where possible;

* ensure that members understand that if assistance is requested, there will not be any adverse professional or personal consequences. An effective leader will

ensure the physical, psychological, and spiritual health of all of their soldiers regardless of the opinions or criticism of others;

* know where to access help for their soldiers. This includes an awareness of pre-deployment, in theatre, and post-deployment services; and

* know the soldiers and be prepared to reach out, listen, and ask subordinates how they can help (i.e., engage in health healing rituals).

As well, leaders need to take steps to look after themselves. They must:

* acknowledge and deal with their own grief. Leaders must process their own grief and then work to facilitate soldiers processing their own. To do so requires that they recognize the feelings of pain, talk about it, forgive them, and take care of their own mental health;

* be given appropriate training in the recognition of both trauma and grief; and

* be encouraged to gather information about the nature of the event, the person's role in the event, the degree of violence, horror, and the sense of personal responsibility and the degree of family and social support following an incident or deployment.

As there is no simple way to deal with the death of a unit member, leaders should enlist help from chaplains, mental health professionals, clinical psychologists, Family Resource Centre specialists, and other local community resources. This access needs to be confidential. With regard to post-deployment, units should utilize unit psychologists, chaplains, and medical resources for education, assessment, and treatment. As well, soldiers should be encouraged to talk with their peers about their experiences to recognize that they are not going through their reactions alone.

## SUMMARY

Doug Manning, a social worker and teacher, stated that grief is not the enemy — it is a friend.[26] It is a natural process of walking through hurt and pain, and growing because of the walk. Soldiers will experience their grief in their own way and with their own resources. Leaders can assist soldiers through the process of education, training, communication, and having the ability and experience to know and recognize when the intervention of a specialist is required.

## NOTES

1. David Grossman, interview by Soldier's Heart, *Frontline Bulletin*, Public Broadcasting System, 2004.
2. CBC Indepth News Backgrounder, http://www.cbc.ca/news/background (accessed 20 June 2007).
3. Brett Litz, Susan Orsillo, Matthew Friedman, Peter Ehlich, and Alfonso Batres, "Post Traumatic Stress Disorder Associated with Peacekeeping Duty in Somalia for U.S. Military Personnel," *American Journal of Psychiatry*, Vol. 154, No. 2 (1997): 178–184.
4. Helen Fitzgerald, *The Mourning Handbook* (New York: Simon & Schuster, 1995), 24.
5. Donald Meichenbaum, *Principles and Practice of Stress Management*, 3rd ed. (New York: Guilford Press, 2005).
6. Kathleen Dunn, "Grief and its Manifestations," *Nursing Standard*, 21–27 July 2004, 45–51.
7. *Ibid.*

8. Kathleen Cowles and Beth Rogers, "The Concept of Grief: A Foundation for Nursing Research and Practice," *Research in Nursing and Health* 14 (1990): 119–127.
9. *Webster's New World Dictionary*, http://www.m-w.com/dictionary.
10. John Bowlby, *Attachment and Loss, Volume 3: Sadness and Depression* (London: Penguin Books, 1980), 17.
11. Naomi Breslau and Glenn Davis, "Post Traumatic Stress Disorder? The Stressor Criterion," *Journal of Nervous and Mental Disease* 175 (1987): 255–264; Joseph Currier, Jason Holland, and Robert Niemeyer, "Sense-Making, Grief, and the Experience of Violent Loss: Toward a Mediational Model," *Death Studies* 30 (2006): 403–428; Jacob Lindy, Bonnie Green, and Mary Grace, "The Stressor Criterion and Post Traumatic Stress Disorder," *Journal of Nervous and Mental Disease* 175 (1987): 269–272; James Rundell, Robert Ursano, Harry Holloway, and Edward Siberman, "Psychiatric Responses to Trauma," *Hospital and Community Psychiatry* 40 (1989): 68–74; Robert Ursano, "Commentary: Posttraumatic Stress Disorder: The Stressor Criterion," *Journal of Nervous and Mental Disease* 17 (1987): 66–75.
12. Howard Gorle, "Knowledge of the Grief," http://www.hospicenet.org.html/knowledge.html.
13. Fitzgerald, 21.
14. Theresa Rando, *Treatment of Complicated Mourning* (Baltimore: Research Press, 1993).
15. Robert Fulton and Deborah Gottesman, "Anticipatory Grief: A Psychological Concept Reconsidered," *The British Journal of Psychiatry* 137 (1980): 45–54.
16. Virginia O'Leary, "Strength in the Face of Adversity: Individual and Social Thriving," *Journal of Social Issues* 54 (1988): 425–446; Virginia O'Leary and Jeanette Ickovics, "Resilience and Thriving in Response to Challenge: An Opportunity for a Paradigm Shift in Women's Health," *Women's Health: Research on Gender, Behavior, and Policy* 1 (1995): 121–142; and Michael Rutter, "Psychosocial Resilience and Protective Mechanisms," *American Journal of Orthopsychiatry* 57 (1987): 316–331.

17. Kathleen Gilbert, Unit 11 — Anticipated Loss and Anticipatory Grief, 2005, http://www.indiana.edu/~famlygrf/units/anticipated.html.
18. Fitzgerald, 25.
19. Albert Smith, *Coping with Death and Grief: A Strategy for Army Leadership* (Carlisle Barracks, PA: U.S. Army War College, 1999), iii.
20. Colin Parkes and Robert Weiss, *Recovery from Bereavement.* (New York: Basic Books, 1983), 16.
21. Mardi Horowitz, "A Model of Mourning: Change in Schemas of Self and Others," *Journal of the American Psychoanalytic Association* 38 (1990): 297–324.
22. Jeffery Brandsmas and Lee Hyer, "Resolution of Traumatic Grief in Combat Veterans," National Centre for Posttraumatic Stress Disorders, 1995, Available from: http://ncptsd.va.gov/ncmain/ncdocs/fact_shts/fs_older_veterans.html, (accessed 15 June 2008).
23. P.C. Bornstien, P.J. Clayton, J.A. Halikas, W.L. Maurice, and E. Robins, "The Depression of Widowhood after Thirteen Months," *British Journal of Psychiatry* 122 (1973): 561–566; Selby Jacobs, Carolyn Mazure, and Holly Prigerson, "Diagnostic Criteria for Traumatic Grief," *Death Studies* 24 (2000): 185–199.
24. Kathleen Nader, "Guilt Following Traumatic Events," 2003, http://www.mental-health-matters.com/articles/articles.php?artID=302.
25. Robert Lifton, "From Hiroshima to Nazi Doctors: The Evolution of Psychoformative Approaches to Understanding Traumatic Stress Syndrome," In *International Handbook of Traumatic Stress Syndrome,* eds. J. P. Wilson and B. Raphael (New York: Plenum Press, 1993), 11–23.
26. Doug Manning, *Do Take My Grief Away* (San Francisco: Harper & Row, 1979).

## SELECTED READINGS

Archer, John. *Nature of Grief: The Evolution and Psychology of Reactions to Loss.* London: Routledge, 1999.

Fitzgerald, H. *The Mourning Handbook*. New York: Simon & Schuster, 1995.

Grossman, Dave. *On Killing: The Psychological Cost of Learning to Kill in War and Society*. Boston: Little Brown, 1995.

Kauffman, Jeffrey. *Loss of the Assumptive World*. New York: Psychological Press, 2005.

Kubler-Ross, Elisabeth, M.D. and David Kessler. *Life Lessons: Two Experts on Death and Dying Teach Us about the Mysteries of Life and Living*. New York: Springer Press, 2000.

Levang, Elizabeth. *When Men Grieve: Why Men Grieve Differently and How You Can Help*. Minnesota: Fairview Publishers, 1997.

Mothers Against Drunk Drivers (MADD). *Death Notification Training*. Oakville, ON: MADD Canada National Office, 2005.

Rando, Theresa. *Treatment of Complicated Mourning*. Baltimore: Research Press, 1993.

Rinpoche, Sogyal. *Tibetan Book of Living and Dying*. San Francisco: Harper, 1992.

Sherman, N., J. Myers, and J. Rosenthal. *Stoic Warriors: The Ancient Philosophy Behind the Military Mind*. London: Oxford University Press, 2005.

# 24

# INFLUENCE

by Bill Bentley and Robert W. Walker

"Influence" originally was defined by ancient astronomers as "the supposed flowing of an ethereal fluid or power from the stars thought to affect the characters and actions of people."[1] The concept of influence has been researched over centuries. Around 5000 BC, the Greeks wrote scholarly essays on rhetorical communications as a means of influence. They argued about speaking styles and emotions, communicator traits, morals and logic, and combinations of these variables.[2] The Romans had similar interests during the first century AD, followed by still similar and subsequent interests by Europeans during and after the Renaissance. By the middle of the 20th century, social scientific methods with grounded theory and empirical approaches were increasingly prevalent for the study of communication, influence, and persuasion. During the Second World War, "applied" research increased as more sophisticated communications technology, interacting with more effective distribution of war propaganda, was used to influence large and geographically dispersed audiences. The depth of valid research on the elements and components of influence and persuasion is reflected in the scope of book chapters and handbooks that cover such ideas as resistance and inoculation to influence, gaining compliance, obedience to authority, gender effects, aggression, fear, deception, superior-subordinate factors, personality, attitudes, changing prejudices, and the media.[3]

All of the above, as background, is important as knowledge about influence and, therefore, for a comprehension of leadership. Leaders are responsible for engendering commitment to an institution's tasks, direction, development, and strategy, regardless of the levels in the organizations at which leaders toil. Influence is defined broadly as the power of persons or groups or events to affect others, either overtly or inadvertently, as a consequence of the presence of such characteristics as expertise, position, authority, abilities, charisma, or prestige, among others.

This chapter will address influence as it is integrated variously into the literature on leadership, first briefly addressing the aspects of value-free "generic" leadership, then, particularly, the components of effective leadership in the Canadian Forces (CF). Next, it will discuss the major leadership "pathways or conduits" (e.g., direct leadership or indirect leadership, leading people or leading the institution, position-based or emergent leadership, etc.) through which influence can be asserted. The chapter will end with a review of the broad aspects and general patterns of influence evident in the "applied" spectrum of leader influence behaviours (e.g., directive, persuasive, supportive, participative, etc.).[4]

## **LEADERSHIP AND INFLUENCE — TOWARD A DEFINITION**

The concepts of influence and leadership cannot be easily separated. For eons, leadership in the scholarly literature has been treated in terms of four essential components: leadership is a process, leadership involves *influence*, leadership occurs within a group context, and leadership involves goal attainment. Despite general agreement on these four components, there are myriad definitions and as many as 65 different classification systems to define the dimension of leadership.[5] However, based on these components, and generally reflecting most of the scholarly research done, one generic and succinct definition provided is: "Leadership is a process whereby an individual influences a group of individuals to achieve a common goal."[6] Another definition, derived from Canadian research, is: "Leadership is directly or indirectly influencing others, by means of formal authority or personal attributes, to act in accordance with one's intent or a shared purpose."[7] This latter leadership definition is a generic and value-neutral example that makes no statements of good or bad behaviour or influence, of effective or ineffective action, or what is required in order to lead. In being both generic and value-neutral, the definition is vague and falls short of describing what is actually requisite for our desired effective, professional military leader.

An "applied" definition for an effective military leader would need to refer to the values at the core of the profession of arms and, in the military ethos, the military's purposes, the means of influence, the importance of followers and their welfare and stability, and leader flexibility. Taking all of these into account, effective leadership for the Canadian Forces has been defined as, "directing, motivating and enabling others to accomplish the mission professionally and ethically, while developing or improving capabilities that contribute to mission success."[8] That is to say, colloquially, that military leaders, through directing, motivating, enabling (i.e., these means of influence) get the job done, look after their people, think and act in terms of the larger team, anticipate and adapt to change, and exemplify the military ethos in all that they do. The CF leadership model is a value-expressive model — one that gives shape to the professional ideal of duty with honour.

This definition of effective military leadership can be applied through several pathways or conduits, the ones most frequently referred to being direct or indirect leadership, leading people or leading the institution, those using either the transactional approach or the transformational leadership approach, plus others identified below. These conduits, like a nest of snakes, are intertwined, overlapping, and competitive in such a way that their interconnectivity contributes to and determines the effectiveness of the specific influence behaviour or behaviours chosen by the leader from the leader-influence spectrum.

## **DIRECT INFLUENCE**

Historically, leadership research, theory, and practical advice have emphasized the personal face-to-face nature of leadership and the associated techniques for directly influencing subordinate performance, or one or more of its behavioural elements, in a fairly immediate way. Accordingly, direct influence incorporates face-to-face influence on others with an immediate effect on their ability, motivation, behaviour, performance, attitudes, or related psychological states, or which progressively modifies the slow-growth attributes of individuals and groups. Verbal direction, goal-setting, practice training, coaching, contingent reward and discipline, performance monitoring, and feedback all are examples of direct-influence behaviours which have immediate effects; intellectual development through education and value development through professional socialization tend to have incremental and delayed effects.

It has been increasingly recognized, however, that leaders also significantly affect behaviour and performance over the longer term by modifying situational conditions to make them more mission-favourable. These propositions logically follow from the observation that behaviour is a joint function of what attributes a person brings to a situation and conditions at play in the organization and external environment. Viewed in this way, leadership is obviously about influencing people, but it is also about shaping the task environment. Leaders make a direct contribution to effectiveness through the immediate effects they have on people and their performance — for example, by clarifying individual and group roles and tasks, developing skills, sharing risks and hardships, maintaining discipline and morale, and encouraging high levels of effort and persistence.

**INDIRECT INFLUENCE**

Leaders contribute to military effectiveness indirectly by designing and creating the group, organizational, and environmental conditions that enhance individual and collective performance, such as a professional culture and identity, cohesion, advanced doctrine, force structure, equipment, and human resource programs and services that support members and ensure their fair treatment. Indirect influence refers to influence on others mediated by purposeful alterations in the task, group, system, institutional, or environmental conditions that affect behaviour and performance. Changes in the content and delivery of training programs, technology, organizational structures and procedures, administrative policies and services, and organizational culture are examples of indirect influence.[9] More than one leader is required to master and control all aspects of a large and complex organization through a collective indirect leadership and, over the long haul, the institution's effectiveness will depend on the "breadth and depth of a strong ... leadership team with a shared sense of responsibility, professional identity, values and purpose."[10]

## **LEADERSHIP, COMMAND, MANAGEMENT, AND INFLUENCE**

A leader's influence "is potentially without limit, and a great leader may transform a nation, the world ..."[11] — particularly if that influence is integrated throughout command, leadership, and management, a circumstance thoroughly addressed elsewhere.[12] The intent here is simply to highlight that influence is not confined only to the conduit of leadership. Command is based on formally delegated authority, usually from a specific position and/or a specific set of activities, enacted downward to subordinates, and with an expression of human will captured in the concept of a commander's intent and influence. Leadership is a role requirement of commanders, as are activities perceived as "management" — planning, organizing, coordinating, controlling, etc. — essentially functions complementary to leadership and subsidiary functions of command. Military command, however, is set apart from management in an important manner, in that military commanders are authorized the use of large-scale lethal force to place military members in harm's way, and to use a distinct military justice structure with powers of punishment. Nonetheless, commanders and managers are expected to lead and to lead well. As for "pure" leadership influence, it is not constrained by formal authority and, unlike command, may be exercised upward and laterally, as well as downward. As a bottom line, it can be stated that, "The interrelationships and interconnectedness of command, management, and leadership functions often make it difficult to disentangle the command, management and leadership effects achieved by individuals in positions of authority."[13] As well, favourable outcomes tend to be attributed to extraordinary leadership, seemingly part of the "romance of heroic leadership" phenomenon of most societies and nations, when those effects may have been the result of strong command or relevant management skills, or a combination of all three, or just pure good fortune.

## LEADING PEOPLE

Leading people is a conduit for effective leadership accomplished primarily at the tactical level of military activities, mostly in a direct manner, face to face and, when on operations, in conditions of uncertainty and danger. Naturally, leading people also occurs at operational and strategic levels in staffs and organizations, large and small, however, the focus at these levels is more on institutional leadership.[14] The doctrine and practices for effective leading of people are applicable to all CF officers and non-commissioned members occupying leadership appointments or exercising situationally determined leader functions such as under circumstances where emergent, non-positional leadership arises.[15]

## LEADING THE INSTITUTION

Although the definition for effective CF leadership, above, remains the same for all CF leaders, the responsibility for leading the institution is different from that for leading people at different levels of military functions, be they tactical, operational, or strategic. Those leading at the institutional level continue to lead people but, predominantly, are involved in activities, and contribute to outcomes, that have impact on all members of the CF at the strategic level and on others internal and external to the National Defence department. Leaders of the institution, whether in staff positions or operational settings, are responsible for the creation of the overall conditions for the success of the CF through development and implementation of internal strategies and policies, stewardship of the profession of arms itself, member well-being and commitment, as well as the establishment of strong and effective external partnerships. It is at this level that leaders exert influence by ensuring a synergy of organizational effectiveness and professional effectiveness, resulting in overall institutional effectiveness.[16]

## **TRANSACTIONAL LEADERSHIP**

The concept of transactional leadership dominated leadership research up until the 1980s when a much different approach — that of transformational leadership — emerged, and continues to influence leadership theory to a much greater extent than the older transactional approach. Nonetheless, transformational leadership has not completely displaced the older theory; they are, to a large extent, complementary. The approach most appropriate to achieve maximum influence depends on the situation, the nature of the group (followers), and the individual leader's characteristics. Transactional leadership occurs when the leader influence is controlled through rewards or disciplines for the follower, depending on the adequacy of the follower's performance. This transactional leadership depends on contingent reinforcement, or on the positive or the more negative forms of management-by-exception, active or passive:

* Contingent Reinforcement: The leader assigns or gets agreement on what needs to be done and promises rewards or actually rewards others in exchange for satisfactorily carrying out the assignment.

* Management by Exception — Active: The leader arranges to actively monitor deviance from standards, mistakes, and errors in the follower's assignments and to take corrective action as necessary.

* Management by Exception — Passive: The leader waits passively for deviances, mistakes, or errors to take place and then takes corrective action.

## TRANSFORMATIONAL LEADERSHIP

The transformational leadership approach to leadership influence refers to the process whereby an individual engages with others and creates a connection that raises the level of motivation and morality in both the leader and the follower. Transformational leaders motivate others to do more than they thought possible. They set more challenging expectations and typically achieve higher performance.[17] Transformational leaders do more with followers then set up simple exchanges or agreements. They behave in ways to achieve superior results by employing one or more of the four components of transformational leadership:[18]

* Idealized Influence: Leaders behave in ways that result in their being role models for their followers. Leaders are willing to take risks and are consistent rather than arbitrary.

* Inspirational Motivation: Transformational leaders communicate high expectations to followers, inspiring them through motivation to become committed to and part of the shared vision in the organization. In practice, leaders use symbols and emotional appeals to focus group members' efforts to achieve more than they could in their own self-interest.

* Intellectual Stimulation: Transformational leaders stimulate their followers to be creative and innovative, and to challenge their own beliefs and values as well as those of the leader and the organization. This type of leadership supports followers as they try new approaches and develop innovative ways of dealing with organizational issues.

* Individualized Consideration: Transformational leaders provide a supportive climate in which they listen carefully to the individual needs of followers. Leaders act as coaches and advisors while trying to assist individuals in becoming fully actualized. These leaders may use delegation as a means to help followers grow through personal challenges.

## **POSITION-BASED VERSUS EMERGENT LEADERSHIP**

Some people are leaders because of their formal position within an organization, whereas others are leaders because of the way other group members respond to them. These two common forms of leadership are called assigned (or position-based) leadership and emergent leadership. The person assigned to a leadership position does not always become the real leader in a particular setting. When an individual is perceived by others as the most influential member of a group or organization, and regardless of the individual's title, the person is exhibiting emergent leadership. The individual acquires emergent leadership through other people in the organization who support and accept that individual's behaviour. This type of leadership emerges over a period of time through communication and group dynamics. Some of the positive communication behaviours that account for successful leader emergence include being verbally involved, being informed, seeking other's opinions, initiating new ideas, and being firm but not rigid.

In addition to effective communication, personality plays a role in emergent leadership. Those individuals who are more dominant, more intelligent, and more confident about their own performance in general are frequently identified as leaders by other members of the group. Transactional and transformational leadership approaches apply to both assigned and emergent leaders. Leader influence behaviours also apply to both assigned and emergent leaders.

In the CF, the transactional and transformational approaches are always, potentially, applicable, depending on the situation. Generally, however, except in the most extreme circumstances — dire emergencies, imminent danger, combat — the most effective approach is the transformational one. Emergent leadership should always be recognized and utilized, but, in the military context, position-based or assigned leaders must demonstrate constantly the effective and professional leadership that is not sidelined by emergent leadership.

## **THE SPECTRUM OF LEADER INFLUENCE BEHAVIOURS**

The Spectrum of Leader Influence Behaviours[19] at Figure 1 provides a comprehensive representation of leader influence behaviours. As illustrated, leader influence behaviours may be differentiated and roughly ordered by the amount of control employed by the leader. This control can range from the total control that characterizes authoritarian leadership through to the complete absence of control typical in laissez-faire leadership. A follower's latitude for discretion generally increases from left to right across the spectrum or continuum of influence behaviours. The appropriateness of an approach represented on the spectrum depends on the level of development and training of those led. Followers who lack necessary skills, knowledge, attitudes, and motivation for the task at hand will need to be led in accordance with the influence behaviours that are more to the left on the spectrum. The most effective leaders are those who can shift between the two approaches to leadership, transactional and transformational, as required by circumstances and characteristics of followers.

These broad patterns of influence commonly are termed "styles of leadership," with the early 20th century studies of leadership styles resulting in the labels *laissez-faire, democratic,* and *authoritarian*. More recent research has resulted in a greater differentiation of leader influence behaviours such that a reasonably more comprehensive inventory can be articulated. Effective influence behaviours can be distributed across a continuum that, appropriately, is stretched between the extremes of authoritarian leadership and laissez-faire leadership. Brief descriptions follow, while discourse on effective use of each leader influence behaviour in the spectrum can be found in the original source:[20]

**Figure 1: A Spectrum of Leader Influence Behaviours**

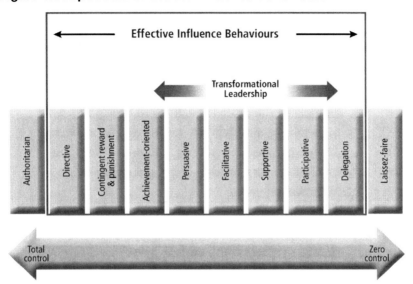

* <u>Directive</u> involves telling followers what they are to do, when, how and to what standard.

* <u>Contingent Reward and Punishment</u> is a style based on tangible rewards and punishment, such as fines or social praise and awards.

* <u>Achievement-Oriented</u> influence is primarily concerned with developing the competence, confidence and independence of subordinates through goal attainment.

* <u>Persuasive</u> influence is primarily intended to affect decision-making and motivation by explaining to, or convincing, others why a certain course of action is necessary.

* <u>Facilitative</u> influence involves, modelling, coaching, mentoring, guiding, and other types of behaviour that either demonstrates a desired behaviour to others or enables that behaviour to be performed by others.

* <u>Supportive</u> influence is intended to assist followers in resolving personal or work related problems or to improve their morale and well-being. This involves concern for well-being and responsiveness to individual needs.

* <u>Participative</u> influence involves timely sharing of decision authority with others.

* <u>Delegation</u> involves a transfer of specific authorities from the leader to one or more followers; thus, enhancing autonomy, motivation, satisfaction, and the increased meaningfulness of role or assignment.

An example, to reflect the integration of these influence behaviours with one or more of the pathways or conduits for achieving influence, is appropriate. Transformational leadership incorporates and combines several of the influence behaviours in the above list, as per Figure 1. Broad patterns of influence can, commonly, be termed "leadership styles." As addressed above, transformational leadership is usually intended to alter the characteristics of individuals, teams, organizations, even societies, in dramatic and substantial ways, to make them more complete or better equipped. Idealized behaviour, one component of transformational leadership, matches up with facilitative behaviour, above; the transformational leadership components of intellectual stimulation and inspirational motivation match up with achievement-oriented behaviour; and individualized consideration in transformational leadership corresponds with elements of supportive and participative behaviours. In military organizations, transformational leaders work to bring together a combination of insight, values, inspiration, imagination, rationality, concern for colleagues, and increased trust, hope, and dedication. The influence behaviours across the spectrum of leadership styles, reflecting transformational leadership, have the potential to achieve the tasks and acquire mission success, as well as to generate respect, trust, and commitment among people.

## SUMMARY

This chapter has dealt with influence in leadership, including military leadership, exploring and defining the aspects of leader influence, the circumstances and variables of leader-follower interactions, and the dynamics of influence in leader approaches and purposes. It has presented and explained the continuum of leader influence behaviours put forth in the CF leadership manuals, and has cross-referenced the alternative influence behaviours with transactional and transformational leadership, direct and indirect influence practices, the challenges of leading people and leading the institution, and the challenging comprehension of these influence behaviours when held up against leadership, command and management. This chapter underscores the profound need among CF leaders for exceptional social capacities, with strong attributes like behavioural flexibility, various and effective communication skills, proficient interpersonal skills, generation of strong group relationships, and establishment of authentic partnering relationships. With such capabilities, influence behaviours can be mastered.

## NOTES

1. *Webster's New World Dictionary*, College Edition (Toronto: Nelson, Foster & Scott Ltd., 1960), 749.
2. James Price Dillard and Michael Pfau, eds., *The Persuasion Handbook: Developments in Theory and Practice* (London: Sage Publications, 2002), ix.

3. Richard M. Perloff, *The Dynamics of Persuasion: Communication and Attitudes in the 21st Century* (London: Lawrence Erlbaum Associates, Publishers, 2003); Timothy C. Brock and Melanie C. Green, eds., *Persuasion: Psychological Insights and Perspectives,* 2nd ed. (London: Sage Publications, 2005); and Eric S. Knowles and Jay A. Linn, eds., *Resistance and Persuasion* (London: Lawrence Erlbaum Associates, Publishers, 2003).
4. It must be emphasized that much of this chapter's content on leader influence behaviours has been extracted from current CF leadership manuals: Canada. *Leadership in the Canadian Forces: Conceptual Foundations* (Kingston: DND, 2005), 4–8, 58–74; Canada. *Leadership in the Canadian Forces: Leading People* (Kingston: DND, 2007), 31–35; Canada. *Leadership in the Canadian Forces: Leading the Institution* (Kingston: DND, 2007). All manuals are available at http://www.cda-acd.forces.gc.ca/cfli.
5. E.A. Fleishman et al., "Taxonomic Efforts in the Description of Leader Behavior," *Leadership Quarterly*, Vol. 2, No.4 (1991): 245–287.
6. *Ibid.*, 246.
7. *Conceptual Foundations*, 7.
8. *Ibid.*, 30.
9. *Ibid.*, 6.
10. *Ibid.*, 6.
11. *Ibid.*, 4.
12. Bernd Horn, "Institutional Leadership: Understanding the Command, Management and Leadership Nexus," in *Institutional Leadership in the Canadian Forces: Contemporary Issues,* ed. Robert W. Walker (Kingston: Canadian Defence Academy Press, 2007), 99–107; and *Conceptual Foundations*, 8–13.
13. *Conceptual Foundations*, 10.
14. See *Leading the Institution*.
15. *Leading People*, 2.
16. *Leading the Institution*, viii.
17. Bernard Bass, *Transformational Leadership: Industrial, Military, and Educational Impact* (Mahwah, NJ: Lawrence Erlbaum and Associates, 1998), 4.

18. Peter Northouse, *Leadership Theory and Practice* (London: Sage Publications, 2001), 137.
19. *Conceptual Foundations*, 64.
20. *Ibid.*, 64–67.


**SELECTED READINGS**

Bass, Bernard. *Transformational Leadership: Industrial, Military, and Educational Impact.* Mahwah, NJ: Lawrence Erlbaum and Associates, 1998.

Burns, James MacGregor. *Leadership.* New York: Harper and Row, 1978.

Canada. *Leadership in the Canadian Forces: Conceptual Foundations.* Kingston: DND, 2005.

Canada. *Leadership in the Canadian Forces: Leading the Institution*, Kingston: DND, 2007.

Canada. *Leadership in the Canadian Forces: Leading People*, Kingston: DND, 2007.

Hughes, Richard L. and Katherine C. Beatty. *Becoming a Strategic Leader.* San Francisco: Jossey-Bass Publishers and Centre for Creative Leadership, 2005.

Northouse, Peter. *Leadership: Theory and Practice.* London: Sage Publications, 2001.

Walker, Robert W., ed. *Institutional Leadership in the Canadian Forces: Contemporary Issues.* Kingston: Canadian Defence Academy Press, 2007.

Yukl, G. *Leadership in Organizations.* 5th ed. Upper Saddle River, NJ: Prentice Hall, 2002.

Zaccaro, Stephen J. and Richard J. Klimoski, eds. *The Nature of Organizational Leadership.* San Francisco: Jossey-Bass Publishers, 2001.

# 25
# INTERPERSONAL FEEDBACK
by Danielle Charbonneau and Allister MacIntyre

One of the essential characteristics of transformational leaders resides in their ability to induce an alignment of values between those of the followers and those of the organization they represent.[1] In the military, this often involves a fundamental change as followers increasingly endorse and internalize military values. Such profound change seldom happens spontaneously. Rather, it is more likely that leaders will encourage change by providing guidance on what individuals need to change and offering feedback on their progress. An understanding of the process of individual change can help leaders to become more effective in transforming their followers.

**PROCESS OF INDIVIDUAL CHANGE**

The process of individual change has been thoroughly examined by psychologists who help individuals to overcome addictions or bad habits.[2] However, this process can be generalized to other situations as well. The change process usually occurs in six steps. The best chances of success occur when individuals complete the transformations associated with each stage before moving on to the subsequent stage. Rushing individuals during the change process almost guarantees an eventual failure. In Stage 1, precontemplation, individuals deny having bad habits and actively resist change. Leaders can promote change by helping their followers to become aware of the disadvantages or negative consequences of their problematic behaviours and by encouraging them to seek objective information about their bad habits.

Individuals in the contemplation stage, Stage 2, admit to having a problem, but they are not quite ready to change. At this stage, leaders can focus on the advantages of changing and helping followers to imagine the end results of change as well as its benefits to them. In addition, leaders can help by identifying specific behaviours that need change in terms of frequency, intensity, and duration.

In Stage 3, the change needs to be planned. Leaders can assist followers in preparing a detailed plan that identifies the most suitable actions that could be taken by a follower. For example, what actions might the follower engage in instead of the problematic behaviour? How will temptations be avoided? What are the high-risk situations and how can they be effectively dealt with? The objective here is for followers to build a considerable amount of confidence in their ability to handle the upcoming change.

During the action step, Stage 4, support and encouragement cannot be given in excess. Followers' self-confidence should be maintained, if not increased. This is the time of active effort and sacrifices on the part of followers. Furthermore, there is a need to appreciate that this is a period that may stretch into several months. Leaders should be alert to any signs of possible discouragement and may want to frequently check how their followers are coping with the change.

Maintaining the change, Stage 5, is often very challenging for individuals. This is partly because the encouragement and support they initially received from others tend to erode with time, whereas their effort in maintaining the change tends to remain relatively constant. Relapse is the rule rather than the exception. Leaders can help by treating the slip as a step back, rather than as a failure, and by remaining supportive. Followers can learn from their slips and refine their action plan to minimize their reoccurrence.

Finally, in Stage 6, it is not uncommon for individuals to recycle back to an earlier stage, especially when the change affects many areas of their life. The key here, for both leaders and followers, is to better plan alternative behaviours for complicated situations and to review the benefits of changing. It is noteworthy that distress precipitates relapse. Hence, during distressing times, leaders may help followers to develop alternative coping strategies.

In sum, knowing an individual's stage of change can be helpful in providing guidance on how to encourage change and improvement. For instance, to foster change up the chain of command, the evidence may be indicative that one's superior is in the precontemplation or contemplation stage. Hence, one may discuss both the advantages and disadvantages of a decision with the superior, in that order. The advantages represent a point of agreement, which is likely to maintain the recipient's interest. Presentation of the disadvantages follows, with the hope that the superior will seriously think about the negative characteristics as well as the more favourable aspects. This discussion would be followed immediately, or a few days later, by another that focuses on the advantages of the proposed alternative solution.

## **FEEDBACK DEFINED**

Several definitions of feedback and criticism are available in the literature, but the one of interest here is Hendrie Weisinger's definition of criticism.[3] According to Weisinger, criticism aims at teaching appropriate skills and knowledge and, therefore, is improvement-oriented. There are several advantages to this perspective. First, it focuses on behaviour, not on the person. Second, the focus is on the current problematic behaviour and how it can be improved in the future. Little energy is spent attacking or defending past situations. Overall, Weisinger recommends a positive approach that assumes that people want to improve in tasks that are meaningful to them.

Many readers will be familiar with the "sandwich" method, one of which alternates between praise and criticism (e.g., "You are doing this well BUT this needs improvement"). This technique is not recommended because (a) if the approach is familiar to the recipient, praise is unheard as the recipient worries about what is coming next;[4] and (b) statements that describe what the recipient can improve and are followed by the recipient's strengths (e.g., "This needs improvement AND you are already good at doing that") generate less negative emotions in the recipient.[5]

## **PREPARING TO GIVE FEEDBACK**

Giving feedback can generally be divided into three steps: the preparation, the delivery, and the follow-up. Each step includes several key components that, if given careful consideration, can greatly increase the chance of positive outcomes. First, the preparation stage can be divided into six stages:

- collecting facts;
- reflecting on our expectations;
- hypothesizing about the causes of poor performance;
- becoming sensitive to the context;
- figuring out how to facilitate change; and
- taking responsibility for the delivery.

*Collecting Facts*

When collecting facts, keeping in mind that criticism aims at improving behaviour, leaders should focus on noting several specific examples of the problematic behaviour. Indeed, two or three specific examples of undesirable behaviour provide more information to the recipient than comments that refer to behaviour in general or to attitudes.

*Expectations*

Many people tend to expect others to adhere to their own personal standards.[6] Such expectations are very subjective. Hence, leaders should be clear about the criteria used to assess performance, as several criteria may be relevant and the recipient may be using a different criterion. For instance, performance may be compared to that of others, and hence be relative. Alternatively, performance may be evaluated on the basis of an established base line, an agreed-upon objective, or past performance.[7]

*Possible Causes*

While preparing to give feedback, leaders should speculate about possible causes of performance problems.[8] Several factors can negatively affect performance, including skill and knowledge deficiencies, insufficient guidance and/or incentive, task difficulty and/or interest, vague or unrealistic objectives, conflicts, low motivation, and personal issues. These factors can operate singly or jointly in impairing performance. It is important to treat one's perception of the causes as hypothetical because it is very easy to focus on one contributor at the expense of others. Adler and Towne recommend that at least two possible interpretations of the problematic behaviour be retained.[9]

*Context*

Consideration should be given to the recipients' context and personal circumstances. What motivates people varies from one individual to another. A recipient's behaviour may be motivated by non-obvious factors. Criticism may be based on inaccurate impressions, second-hand information, and/or a downplaying of the recipient's qualities.[10] Leaders should be aware of the possibility of having some gaps in the knowledge and/or understanding of the recipient's context. Every effort should be made to fill in these gaps as completely as possible.

*Facilitating Change*

Prior to giving feedback, the definition and assessment of a successful change or end result should be clear. To this end, leaders should reflect not only on the desired specific end results or goals, but also on how leaders can help facilitate the change.[11] The use of a team approach to implement the change usually increases recipients' involvement.

*Delivery*

Finally, taking responsibility for the delivery of the feedback includes having a good grasp on the criteria used to evaluate the poor performance and the desired end results. Although a superior often has the authority to impose a solution, eliciting a decision from the recipient will normally increase the recipients' involvement in implementing the solution. To maximize the recipient's collaboration, not only is it important to plan what to say, one must also think about how to say it.[12] For instance, most people enjoy answering questions. Harry Mills suggests the use of questions beginning with "What effect …?", "How do you …?", "What would be different if …?", "What needs to happen for … to increase?"[13] or "If your unit were operating effectively, what would it look like?", "What are the disadvantages in doing it this way?", and "What would be the advantages of doing it this way instead?" Such a coaching approach increases recipients' awareness and assures them that their leader is interested in their personal development. It is also important to understand that these question stems are all open-ended. The avoidance of closed questions (e.g., easily answered with a yes or no) will lead to a more comprehensive understanding of a follower's perspective and help generate more meaningful solutions.

Another aspect of taking the responsibility for the delivery of feedback includes the selection of an appropriate setting and timing. Under most circumstances, negative feedback should be delivered in private, and at a time that is convenient to both the deliverer and the recipient. There are very few exceptions to this rule and public negative feedback should only be considered in urgent situations where issues like safety are at stake. Assurances of confidentiality should be provided and deliverers should block sufficient time to listen to recipients' perspectives and responses. Finally, interruptions can be very disruptive and are strongly discouraged.

## **FEEDBACK DELIVERY**

Essential elements of feedback delivery comprise listening to recipients' perspective, presenting your own evaluation, allowing recipients to explain, resolving disagreements, and establishing goals for improvement.[14] Throughout the conversation, the use of active listening increases the likelihood that the recipient understands the message and the feedback giver understands the recipient's perspective. (See Chapter 10 on Counselling.) It is important to obtain the recipients' perspective early in the conversation for two main reasons: (a) the recipient may possess information that is unknown to the deliverer and (b) the deliverer will be able to gauge recipients' stage of change and adjust their conversation focus accordingly.

Following this, deliverers can present their own descriptions of specific undesirable behaviours along with their negative impact on work, without forgetting to mention the criteria used for classifying the behaviours as undesirable. Interpretations of behaviours may be offered, as long as they are presented in a tentative way. The recipient should then be allowed to comment on what was just said. Disagreement in interpretation or opinion should be resolved; otherwise there is little chance of agreement on setting objectives. Without agreement on the objectives, the anticipated change may not happen.

Objectives may be personal, task-related, social, or group-related. They may range from reflecting on the negative consequences of the problematic behaviour (when the recipient is in the precontemplation stage), to looking for information on the benefits of changing (contemplation stage), to identifying specific actions to be taken to improve or change the behaviour (preparation stage), and to problem solving when a relapse occurs (maintenance of change). Improvement may not happen as quickly as one would hope for. However, as mentioned earlier, trying to accelerate change will likely lead to some relapse in the future. So patience may pay off.

Word selection can contribute greatly in diffusing possible tensions by protecting the recipient's self-esteem. The application of three simple rules may facilitate the discussion.[15] First, leaders should clearly state that comments are based on observations and begin sentences with "I think …" or "In my opinion …" as a way of acknowledging the subjective nature of giving feedback. Second, generalizations and absolute statements, such as "always" and "never" should be avoided as they can easily be refuted with a single counter-example. The word "sometimes" is viewed as less offensive. Similarly, "should have done" can be replaced by "could have done."[16] Finally, positively charged words are preferred. For instance, "below standards" becomes "not quite up to our standards,"[17] or "almost up to our standards," or "sometimes your work needs to be more thorough."[18]

## FOLLOW-UP

There is a strong tendency to underestimate the importance of follow-up. Change is difficult and can take time. It is important to communicate any improvements that are noticed, even very small improvements. Alternatively, efforts toward change can also be praised.[19] One can never provide too much encouragement in the follow-up period, when the change actually occurs.

## RECEIVING NEGATIVE FEEDBACK

It can be very challenging to be at the receiving end of negative feedback. The key is to remain calm and non-defensive and to remember to breathe deeply. Weisinger suggests that someone receiving negative feedback can often refocus the conversation by asking the other person how he or she thinks you can be more effective.[20] One way to do this is to request some specific instances of behaviours that form the basis of the other person's feedback (e.g., "What did I do that gave you the impression that …?"). Because intentions are invisible to others, behaviours may be wrongly interpreted, thereby allowing others to draw inappropriate conclusions. Alternatively, clarification of the other person's expectations may be in order (e.g., "What do you think would have been a better way for me to deal with the situation?"). Indeed, the criticism may be based on unmet expectations that were never shared. Such misunderstandings often can be resolved.

## SUMMARY

Taking care of the subordinates' personal development is an essential aspect of military leadership. Change takes effort, time, and support but it will be beneficial for both the individual and the organization. Encouraging someone to change for the better is more likely to succeed when that person's stage of change is known and taken into consideration when the feedback is provided. It always pays off to protect the recipient's self-esteem, especially when giving negative feedback. Not only does the content of the feedback matter, but also the manner of delivery is crucial to ensure success. Finally, leaders must understand that there is more to the change process than simply initiating some sort of change. The importance of follow-up actions and ongoing guidance cannot be overemphasized. When the commitment to change is encouraged, monitored, nurtured, and maintained, positive results will be naturally produced.

## NOTES

1. B.M. Bass, *Transformational Leadership: Industrial, Military, and Educational Impact* (Mahwah, NJ: Lawrence Erlbaum Associates, 1998).
2. J.O. Prochaska, J.C. Norcross, and C. C. Diclemente, *Changing for Good* (New York: HarperCollins, 1994).
3. H. Weisinger, *The Power of Positive Criticism* (New York: Amacom, 2000).
4. J.P. Zima, *Interviewing: Key to Effective Management* (Chicago: Science Research Associates, 1983), 266.

5. Weisinger, 38–39.
6. G.T. Fairhurst and R.A. Sarr, *The Art of Framing: Managing the Language of Leadership* (San Francisco: Jossey-Bass Publishers, 1996).
7. H. Weisinger, *The Critical Edge: How to Criticize Up and Down Your Organization and Make It Pay Off* (Boston: Little, Brown and Company, 1989).
8. Zima, 258.
9. R.B. Adler and N. Towne, *Communication et Interactions, 3ième Édition* (Montréal: Groupe Beauchemin Éditeur, 2005).
10. *Ibid.*, 77.
11. Weisinger, 2000.
12. *Ibid.*
13. H. Mills, *Artful Persuasion* (New York: Amacom, 2000), 191–194.
14. Zima, 268.
15. Weisinger, 2000.
16. *Ibid.*, 23.
17. *Ibid.*, 21.
18. *Ibid.*, 23.
19. *Ibid.*
20. *Ibid.*, 7.

**SELECTED READINGS**

Adler, Ronald B. and Neil Towne. *Communication et interactions. 3ième Édition.* Montréal: Beauchemin, 2005.

Fairhurst, Gail T. and Robert A. Sarr. *The Art of Framing.* San Francisco: Jossey-Bass Publishers, 1996.

Mills, Harry. *Artful Persuasion.* New York: Amacom, 2000.

Weisinger, Hendrie. *The Critical Edge: How to Criticize Up and Down Your Organization and Make It Pay Off.* Boston: Little, Brown and Company, 1989.

Weisinger, Hendrie. *The Power of Positive Criticism*. New York: Amacom, 2000.

Zima, Joseph P. *Interviewing: Key to Effective Management*. Chicago: Science Research Associates, 1983.

# 26

# JUDGMENT

by Robert W. Walker

Dictionaries and thesauruses usually integrate definitions, meanings, and synonyms for "decision-making" and "judgment," creating the impression that they are one and the same.[1] Adding more to this perspective, both decision-making and judgment can include similar components such as informed reasoning, intellectual thinking, ethical confidence, and decisive resolve. In contrast, Chapter 15 on Decision-Making provides evidence that, although similarities exist between decision-making and judgment, they are not identical. Accordingly, because military doctrine stipulates that decision-making is one of a military leader's primary responsibilities, but also because understanding the differences is fundamental to a leader's processes of judgment and decision-making, these terms and their definitions require some investigation.

## SOME HISTORY — MILITARY INCOMPETENCE AND POOR JUDGMENT

Two well-known books, written in the last three decades, address military incompetence[2] and military failures related to military incompetence.[3] Their value comes from their exploration of military leader behaviour, particularly their judgment, in warfighting campaigns, and in military institutions. Essentially, the books cover the last 100 years with an emphasis on the First and Second World Wars and focusing particularly on poor judgment and the reasons for it.

British scholar, psychologist, and ex-military officer Dr. Norman Dixon addressed the psychology of military incompetence, and the judgment that contributed to it, from two perspectives. He elaborated substantially on "the psychopathology of individual leaders" and on "the social psychology of military organizations." Dixon explored the human-incapacity perspective from several viewpoints, emphasizing that in fact some senior leaders, comfortably ensconced in positions of authority, were in fact almost totally devoid of cognitive capacities required for judgment. Alternatively, he emphasized, some leaders, in light of their professional circumstances and personal priorities, may have predisposed themselves, "psychopathologically," to not use their cognitive capacities for good judgments in order not to jeopardize their careers by "standing out," having already been institutionally shaped into conformist behaviour and risk-free careers. With respect to the military institution, Dixon's exploration of "the social psychology of military organizations" addressed inflexible hierarchical structures and misguided human resources practices that contributed to the presence of those least able or willing to challenge obsolete practices of war, leadership, and professionalism. Dixon provided innumerable and impressive examples in support of his detailed chronology of military incompetence.

American strategic analyst and author Eliot Cohen, and British professor and historian John Gooch took an innovative look at the factors that undermined armies, leading to military "misfortunes" and the defeat of nations. They emphasized that military misfortunes should not be seen as failures of an individual leader, but of clusters of individuals representing and leading an organization. They also emphasized that the reasons for such failures are not easy to discern. Cohen and Gooch analyzed failures of judgment from military leader groups and institutional perspectives (Passchendaele or Pearl Harbour), from industry (the Edsel), and from government (the U.S. Three Mile Island nuclear disaster), to map out misfortune. Their contribution, engaging an analytical matrix specifically for their anatomy of failures in war, was to identify the strategic judgment necessary, particularly, by learning, anticipating, and adapting, all in order to master organizational learning that would support decisive and effective judgment. Again, like Dixon,

Cohen and Gooch provided strong examples of military misfortune, warfighting failures, and inadequate judgment by senior leaders.

## UNDERSTANDING JUDGMENT

The history of military incompetence and failures in warfare, the vastness of which is inadequately represented by just two above-mentioned reference sources, nonetheless emphasizes the major relevance of sound judgment. Judgment, essentially, should be seen as a component competency that supports the requisite decision-making practices needed to avoid misfortunes. Were decision-making the cake, judgment would be the icing. With respect to this icing:

* judgment, applied during the leaders' cognitive transformation of complex components of knowledge, while they are engaging their intuition and intellect, can generate leaders' high-quality understanding that, when applied to the final outcome, will have increased the probability that the reasoned cognitive outcome is appropriate to the required action;

* judgment involves a clear perception of the developing circumstances, an interpretation of all evidence available, a weighing and balancing of options, an extrication of the essentials from among the non-essentials and the irrelevant, a decisiveness born of objectivity and due consideration, and the delivery of a reasoned opinion to a final and practical application;

* judgment is of particular importance when the information being used is less than perfect, when a rational approach with thorough consideration of

alternatives and the implementation of detailed analyses is next to impossible, when time constraints are severe, and when immediate actions are required.

Chapter 15 on Decision-Making in this handbook describes the Rational Decision-Making Model as an example of an ideal rational process in which, it is hypothesized, perfect rationality can exist, as would a judgment strategy that is perfectly informed and perfectly logical. This model is predicated on formal processes based on normative theories of probability and logic.[4] At the junior leader level, a generous application of common sense, tempered by relevant experience with the leading of people, and supplemented by linear analytical thinking, frequently can generate an adequate judgment.

Importantly, and essentially, this "limited" rational model of decision-making at the junior leader level also emphasizes the differences between decision-making and judgment. Judgment can be a component competency of decision-making, but as we will see, judgment needs to "go beyond" analytic-thinking decision-making as performed at basic, junior leader levels. Particularly in circumstances of senior leaders with increasing responsibilities — a more expansive operating environment, an acquired leading-the-institution focus, a longer-term strategic perspective — such responsibilities would make a rational decision-making approach inadequate, and would cry out for creative and abstract cognitive judgment.

This somewhat dichotomous circumstance of junior leader/analytic cognition and senior leader/creative cognition is foundational for differentiating between decision-making and judgment. A representative part of that foundation is the Canadian Defence Academy's Professional Development Framework (PDF), utilized for oversight of training and development.[5] (The PDF is explained in detail in Chapter 32 of this handbook.) One of the PDF's five leader metacompetencies is Cognitive Capacities, which spreads across the continuum of junior through senior leader levels, evolving from an analytic cognitive focus at the more junior and into the intermediate levels of leadership (i.e., predominantly engaged in such thinking skills as troubleshooting and problem-solving) to a creative and more abstract cognitive focus at the advanced and senior leader levels (i.e., engaged in imaginative, original, inventive, and innovative deliberation). This comparison of the opposing ends of the continuum obviously highlights that dichotomous nature of decision-making and judgment.

With reference once again to Chapter 15, Bill Bentley's essay on decision-making in this handbook, he states:

> The nature of the real world battlespace limits a human decision-maker's ability to implement truly analytical processes. This has led to the development of another approach that comprises what are called naturalistic or intuitive theories of decision-making ... based on the premise that people use informal procedures or heuristics to make decisions within the restrictions of available time, limited information and limited cognitive processing capacity ... [procedures that are based on] three basic principles ... decisions made by sequential, holistic evaluations of courses of action ... [reliance] primarily on recognition-based processes to generate options ... [and, for judgment,] adopt a satisficing criterion, stopping the search when an acceptable course of action is identified.[6]

Bentley described recognition-primed decision-making models, narrative-based models, and, specifically, hard systems thinking, soft systems methodology, and systemic operational design, the latter comprised of seven steps that build upon each other and lead to an articulation of a holistic operational design for detailed planning and implementation. Bentley saw this not as a panacea for contemporary operational design challenges, but as "a viable alternative to current CF operational planning process methodologies, in that its theoretical underpinnings take the complex and indeterminate nature of the contemporary operating environment and emerging science into account."[7] He concluded that modern warfare and conflict required a mix of decision-making models and techniques — structured, linear techniques at tactical levels with junior leaders and, for senior leaders needing techniques at operation and strategic levels, non-linearity, systems-thinking and creativity to address the broad unpredictability of challenges.

## **EXAMPLES OF CREATIVITY-BASED JUDGMENT**

Elsewhere, beyond Bentley's applications of models to warfighting, are two other CF sets of circumstances of unpredictability, or perhaps ambiguity and complexity, which cry out for strong professional judgment. They are professional ideology[8] and leader ethics,[9] related but separate challenges, being practised in the CF.

Practitioners in the profession of arms, as members of a larger group that would include doctors, lawyers, clerics as professionals, develop their professional ideology to serve them in their execution of roles and responsibilities in their profession. The Canadian Forces' profession of arms, with a military ethos embedded in it, is described extensively elsewhere.[10] The Canadian military version of professional ideology embraces institutional/CF values, norms, standards, perceptions, practices. However, in line with the predisposition of each member of the profession of arms, it is individualistic, like any ideology, in that a personal, complex, values-based, knowledge-based, and experience-based conceptualization comes to exist in each mind of each professional. As discussed below, under ethics, an individual's personal identity development is yet one more aspect of a professional ideology, as leaders' identity structures influence how they see themselves and how they make sense of the institution in which they serve as professionals.[11] Central to all of this professional–institution congruence is judgment.

However, importantly, as Bentley articulates, the profession of arms varies greatly from other professions in that, as a collective profession, it is distinguished by the concept of service before self, a very expansive interpretation of duty, the lawful and ordered application of military force, and the acceptance of the concept of unlimited liability, all components demanding exceptional judgment for life-and-death situations.[12] The ideology in the profession of arms also claims a unique body of military knowledge that is developed and imparted differently than in any other profession. This knowledge has a significant level of abstraction resulting from its basis in theory, thereby requiring extensive education, career-long study, and exceptional judgment. As Carl von Clausewitz warned us in his book *On War*: "Theory cannot equip the mind with formulae for solving problems … [b]ut it can give the mind insight into the great mass of phenomena and of their relationships, then leave it free to rise into the higher realm of action."[13]

By virtue of the complexity of the professional and ideological exigencies of military life in the profession of arms, survival of members, particularly leaders of the institution, depends exponentially on strong, relevant, creative, and timely judgment. But, there is more. Another domain that draws out the differences between "routine" decision-making and that strong, relevant, timely judgment is military ethics. It acts just as senior leadership of institutions or professional ideology, with their ambiguities, abstractions, and complexities, have acted. Ethics is about human behaviour, about interpretations of rights and wrongs, or choices between/among two or more rights, or two or more wrongs. Ethics exist against a backdrop of culture, climate, and environmental factors, not a passive backdrop, but sources of considerable influence. Ethics embraces values, norms, standards, perceptions, practices, just as professional ideology does, and military duty challenges military leaders to internalize such values and beliefs that underpin the military ethos.

Additionally, Dr. Lagacé-Roy has explored Robert Kegan's personal [leader] identity development model with respect to its relevance to ethics, and to what Kegan called "meaning-making" for a broader world view. This more complex way of thinking generates a deeper reflection [judgment] before a situation is addressed. Accordingly, the identity structure of professionals in the profession of arms will influence just how these professionals, as leaders, see themselves, and how they interpret and make sense of the institution they serve, and how their ethical conduct will contribute to positive outcomes, member well-being, and mission success.[14] As Lagacé-Roy explained, "Kegan's stages focus on the aspect of oneself that evolves toward an understanding of systems and groups that shaped the person ... and of which the self is part."[15] Kegan's advanced stages, 4 and 5, involve internalized self-reflections, principles, and obligations that leaders will incorporate into their judgments about ethical dilemmas, and eventually leaders can acquire a capacity to recognize a plurality of systems, groups or theories seen as universal and from which they can construct a broader view of society, or the world, in order to recognize their own perspective process when dealing with complex issues. Such a process is a desired outcome of values-based leadership and professional ethics.[16] When leaders inspire, contribute, and create changes valued by the institution, they are transformational leaders influencing the ethical behaviour of members and the ethical climate of the institution. Their component competency to reach this broad capacity very much involves enlightened judgment.

## **SUMMARY**

So, why is sound and effective judgment important to military leadership? Essentially, it's the environment — an environment in which people can be killed! For a military, its surroundings are ones in which mission success is of primary importance, a mission involving unlimited liability in warfighting, a fighting spirit of commitment and morale generated by confidence in leaders to judge and act, and the assurance by such leaders, as much as possible, of the well-being of members. Without strong judgment, and the confidence by followers in that judgment, mission success is jeopardized.

This chapter, in exploring generations of military leader behaviour involving consistently inappropriate judgments, reflects that effective judgment in military circumstances is not a new and recent challenge. This history, however, as a backdrop, also reinforces that, today, "The times, they are a-changin'," and they are rapidly a-changin' into far more complex, demanding, and challenging times. Today's threatening military circumstances include an increasingly and volatile international security environment, the unpredictable natures of societies within it, and the high-speed technological innovations of warfare, all necessitating an even greater informed and effective judgment for those leaders in the profession of arms.

But, what is judgment? Judgment — sound, enlightened, disciplined but, importantly, also innovative and inventive thinking — is the component increasingly important to success. This crucial judgment is what supports the requisite decision-making needed to avoid error and failure in the military. It is part of the conceptual transformation of complex knowledge, through the leader's intuition and intellect, into a high-quality understanding applicable to the final outcome. Judgment is the integration of all evidence available, a weighing of options, a separating out of the essentials from the nonessentials, and the drawing up of a reasoned intention for a final and practical application. Judgment is of particular importance when the information being used is less than perfect and less than complete, when a rational approach with thorough consideration of alternatives and the implementation of detailed analyses is next to impossible, when time constraints are severe, and when imminent action is required. Additionally, judgment is particularly important and even more complex in leadership at the upper echelons of the profession of arms. With seniority and promotion, practical and linear problem-solving and trouble-shooting analytical demands evolve into ambiguous, complex, longer-term, strategic, and institutional responsibilities for which "naturalistic" decision-making theories, systems-thinking, and creative applications of judgment are required. So, judgment is not just one thing — it is a system of creative cognition founded upon a sequence of steps and actions as explained above, but it also is a process that is transformed as senior leader roles expand, consequences become more dire, and the scope of influences broadens.

## NOTES

1. *Webster's New World Dictionary* (Toronto: Nelson, Foster & Scott Ltd., 1960), 380 and 792. Defines making a decision as giving a

judgment, a decision as a judgment reached, and deciding as judging. *Roget's International Thesaurus*, 6th ed. (New York: HarperCollins Publishers, 2001), 598.09 recommends one for the other — decision/judgment.

2. Norman F. Dixon, *On the Psychology of Military Incompetence* (London: Jonathan Cape Ltd., 1976).

3. Eliot A. Cohen and John Gooch, *Military Misfortunes: The Anatomy of Failure in War* (New York: Anchor Books/Random House, 1990).

4. See Chapter 15 on Decision-Making in this handbook, authored by Bill Bentley, specifically the section on the Rational Decision-Making Model and the (Bounded) Rational Decision-Making Model. Bentley's original source was Herbert Simon, *The Sciences of the Artificial* (Cambridge, MA: MIT Press, 1996), 132.

5. Robert W. Walker, *The Professional Development Framework: Generating Effectiveness in Canadian Forces Leadership*, CFLI Technical Report 2006-01. (Kingston: Canadian Forces Leadership Institute, 2006). Available at http://www.cda-acd.forces.gc.ca/cfli.

6. Bill Bentley, "Decision-Making," in *The Military Leadership Handbook,* eds. Colonel Bernd Horn and Dr. Robert Walker (Toronto and Kingston: Dundurn Press/Canadian Defence Academy Press).

7. *Ibid.*

8. Bill Bentley, *The Professional Ideology and the Profession of Arms in Canada* (Toronto: Canadian Institute of Strategic Studies, 2005).

9. See Daniel Lagacé-Roy's "Ethics" chapter in this handbook and his chapter "Institutional Leader Ethics" in *Institutional Leadership in the Canadian Forces: Contemporary Issues,* ed. Robert W. Walker (Kingston: Canadian Defence Academy Press, 2007), 109–121. He also has made reference to the Canadian Forces Defence Ethics program, *Introduction to Defence Ethics,* 2nd ed., 2005. For broader coverage related to Judgment, see also Lagacé-Roy's two publications, *Ethics in the Canadian Forces: Making Tough Choices* (Instructor Manual) (Kingston: DND, 2006), and *Ethics in the*

    *Canadian Forces: Making Tough Choices* (Workbook) (Kingston: DND, 2006).
10. Canada. *Duty with Honour: The Profession of Arms in Canada.* (Kingston: DND, 2003), as well as Bentley's *The Professional Ideology and the Profession of Arms in Canada.*
11. Robert Kegan, *The Evolving Self: Problem and Process in Human Development.* (Cambridge, MA: Harvard University Press, 1982.)
12. Bentley, *The Professional Ideology*, 56–58.
13. Carl von Clausewitz, *On War,* eds. Michael Howard and Peter Paret (Princeton: Princeton University Press, 1976), 578.
14. Daniel Lagacé-Roy, "Institutional Leader Ethics" in *Institutional Leadership in the Canadian Forces: Contemporary Issues,* Robert W. Walker (Kingston: Canadian Defence Academy Press, 2007), 110, 114–117. Original source is Robert Kegan, *The Evolving Self: Problem and Process in Human Development* (Cambridge, MA: Harvard University Press, 1982).
15. Daniel Lagacé-Roy, "Institutional Leader Ethics," 115.
16. *Ibid.*, 116.

## SELECTED READINGS

Canada. *Ethics in the Canadian Forces: Making Tough Choices* (Instructor's Manual). Kingston: DND, 2006.

Canada. *Ethics in the Canadian Forces: Making Tough Choices* (Workbook). Kingston: DND, 2006.

Canada. *Leadership in the Canadian Forces: Conceptual Foundations.* Kingston: DND, 2005. http://www.cda-acd.forces.gc.ca/cfli.

Cohen, Eliot A. and John Gooch. *Military Misfortunes: The Anatomy of Failure in War.* New York: Anchor Books/Random House, 1990.

Dixon, Norman F. *On the Psychology of Military Incompetence*. London: Jonathan Cape Ltd., 1976.

Horn, Bernd and Stephen J. Harris. *Generalship and the Art of the Admiral*. St. Catharines, ON: Vanwell Publishing Ltd., 2001.

Hughes, Richard L. and Katherine C. Beatty. *Becoming a Strategic Leader*. San Francisco: Jossey-Bass and the Center for Creative Leadership, 2005.

Kegan, Robert. *The Evolving Self: Problem and Process in Human Development*. Cambridge, MA: Harvard University Press, 1982.

Walker, Robert W. *The Professional Development Framework: Generating Effectiveness in Canadian Forces Leadership*, CFLI Technical Report 2006-01. Kingston: Canadian Forces Leadership Institute, 2006. http://www.cda-acd.forces.gc.ca/cfli.

Walker, Robert W., ed. *Institutional Leadership in the Canadian Forces: Contemporary Issues*. Kingston: Canadian Defence Academy Press, 2007.

Wong, Leonard, Stephen Gerras, William Kidd, Robert Pricone, and Richard Swengros. *Strategic Leadership Competencies*. Carlisle Barracks, PA: U.S. Army War College, Strategic Studies Institute, 2003. http://www.carlisle.army.mil/research_and_pubs/research_and_publications.shtml.

# 27

# MENTORING[1]

by Daniel Lagacé-Roy

In the current ambiguous, complex and ever-changing world, individuals, particularly military professionals, are searching for mentors to help cope with challenges and difficult situations. Furthermore, the role of a mentor encompasses the desire and responsibility to participate in someone's development, which provides additional tutelage for those navigating through the difficult security environment. As such, mentoring is a learning tool that offers a more holistic approach to development and provides the opportunity to better prepare individuals to meet the challenges and obligations of their appointments.

## BACKGROUND

Historically,[2] the term *mentor* derives from Homer's epic poem, *The Odyssey*, in which the Greek poet introduces a character named Mentor. He was the son of Alcumus and, in his old age, a trusted advisor and old friend of Odysseus, King of Ithaca. Before leaving for the siege of Troy, Odysseus appointed Mentor to be a surrogate father to his son, Telemachus, and was entrusted as a guardian to protect the royal household. It is interesting to note that an assiduous reading of *The Odyssey* reveals that the goddess Pallas Athene took, on some occasions, the Mentor's form. Therefore, the Greek mythology did not limit the role of a mentor to a specific gender.

*Mentor*, as a common noun, entered both French and English languages only in the early 18th century. It is the French author Fénélon in his 1869 work, *Les Aventures de Télémaque*, that brought the term *mentor* into the language. It is argued that today's meaning of the word *mentor* derives from Fénélon's interpretations.[3]

With the passage of time, mentoring has been associated with the concept of apprenticeship as a method for developing the personal and professional skills of employees. Young men were apprentices to master craftsmen, traders, or a ship's captain, who passed down their knowledge. Sometimes personal relationships developed between a master (mentor) and an apprentice.

This mentor-apprentice dyad evolved and was subjected to adaptations in order to satisfy particular needs. For example, after the Second World War, mentoring was primarily applied to develop high potential personnel for managerial positions. It was only after 1960 that the importance of mentoring as a career development strategy benefiting all employees was recognized.

As societies became more complex and impersonal, in part due to the advancement of technology, the need for person-to-person mentoring became increasingly important. Organizations began to use mentoring as an important developmental tool for professional and personal skills and this developmental tool is quite prevalent today.[4]

## WHAT IS MENTORING?

The concept of mentoring is defined by a professional relationship in which a more experienced person (a mentor) voluntarily shares knowledge, insights, and wisdom with a less experienced person (a mentee[5]) who wishes to benefit from that exchange. It is a "dynamic, time-consuming relationship in which the person mentored matures professionally and personally under the mentor's tutelage so he [she] can 'innovate, think, and adapt to the demands of a fast-paced, highly stressful, rapidly changing environment.'"[6] Furthermore, it is a medium to long-term learning relationship founded on respect, honesty, trust, and mutual goals. Mentoring generally focuses on four long-term developmental areas:

    a. Leadership (e.g., enhancing leadership skills);
    b. Professional (e.g., encouraging growth through up-to-date courses);
    c. Career (e.g., identifying possibilities and opportunities for advancement);
    d. Personal (e.g., discussing ways to balance family and work).

It is important to keep in mind that each mentoring relationship is unique. The topics of discussion, such as understanding organizational culture and improving performance, will largely depend on the mentee's needs and the issues that the mentor feels are important for a long-term development. One key ingredient in a mentoring is that the mentor and the mentee are willing to commit time and energy to make the relationship work.

## **WHAT IS A MENTOR?**

A mentoring relationship offers the setting in which a mentor provides guidance in terms of the areas of development mentioned above. In order to provide the appropriate guidance, the mentor should hold a higher position (e.g., at least two levels above) than the mentee. However, the most important factor is that the mentor is a more experienced person (e.g., team leader mentoring a junior employee) and he/she is willing to motivate, encourage, empower, and support the mentee to the best of his/her abilities.

During the development of a mentoring relationship, a mentor may be called to be a counsellor, a teacher, a guide, a motivator, a role model, and a coach.[7] These functions are determined by the mentee's needs,[8] and can be briefly described as follows:

    a. a counsellor listens, encourages, and assists in developing self-awareness;

b. a teacher helps set realistic goals and informs about professional obligations;
c. a guide shares experience and acts as a resource person;
d. a motivator recognizes strengths, areas of development, and empowers to find answers;
e. a role model acts as a person with integrity, one whose actions and values are to be emulated; and
f. a coach focuses on the development of specific skills.

It is important to pay attention to the last function, which is the role of a coach. In mentoring, that function represents only one facet of the relationship. When a mentor undertakes the role of a coach, it is done with the knowledge that the skills (e.g., to overcome shortcomings) learned through coaching are means to achieve the overall goals (e.g., to become a better leader) set by the mentee and the mentor at the outset of their mentoring relationship.

## **WHAT IS A MENTEE?**

A mentee is usually described as a more junior professional with less experience who is highly motivated to learn, develop, and grow personally and professionally. A mentee seeks out and is receptive to feedback and welcomes new challenges and new responsibilities. The success of a mentoring relationship often relies on the mentee's eagerness to learn and to take an active role for development to occur. Another factor for success is that the mentee chooses his or her mentor. This choice will be based on the mentee's needs (e.g., keen to learn from previously successful experiences) and also the desire to seek out for a mentor who is prepared to dedicate time to someone's development.

## WHAT IS THE DIFFERENCE BETWEEN MENTORING AND COACHING?

The use of the terms mentoring and coaching has come to be used interchangeably and their meanings are often confused. Although mentoring, as it was already mentioned, encompasses coaching as one particular function, it is not synonymous with it. "Thus, the scope of mentoring is vastly greater than coaching, which is, itself, a small subset of mentoring."[9]

Coaching, as a single function, is a short-term relationship in which a person (coach) is focused on the development and enhancement of performance, skills, effectiveness, and potential of another person (coachee).[10] To be more precise, coaching is a process in which a coach directs the learning and instruction with the intention "to correct inappropriate behaviour, improve performance, and impart skills that the employee needs to accept news responsibilities."[11] Therefore, the main focus of a coaching role is job-focused in guiding a person to achieve concrete end results (e.g. learn how to delegate tasks) during a short-term relationship.

## TYPES OF MENTORING RELATIONSHIPS

The development and the process of a mentoring relationship usually take shape inside a particular type of relationship. The more suitable type depends on the goals to achieve by the parties (e.g., mentee, mentor, organization) involved. Three types of relationships are briefly described here. The two most referred to are formal and informal mentoring relationships. A third type, called "semi-formal," is also offered as a "compromise" between formal and informal. This last type of mentorship is often the most effective in large organizations.[12]

The first type is called "formal," given that the mentoring relationship is structured and managed by a specific organization and based on the organizational needs and goals. That formal setting calls for some pressures to participate, and mentors and mentees are paired based on compatibility. In addition, training is provided for both and outcomes are measured.

The second type is called "informal" since the mentoring relationship happens in a spontaneous and ad hoc manner. There is a self-select process between a person who believes in someone's potential, and an individual who selects someone whom he/she might view as a role model. The process and benefits are not always recognized.

The third type is called "semi-formal" because it includes all of the attributes of the formal program, with the exception of the matching process. It is more flexible and available to all personnel as part of the organization's developmental programs. The organization provides training and education and facilitates matching by providing opportunities for participants to meet and form their mentoring relationships.

## **BENEFITS OF MENTORING**

The benefits from a mentoring relationship can be easily identified at three different levels: the mentee, the mentor, and the organization (when a formal program is in place).

The benefits, from the mentee's perspective, include, but are not limited to:

a. receiving career guidance;
b. understanding roles and expectations within the organization;
c. increasing organizational knowledge;
d. learning how to deal with difficult situations; and
e. increasing self-confidence and productivity.

The benefit of being involved in the development of subordinates and future leaders is, from the mentor's point of view, a great reward in itself. In addition to this accomplishment, personal and professional benefits play an important role. Benefits include:

a. personal satisfaction;
b. new ideas and fresh perspectives;
c. contribution to the corporate memory;
d. opportunities to reflect on personal and professional achievements; and
e. collaboration and collegiality.

Organizational benefits of mentoring programs are clearly acknowledged in this quotation by the Conference Board of Canada: "leading organizations are considering innovative ways to retain talent, help employees learn and grow, and transfer knowledge from one experienced worker to those less experienced."[13] Other benefits include, but are not limited to:

a. a direct influence on job performance;
b. early career socialization;
c. organizational communication and understanding; and
d. long-range human-resource development planning and leadership succession.

## **MENTORING AND LEADERSHIP**

The importance of creating a learning environment which ensures that members develop professionally and personally through knowledge and experience is increasingly recognized. Such a learning environment allows future leaders to adapt themselves and be receptive to internal and external changes. Furthermore, it allows them to encourage their subordinates to "engage in broad inquiry, to think critically, and to venture and debate new ideas in the interest of contributing to collective effectiveness."[14]

In fact, a developmental learning environment promotes the type of leaders who are responsible for mentoring "people in apprenticeship positions and challenging assignments, and encourage and support subordinate participation in educational, professional, and personal-growth activities over their career span."¹⁵ By participating in their leadership, professional, career, and personal development, leaders, as mentors, embrace mentoring as a learning capacity in enhancing individuals' abilities. Furthermore, by providing the appropriate mentoring relationship, leaders, as mentors, ensure that future leaders possess "capacities" that are needed for an effective organization.

Mentoring is therefore a proven learning process and leadership developmental tool. Indeed, to be effective, mentoring cannot be separated from the critical role of development. Thus, "leadership development is a career-long process and that leadership development training must address how to interact with subordinates and assist their growth and development as well as a focus on personal leadership development."¹⁶ The mentoring process, in terms of development, is suited to address the continual enhancement of subordinates and future leaders. To be more precise, mentoring serves as a way to develop social capacities (e.g., effective communication) that are needed "… to formulate solutions in complex organizational environments …"¹⁷

## **MENTORING AND THE CANADIAN FORCES (CF) AS A CASE STUDY**

The concept of mentoring is not completely foreign to the CF and some forms of mentoring initiatives are established at various levels. In fact, mentoring has always been an important facet of CF leadership. While there is no formal CF mentoring program at the time of writing, a number of mentoring initiatives have already been implemented at different levels of the organization (e.g. unit level, military occupation classification (MOC) level). The CF offers a unique setting for mentoring relationships: less experienced members can learn and develop from more experienced members. Senior officers and senior non-commissioned officers (NCOs) often adopt this role informally by encouraging and challenging subordinates in their development. Mentoring is also beneficial for the CF as an institution since it has an impact on job satisfaction, productivity, and retention.

Most noticeably, the interest in mentoring has increased recently and it is becoming a more integral part of leadership development. As one example: "the Naval Reserve uses the practice of mentoring in preparing its aspiring Ship's Commanding Officers to develop the command thought process and assist them in preparing for their Command Boards."[18] Training and educational establishments in the CF, as another example, have incorporated lectures on the subject of mentoring and coaching. This is the case for the Advanced and Chief Qualification courses at the Professional Development Centre at RMC St-Jean.

It seems that those initiatives echoed the Chief of the Defence Staff's concerns for CF members' development. General Rick Hillier stated that the "CF must develop a far more holistic, integrated approach to its personnel management and support as it creates a human resources strategy for the years to come. Our men and women are the single essential element in everything that we do and we must continue to improve how we recruit, train, educate, integrate, and employ these personnel to ensure we are properly focused on our operational mission."[19] The interest in the development of CF members has two goals: one is to create a "human resources strategy" for the future, and the second is to prepare for operational effectiveness. These two aims were already mentioned in the document *Military HR Strategy 2020: Facing the People Challenges of the Future*: "The operational capacity of the Canadian Forces is ultimately derived from its people.... Look after our people, invest in them and give them confidence in the future."[20] These two aims are the foundations on which a mentoring program should be based. If the CF has to invest in its people, mentoring should be a key component in developing this important investment.

## **SUMMARY**

This chapter introduced concepts and different aspects of a mentoring relationship. Mentoring is presented as a more holistic approach to learning and, as such, can benefit the mentee, the mentor, and the organization. Furthermore, the mentoring process is articulated as an integral part of leadership. It is also suggested that mentoring is a good tool for developing subordinates and preparing future leaders. It is the duty of leaders to improve performance and effectiveness of the organization as a whole by emphasizing individual characteristics of their members (e.g., skills and abilities; attitudes and motivation; needs and values). If mentoring is ideally suited to address the needs unique to each individual, especially given the fact that it is a long-term professional relationship as opposed to a one-time meeting, it might indicates that mentoring should be supported by any organization.

## NOTES

1. This chapter is based on the *Mentoring Handbook*, published by CFLI and written by the author and Lieutenant-Colonel Janine Knackstedt, PhD.
2. James Donovan, "The Concept and Role of Mentor," *Nurse Education Today* 10 (1990): 294–298; Andy Roberts, "Homer's Mentor: Duties Fulfilled or Misconstrued," http://home.att.net/~nickols/homers_mentor.htm (accessed February 12, 2006).
3. Priscilla P. Clark, "The Metamorphoses of Mentor: Fénélon to Balzac," *Romanic Review* 75 (1984): 200–215.
4. J. Monaghan and N. Lunt, "Mentoring: Person, Process, Practice and Problems," *British Journal of Educational Studies*, Vol. 40, No. 3 (August 1992): 248–263.
5. "Associate" and "protege" are terms also used to describe the less-experienced person.
6. Nate Hunsinger, "Mentoring: Growing Company Grade Officers," in *Military Review*, September–October 2004, 6.
7. Kathy E. Kram, "Phases of the Mentor Relationship," *Academy of Management Journal*, (December 1983): 613–614.
8. Janine Knackstedt, "Organizational Mentoring: What about Protégé Needs?," paper presented at the 16th Annual Conference, Society for Industrial and Organizational Psychology, San Diego,

California, April 2001.
9. Harvard Business Essentials, *Coaching and Mentoring* (Boston: Harvard Business School Press, 2004), 77.
10. *Coaching & Facilitation for CFC Directing Staff*, document prepared by Wayne Chamney and Andrea Legros, Canadian Forces Learning and Development Centre, August 2006, B1.3.
11. Harvard Business Essentials, *Coaching and Mentoring,* Op cit., 79.
12. Janine Knackstedt, "A Proposed Military Executive Mentoring Model: The Way Ahead for Executive Leadership Development," Draft paper, 2001, 10–29.
13. Janice Cooney, "Mentoring: Finding a Perfect Match for People Development," *The Conference Board of Canada*, June 2003, http://www.conferenceboard.ca/documents.asp (accessed 17 August 2006).
14. Canada. *Leadership in the Canadian Forces: Conceptual Foundations*, (Kingston: DND, 2005), 126.
15. *Ibid.*, 50.
16. Mark White, "Mentoring as a Mainstay for the Canadian Forces" (master's thesis, Canadian Forces College, 2004), 24–54.
17. Robert W. Walker, "The Professional Development Framework: Generating Effectiveness in Canadian Forces Leadership," CFLI Technical Report 2006-01 (Kingston: Canadian Forces Leadership Institute, 2006), 34.
18. White, 36–54.
19. CDS-TRANSFORMATION, SITREP, 03/05. http://cds.mil.ca/cft-tfc/pubs/sitrep_e.asp (accessed 17 August 2006).
20. Department of National Defence, *Military HR Strategy 2020: Facing the People Challenges of the Future* (Ottawa: Minister of National Defence, 2002), i, 1.

**SELECTED READINGS**

Canada. *Military HR Strategy 2020: Facing the People Challenges of the Future*. Ottawa: DND, 2002.

Ensher, Ellen A. and Susan Elaine Murphy. *Power Mentoring: How Successful Mentors and Protégés Get the Most out of Their Relationships*. San Francisco: Jossey-Bass Publishers, 2005.

Harvard Business Essentials. *Coaching and Mentoring*. Boston: Harvard Business School Press, 2004.

Hunsinger, Nate. "Mentorship: Growing Company Grade Officers," *Military Review*, September/October 2004, 78–85.

Kram, Kathy E. "Mentoring in the Workplace," in *Career Development in Organizations*. Edited by Douglas T. Hall and Associates. San Francisco: Jossey-Bass Publishers, 1986, 160–201.

———. "Phases of the Mentor Relationship." *Academy of Management Journal*, December 1983, 613–614.

Lagacé-Roy, Daniel and Janine Knackstedt. *Mentoring Handbook*. Kingston: Canadian Forces Leadership Institute, 2007.

McDermott, Ian and Wendy Jago. *The Coaching Bible: The Essential Handbook*. London: Piatkus, 2005.

Walker, Robert W. *The Professional Development Framework: Generating Effectiveness in Canadian Forces Leadership*. CFLI Technical Report 2006-01. Kingston: Canadian Forces Leadership Institute, 2006.

White, Mark. *Mentoring as a Mainstay for the Canadian Forces*. Master's thesis, Canadian Forces College, 2004.

Wild, Bill. *Understanding Mentoring: Implications for the Canadian Forces*. Kingston: Canadian Forces Leadership Institute, May 2002.

# 28
# THE MILITARY PROFESSIONAL
by Bill Bentley

Samuel Huntington, the author of the seminal work on the military as a profession, noted in 1957 that prior to 1800 there was no such thing as a professional officer. However, he noted that in 1900 such bodies existed in virtually all major Western countries.[1] Arguing along the same lines, the military sociologist Morris Janowitz observed in 1960 that "the officer corps can now also be analyzed as a professional group by means of sociological concepts."[2] The emergence of a professional military, specifically a professional officer corps, had, according to these two students of military professionalism, been a slow and gradual process with many interruptions and reversals. Mercenary officers were ubiquitous in the 16th century but the outline forms of professionalism were not clearly discernible until the beginning of the 18th century. It took another 100 years or so before one could speak of the emergence of an integrated military profession.

The parallels with the evolution of professions in general (e.g., medicine, law, engineering) in the 19th century are inescapable, with many of the same factors that account for the professionalization of select occupations in the latter half of that century at play in Western militaries. It is generally the case that the military profession, as we know it, emerged out of the European state system, the development of modern society and the European wars of the last 200 years. In support of this contention, a British authority on the subject, General Sir John Hackett, observed that:

> Service under arms has evolved into a profession, not only in the wider sense of what is professed, but in the

narrower sense of an occupation with a distinguishable corpus of specific technical knowledge and doctrine, a more or less group coherence, a complex of institutions peculiar to itself, an education pattern adopted to its own specific needs, a career structure of its own and a distinct place in society which has brought it forth. In all these aspects it has strong points of resemblance to medicine and law, as well as to holy orders. Though service under arms has strongly marked vocational elements and some appearances of an occupation, it is probably as a profession that it can be most profitably studied.[3]

## **PROFESSIONS DEFINED**

What then are the generic attributes or characteristics of a profession that can also characterize the profession of arms? The form that modern professions takes in the course of the Industrial Revolution and beyond is that of corporate groups attempting, first of all, to organize production for a special type of market and to gain quasi-monopolistic control within that market. Given the singular nature of the commodity to be exchanged, the organization of production is concerned not with an inanimate product, but with the selection of producers or providers of services based on specialized knowledge. The end point of this primary aspect of professional organization is, therefore, the monopoly of relatively standardized education. Thus, two of the most generalized ideas underlying professionalism are first, the belief that certain work is so specialized as to be inaccessible to those lacking the required training and experience, and second, the belief that it cannot be standardized, rationalized or commodified. At the core of professionalism is its claim to discretionary specialization.

Drawing upon these themes the noted sociologist Talcott Parsons enumerated three core criteria by which to assess an occupation:

- \* The requirement of formal technical training accompanied by some institutionalized mode of validating both the adequacy of training and the competence of trained individuals. Among other things, the training must lead to some order of mastery of a generalized cultural tradition, and do so in a manner giving prominence to an intellectual component — that is, it must give primacy to the validation of cognitive rationality as applied to a given field.

* Not only must the cultural tradition be mastered, in the sense of being understood, but skills in some form of its use must also be developed.

* A full-fledged profession must have some institutional means of making sure that such competence will be put to socially responsible uses.[4]

A more recent and even more useful definition of professionalism is that provided by Professor Eliot Freidson, who describes the five interrelated attributes of ideal-type professionalism as follows:

* Specialized work in an officially recognized economy that is believed to be grounded in a body of theoretically based, discretionary knowledge and skills and that is accordingly given special high status in the work force.

* Exclusive jurisdiction in a particular division of labour created and controlled by occupational negotiation.

* A sheltered position in both internal and external labour markets, based on qualifying credentials created by the occupation.

* A formal training program lying outside the labour market that produces the qualifying credentials, which is controlled by the occupation, and associated with higher education.

* An ethic that asserts greater commitment to doing good work than to economic gain and to the quality rather than the economic efficiency of work.[5]

Professions, therefore, claim both specialized knowledge that is authoritative in a functional and cognitive sense, and a commitment to a transcendental value that guides and adjudicates how that knowledge is employed. In this conceptualization, the central role of knowledge and trust is readily identified and, in conjunction with the professional claim, forms the content of professional ideology.

## **THE MILITARY PROFESSION DEFINED**

The American political scientist Samuel Huntington was one of the first scholars to discuss military professionalism explicitly in terms of the theoretical construct outlined above. Huntington acknowledged that bureaucracy was, indeed, a characteristic of the military but argued that it was a secondary, not essential characteristic. He noted further that the nature and history of professions in general had received considerable scholarly attention, but that the public, as well as the scholars, hardly conceived of the military in the same way that they did medicine or law; withholding from the military the deference they accorded other professionals.

Huntington rejected this exclusion of the military from the professional construct both in theory and practice. He stated explicitly that the overriding theme of his seminal ground-breaking book, *The Soldier and the State*, was "that the modern officer corps is a professional body and the modern military officer a professional man. Professionalism was what distinguished the military officer of today from warriors of previous ages."

Huntington identified expertise, responsibility, and corporateness as the distinguishing attributes of that special type of vocation known as a profession. He maintained that according to these criteria, the military qualified fully as a profession. However, writing, as he did in 1957, Huntington invoked the sociological theory of closure to argue that only officers, and basically operational officers at that, were true professionals. Non-Commissioned Officers (NCOs), according to Huntington, were part of the organizational bureaucracy but not of the professional bureaucracy. Enlisted personnel, in Huntington's view, had neither the intellectual skills nor the professional responsibility of the operational officer. They were specialists in the application of violence, not in the management of violence. Until recently, this differentiation constituted the basis for a decidedly exclusionary model of the profession of arms common throughout Western militaries.

The decades between 1960 and 2000 saw significant, even profound, changes in both the geopolitical and politico-strategic environments as well as society in general. Socio-demographic changes, as they relate to the military profession, included increasing educational levels, the rise of individualism, cosmopolitanism, decline in deference to authority, and multiculturalism. Each of these phenomena influences the attributes of professionalism, affecting how education is acquired and distributed, and especially how military ethos structures the military community and its professional identity.

All of these changes strongly informed the re-assessment of the nature of military professionalism and its hitherto exclusionary bias that took place in Canada and the U.S. in the 1990s and into the new century. In the latter case, the work of the military sociologists Charles Moskos and Sam Sarkesian reflects two significant developments in the theory of military professionalism. First, the modifications they proposed to the conventional model resulted in a more inclusive construct in which NCOs are considered members of the profession of arms. Second, there is an acute awareness of the changing roles and tasks of military forces and the consequent impact of this phenomenon on all the attributes of the profession.

## MILITARY PROFESSIONALISM IN THE CANADIAN CONTEXT

In Canada, the work on the reformulation and codification of the Canadian military profession began in earnest in the late 1990s. The basis of this work initially rested on the Huntington model and the work of Moskos, Sarkesian, and especially Freidson, cited above, and on Professor Andrew Abbott's book, *The System of Professions*. Huntington's model was judged to be deficient in two major respects — its exclusionary nature and the inadequate treatment of the concept of professional ideology.

The Canadian model established four attributes of military professionalism — Responsibility, Expertise, Identity, and Professional Ideology. The profession of arms, therefore, exists within a complex formal structure that requires that the attributes of responsibility, expertise, and identity be differentiated and distributed within the profession yet coordinated and synchronized to ensure effectiveness. Their relationship is depicted in Figure 1.

The pivotal role of professional ideology is clearly seen in this construct. It is defined as the claim to a systematic, specialized, theory-based, and discretionary body of knowledge, authoritative in both a functional and cognitive sense; and, a transcendental value that adjudicates how that knowledge is applied. In the case of the military profession the body of knowledge in question is the General System of War and Conflict and the transcendental value is the Canadian military ethos. (Professional Ideology is addressed in Chapter 33 in this handbook.)

**Figure 1: Canadian Forces Military Professionalism**

**Jurisdiction**
(Government Direction and Control)

*Responsibility*

Collectively, members of the Canadian profession of arms have a core responsibility to the government and the people of Canada to defend the nation and its interests as well as projecting its values. This involves keeping the Canadian Forces effective. To manage this responsibility, members of the profession must fully understand the process of civil control of the military within the Canadian political system and demonstrate administrative competence and accountability. All Canadian military professionals share a broad responsibility to maintain the integrity and reputation of the profession of arms by ensuring that Canadian values and the Canadian military professional ideology guide their actions.

*Expertise*

Within the profession, expertise derives from a deep and comprehensive understanding of the theory and practice of armed conflict in its many forms, ranging from warfighting to humanitarian missions. Increasingly, this means mastering joint, combined, and multi-agency operations, national security issues and the law of armed conflict. Such expertise must be tempered with critical judgment skills on the appropriate use of force. Expertise and critical judgment are increasingly required from more junior ranks, because in an organizationally sophisticated, fast-paced, highly technical conflict, environment decisions taken at the lowest levels can have immediate and far-reaching implications.

*Identity*

Military members are indeed part of Canadian society, a fact reflected in the Canadian values incorporated into the military ethos, part of professional ideology. At the same time, they have a sense of a separate, distinct identity. Thus, while they are not civilians, military members are always citizens. Internally, the professional identity is differentiated in a number of ways, and the realities of combat and other operations at sea, on land, and in the air mean there will always be differences in the way military culture is expressed in the Navy, the Army, and the Air Force. These many separate identities coalesce around a hierarchy of loyalties that operates in descending order from law and government to the CF and thereafter through individual services to unit and branch. All Canadian military professionals share a common understanding of the Canadian professional ideology that unifies the CF around the concept of duty with honour.

The professional construct described above was codified in the capstone doctrinal manual, *Duty with Honour: The Profession of Arms in Canada*, signed by the Governor General, the Minister of National Defence, and the Chief of the Defence Staff in 2003. The manual has been distributed widely across the CF and forms the basis of all professional development in the areas of leadership and professionalism throughout the CF's Professional Development System.

In brief, the Canadian Statement of Military Professionalism is as follows:

> The profession of arms in Canada is comprised of military members who are dedicated to the defence of Canada and its interests, as directed by the Government of Canada. The profession of arms is distinguished by the concept of service before self, the lawful, ordered application of military force, and the acceptance of the concept of unlimited liability. Its members possess a systematic and specialized body of military knowledge and skills acquired through education, training and experience, and they apply this expertise competently and objectively in the accomplishment of their missions. Members of the Canadian profession of arms share a set of core values and beliefs found in the military ethos component of their professional ideology. This guides them in the performance of their duty and allows a

special relationship of trust to be maintained with Canadian society.

## SUMMARY

The concept of a profession, and the military profession in particular, is not a static one. As described above, the military profession has evolved in close parallel with other professions but has been uniquely shaped by its loyalty to only one client — the government — and its special function of the ordered application of military force in accordance with government direction. The evolution of the military profession has also been specifically shaped by the history, governance, and society of the particular nation in which it is embedded.

In the case of Canada, the continued dynamism and effectiveness of the military profession is guided by four principles: Relevance, Openness, Consistency, and Reciprocity.

*Relevance*

The principle of relevance speaks to the need to ensure that the profession continues to meet Canadians' expectations. To do this, it must demonstrably be capable of succeeding across the full range of missions that could be assigned. To be relevant, the profession must be accorded full legitimacy by Canadians because of its operational effectiveness, combat capability, reflection of Canadian values, and adherence to the core military values of duty, loyalty, integrity, and courage.

*Openness*

The principle of openness speaks to the profession's need to ensure that professional knowledge and practice are current and germane. Consequently, the profession must incorporate a philosophy of openness to new ideas and anticipate changes to meet future challenges. New responsibilities and different ways of doing things must be welcome if they strengthen professionalism.

*Consistency*

The principle of consistency speaks to the need to ensure that assigned responsibilities, expertise, and identity, as well as the manifestation of professional ideology across the profession, are integrated, coordinated, and aligned so that the Canadian Forces maintains its ability to achieve missions rapidly and decisively. New responsibilities may be acquired, but the fundamental ones to the country, government, and professional colleagues will continue. Expertise will remain differentiated by rank and function, but integrated by mission requirements. Responsibilities will continue to differentiate identity among members of the profession, but these will continue to be integrated through the strength of the Canadian professional ideology held in common.

*Reciprocity*

The principle of reciprocity speaks to the need to ensure an appropriate, principles-based balance of expectations and obligations both between the profession and Canadian society, and between the profession as a whole and its members. Internally, this principle addresses the responsibility for the care of all members, based on the recognition that membership will necessarily infringe on the full range of rights and freedoms enjoyed by other citizens, and therefore incurs an added moral obligation to address members' requirements. This obligation applies to all preparations before operations, the conduct of such operations, and the continued care of members and their families upon return from operations.

## NOTES

1. Samuel Huntington, *The Soldier and the State* (New York: Vintage, 1957), 19.
2. Morris Janowitz, *The Professional Soldier* (New York: The Free Press, 1960), 6.
3. General Sir John Hackett, *The Profession of Arms* (London: Times Publishers, 1963), 2.
4. T. Parsons, "Professions," in *International Encyclopedia of the Social Sciences,* Volume 12, ed. David Sils (New York: McMillan, 1968), 545.
5. Eliot Freidson, *Professionalism* (Chicago: University of Chicago Press, 2001), 127.

## SELECTED READINGS

Abbott, Andrew. *The System of Professions.* Chicago: University of Chicago Press, 1988.

Bentley, Bill. *Professional Ideology and the Profession of Arms in Canada.* Toronto: Canadian Institute of Strategic Studies, 2005.

Freidson, Eliot. *Professionalism: The Third Logic.* Chicago: University of Chicago Press, 2001.

Horn, Bernd and Stephen J. Harris. *Generalship and the Art of the Admiral.* St. Catharines: Vanwell Press, 2001.

Huntington, Samuel. *The Soldier and the State.* New York: Vintage, 1957.

Janowitz, Morris. *The Professional Soldier.* New York: The Free Press, 1960.

Moskos, Charles. "From Institution to Occupation: Trends in Military Organization." *Armed Forces and Society*, Vol. 4, No. 1 (Fall 1977).

Moskos, Charles, John Williams, and David Segal. *The Post-Modern Military*. New York: Oxford University Press, 2000.

Sarkesian, S. *The U.S. Military Profession into the 21st Century*. London: Frank Cass, 1999.

# 29

# MORALE

by Bernd Horn and Daniel Lagacé-Roy

Arguably, morale is one of, if not the greatest, combat multipliers. Napoleon Bonaparte, the great commander and strategist, insisted that "morale was to the physical as three is to one."[1] After all, morale is critical to achieving superior performance, meeting organizational objectives, building strong teams, and contributing to effectiveness and readiness, as well as minimizing combat stress casualties. In the military profession, morale is a vital element that joins together members of a unit or section in order to attain the objectives, particularly during crisis and danger. One study determined that "High morale can sometimes outweigh the effects of sleep loss on performance."[2] "During a sticky battle," explained Lieutenant-Colonel Syd Thomson, a commanding officer of the Seaforth Highlanders during the Second World War, "morale is as important, if not more important, than good tactics."[3] Similarly, Field-Marshal Montgomery concluded, "the morale of the soldier is the greatest single factor in war."[4]

## WHAT IS MORALE?

Morale, much like leadership, means many things to many people. According to the *Encyclopedia of Psychology*, the term "morale" is defined by an individual's "positive attitude to a group, its aims, and leadership." It is also a "synonym for satisfaction in the working group with the result of a high readiness to work."[5] This element of satisfaction in the working environment is similarly emphasized by the *Concise Encyclopedia of Psychology*, which states: "The study of morale in organizations often concentrates on two major dimensions: work motivation and job satisfaction."[6]

The *Handbook of Military Psychology* defines morale as a "function of cohesion and esprit de corps."[7] The U.S. Army identifies morale as "the mental, emotional, and spiritual state of the individual. It is how [s/]he feels — happy, hopeful, confident, appreciated, worthless, sad, unrecognized, or depressed."[8] However, a more comprehensive explanation was written by Edward Munson, an American staff officer, in 1921. He explained that morale was the "confident, resolute, willing, often self-sacrificing and courageous attitude of an individual to the functions or tasks demanded or expected of him by a group of which he is a part that is based upon such factors as pride in the achievements and aims of the group, faith in its leadership and ultimate success, a sense of fruitful participation in its work, and a devotion and loyalty to the other members of the group."[9]

More succinct definitions have also been developed. A NATO working group concluded, "One way to define it [morale] is a service member's level of motivation and enthusiasm for accomplishing mission objectives."[10] Lord Moran, a medical officer in the First World War, simply stated that morale "is the ability to do a job under any circumstances to the limit of one's capacity."[11] Similarly, scholar J.C. Baynes explained that, "Morale is the enthusiasm and persistence with which a member of a group engages in the prescribed activities of that group."[12]

Once the various definitions are distilled down to their components, it becomes clear that morale describes the spirit, determination, and confidence within a group to overcome challenges, dangers, and obstacles to achieve an assigned task, self-imposed goal, or situation in which they may find themselves.

## THE IMPORTANCE OF MORALE

Research on morale has shown unequivocally that high morale positively affects performance. In addition, high morale is associated with fewer stress causalities.[13] These two findings indicate that morale has a great impact on group effectiveness and on the well-being of individuals. The nature of a military operation, for example, relies on the group's capacity to achieve its mission. This capacity can be described as being able to:

* work together (e.g., cohesion);
* embrace the same cause (e.g., sense of purpose); and
* fight for (e.g., esprit de corps), and protect each other (e.g., loyalty).[14]

As for the members' well-being, morale plays a crucial role as an indicator that individuals trust and believe that their needs (e.g., biological, psychological, spiritual) are looked after.[15] High morale is usually a sign that individuals know that leaders "care" for them and that their superiors will defend their interests.

Robert Greene, in his book *The 33 Strategies of War*, reinforced the importance of morale when he wrote, "Morale is contagious: put people in a cohesive, animated group and they naturally catch that spirit."[16] While high morale is contagious in a very positive way as noted above, low morale can be detrimental to individuals and groups alike. "Low morale develops more rapidly and is even more difficult to overcome."[17] Low morale is often the result of a number of causes, such as:

* unresolved conflict between colleagues;
* individual agendas;
* competing strong personalities;
* negative team dynamics;
* rumour-mongering;
* poor leadership;
* poor communications;
* task/mission failure;
* adversity (e.g. harsh climatic or environmental conditions); and
* hardship (e.g. lack of food, water, sleep).

## INDICATORS OF MORALE

Although it is difficult to observe morale directly, its state (i.e., whether high or low) can be gauged by the behaviour, mood, and interaction of personnel in a unit or organization. Units or organizations with high morale are usually characterized by a number of indicators:

* lower rates of disciplinary problems;
* high levels of bearing, dress, and deportment;
* good care and maintenance of equipment and vehicles;
* high levels of teamwork;
* proficient drills and a high level of training;
* transparency and clear lines of communication; and
* pride in unit accomplishments and reputation.

Conversely, units or organizations with low morale are normally characterized by:

* poor communication;
* poor dress and deportment;
* disciplinary problems;
* restlessness;
* dissatisfaction, griping, and pessimism;
* oversensitivity to criticism;
* disobedience;
* malingering;
* relatively higher levels of individuals reporting sick; and
* higher levels of absence without leave (AWOL).

## BUILDING MORALE

In his book titled *War*, the military historian Gwynne Dyer argued, "Nothing is quite so effective in building up a group's morale and solidarity ... as a steady diet of small triumphs."[18] This quotation refers to one aspect of building morale, which is recognizing and sharing accomplishments by taking pride in them. Furthermore, this quotation implies that the focus of building morale is to inculcate "confidence in one's self and the team," encourage "group cohesion," and emphasize a "sense of contribution"[19] by achieving desired results that matter to the members.

Overall, the components of good morale are basically consistent over time and organizations:

* good leadership — sharing risk and hardship, leading by example, demonstrating competence and compassion;
* self-confidence — believing in own abilities and skills to accomplish task/perform duties;
* cohesion — bonding of personnel, sharing hardship and experience, confidence in peers and superiors;
* esprit de corps — pride in unit or organization, loyalty and confidence beyond the immediate work group;
* provision of basic needs — ensuring personnel receive adequate food, water, sleep, hygienic considerations;
* provision of adequate equipment, weapons, vehicles and logistical support — ensuring timely, reliable, and sufficient material support and equipment; and
* meaningful purpose – belief that unit/organizational role, objectives, and missions contribute to an accepted and meaningful cause.

## **MORALE AND LEADERSHIP**

Leadership is a key component of high morale. Military personnel expect leaders to watch out for their best interests. If military personnel understand the mission and feel supported professionally by their leaders, they will be willing to withstand the rigours of deployment.[20] After all, leaders set the conditions for their subordinates. They cultivate a conducive environment that allows individuals to build confidence and cohesion through empowerment, good training, effective group dynamics, competent direction, and leader concern. The leader must create an atmosphere that ensures "increased communication, a friendly atmosphere, loyalty, and member participation in decisions and activities."[21] In short, the leader's responsibility in building morale consists of:

* defining a clear vision;
* promoting teamwork and collegiality;
* engaging in team-building;
* practising fairness and justice;
* encouraging open communication;
* instilling discipline;
* eliminating ambiguity and confusion;
* recognizing mistakes; and
* managing conflicts.

In essence, "Leaders make a direct contribution to morale by clarifying individual and group roles and tasks, developing skills, sharing risks and hardships, maintaining discipline, and encouraging high levels of effort and persistence."[22] To achieve this direct contribution, leaders must adopt a supportive influence intended to assist followers in resolving personal or work related problems and to improve morale and well-being. This supportive influence includes:

* recognition of and response to individual needs;
* demonstrations of understanding and empathy;
* offers of help or collaboration;
* representation of subordinate interests to administrative authorities; and
* efforts to improve unit climate.[23]

Research in the areas of morale, cohesion, and leadership indicates that there often are discrepancies between leaders and subordinates in regard to their perception of unit/organizational morale.[24] Leaders usually are unaware of the gaps and "tend to have unrealistically optimistic perceptions of morale in their organizations. If morale is suffering, senior [leadership/] management is often the last to know."[25] To prevent such gaps between perception and reality, leaders must continually and realistically assess unit morale using all available tools (e.g. unit climate survey).

In essence, leaders must attempt to take the "pulse" of their personnel. This can be done informally or formally. The informal assessment is the most common one because it is done through informal moments such as listening to subordinates during professional development days, sporting events, social gatherings, and "down time" during and after the duty day. While these informal moments provide a general "feeling" about the group's morale, it may not reflect the views or the concerns of the majority of the group. In addition, some members may be afraid to voice their opinions because of bad experiences with being open in the past, or they may feel it is not in their best interests to open up. In that case, a more formal assessment is required.

Formal assessment is a more objective method to evaluate morale and to track changes over time. For example, formal assessment may examine problematic behaviours such as disciplinary violations, accidents, injuries, unauthorized absences, and sick leave. Several methods are available for diagnosing and maintaining morale, including focus groups and surveys.[26]

Focus groups give a quick assessment on possible issues in a unit or organization. A focus group is a special type of group in terms of purpose (i.e. to gather information), size (i.e., 8–10 members), composition (i.e. members have something in common), and procedures (i.e., a structured discussion). The members voice their concerns, raise questions, and provide suggestions to resolve specific issues. Successful focus groups use:

* a skilled moderator that is not in the chain-of-command;

* structured questions prepared ahead of time to emphasize particular issues; and
* participants who are representative of the unit or the section.

Finally, the morale survey is the most comprehensive way in which to study morale. In a survey, members are asked to respond to a set of questions, which, when completed, provides leaders with a more accurate perception of the level of morale. Surveys (normally by questionnaire) are usually developed by leaders in collaboration with professionals trained in survey methodology. NATO-sponsored research has developed a core list of important items that should be included in a morale survey:

* climate — perceptions that members are treated and cared for;
* cohesion — the degree to which members feel connected with their unit or section;
* leadership behaviours — confidence that leaders are effective;
* self-efficacy — members' confidence in their skills and abilities;
* stressors — environmental and personal factors such as living conditions, boredom;
* deployment events — exposure to snipers, brutal deaths; and
* psychological health — overall unit mental health.

In the military, morale surveys could be conducted prior to a deployment (e.g., determine how the unit is doing); during deployment (e.g., identify changes in level of morale); and after deployment (e.g., how members are doing after returning home).[27] According to Douglas A. Benton, "… a morale survey tends to boost morale because it indicates to employees that the company is interested in what they think." Benton reveals that this increase in morale will remain "… as long as employees believe [that] their opinions are contributing to change."[28]

## SUMMARY

Morale is unquestionably a key combat multiplier. However, whether in war, conflict, or peace, morale is a critical component in the effectiveness of any unit or organization. High morale ensures an efficient, harmonious, and highly effective group. Conversely, low morale breeds dissention, ill-discipline, and discontent. In the end, leadership is key to creating high morale within groups, for it is leaders who are instrumental in ensuring a conducive, dynamic, open, and team-orientated environment. Investment in creating high morale never will be wasted. "Morale is a state of mind," acknowledged Field Marshal William Slim, "It is that intangible force which will move a whole group of men to give their last ounce to achieve something, without counting the cost to themselves; that makes them feel they are part of something greater than themselves."[29]

## NOTES

1. http://www.militaryquotes.com/Napoleon.htm (accessed 24 March 2007).
2. Major P.J. Murphy and Captain C.J. Mombourquette, "Fatigue, Sleep Loss, and Operational Performance," Personnel Research Team, *Operational Effectiveness Guide 98-1* (April 1998), 9.
3. Cited in Doug Delaney, "When Leadership Really Mattered: Bert Hoffmeister and Morale During the Battle for Ortona, December 1943," in *Intrepid Warriors,* ed. Bernd Horn (Toronto and Kingston: Dundurn Press and the Canadian Defence Academy, 2007), 139.
4. Field-Marshal The Viscount Montgomery, *The Memoirs of Field-Marshal the Viscount Montgomery of Alamein* (London: Collins, 1958), 83.
5. H.J. Eysenck, W. Arnold, and R. Meili, eds., *Encyclopedia of Psychology* (New York: Continuum Publishing Company, 1982), 680.

6. Raymond J. Corsini and Alan J. Auerbach, eds., *Concise Encyclopedia of Psychology* (New York: John Wiley & Sons, 1996), 577.
7. Frederick J. Manning, "Morale, Cohesion and Esprit de Corps," in *Handbook of Military Psychology*, eds. R. Gal and D. Mangelsdorff (New York: John Wiley & Sons, 1991), 453.
8. *Ibid.*, 454.
9. *Ibid.*, 455.
10. NATO Task Group HFM, *A Leader's Guide to Psychological Support Across the Deployment Cycle*, draft document, (19 January 2007), 32.
11. Frederick J. Manning, 455.
12. Samantha K. Brooks, Don G. Byrne, and Stephanie E. Hodson, "Non-Combat Occupational Stress and Fatigue: A Review of Factors and Measurement Issues for the Australian Defence Force," *Australian Defence Force Journal*, No. 145, (November/December 2000): 38.
13. *Ibid.*
14. Hew Strachan, "Training, Morale and Modern War," *Journal of Contemporary History*, Vol. 41, No. 2 (2006): 211.
15. Frederick J. Manning, 459.
16. Robert Green, *The 33 Strategies of War* (London: Viking, 2006), 80.
17. Douglas A. Benton, *Applied Human Relations: An Organizational and Skill Development Approach* (Upper Saddle River, NJ: Prentice Hall, 1998), 168.
18. Gwynne Dyer, *War* (New York: Crown Publishers Inc., 1985), 47.
19. Reuven Gal and Frederick J. Manning, "Morale and its Components: A Cross-National Comparison," in *Journal of Applied Social Psychology*, Vol. 17, No. 4 (1987): 370.
20. NATO Task Group HFM, *A Leader's Guide*, 33.
21. Richard L. Daft, *Leadership: Theory and Practice* (Fort Worth, TX: The Dryden Press, 1999), 285.
22. Canada, *Leadership in the Canadian Forces: Conceptual Foundations* (Kingston: DND, 2005), 6.
23. Canada, *Leadership in the Canadian Forces: Leading People* (Kingston: DND, 2007), 38.

24. Kelly M.J. Farley and Jennifer A. Veitch, "Measuring Morale, Cohesion and Confidence in Leadership: What Are the Implications for Leaders," *The Canadian Journal of Police & Security Services*, Vol. 1, No. 4 (Winter 2003): 355. See also Karen D. Davis, "Culture, Climate and Leadership in the Canadian Forces: Approaches to Measurement and Analysis," in *Dimensions of Leadership*, eds. Allister MacIntyre and Karen D. Davis (Kingston: Canadian Defence Academy Press, 2006), 311–336.
25. Glen Klann, "Morale Victories: How Leaders Can Build Positive Energy," *Leadership in Action*, Vol. 24, No. 4 (September/October 2004): 7.
26. This section on formal assessment stems mainly from the chapter on "Morale" in NATO Task Group HFM, *A Leader's Guide*, 34–36. For more information on focus groups and surveys, see Richard A. Krueger and Mary Anne Casey, *Focus Groups: A Practical Guide for Applied Research*, 3rd ed. (Thousand Oaks, CA: Sage Publications, 2000) and Ronald Czaja and Johnny Blair, *Designing Surveys: A Guide to Decisions and Procedures*, (Thousand Oaks, CA: Pine Forge Press, 2005).
27. For a good example of a morale survey in operations, see Peter J. Murphy and Kelly M.J. Farley, "Morale, Cohesion, and Confidence in Leadership: Unit Climate Dimensions for Canadian Soldiers on Operations," in *The Human in Command: Exploring the Modern Military Experience*, eds. Carol McCann and Ross Pigeau (New York: Kluwer Academic/Plenum Publishers, 2000), 311–331.
28. Douglas A. Benton, *Applied Human Relations*, 181.
29. Field-Marshal Viscount William Slim, *Defeat into Victory, Battling Japan in Burma and India, 1942–1945* (New York: Cooper Square Press, 1956), 182.

## SELECTED READINGS

Biehl, Heiko and Gerhard Kümmel. "Morale and Cohesion." In *Armed Forces and International Society: Global Trends and Issues*. Edited by Jean Callaghan and Franz Kernic. Münster: Lit Verlag, 2003, 249–253.

Green, Robert. *The 33 Strategies of War*. London: Viking, 2006.

Holmes, Richard. *Acts of War: The Behavior of Men in Battle*. New York: The Free Press, 1985.

Kellett, Anthony. *Combat Motivation*. Operational Research and Analysis Establishment (ORAE) Report No. R77. Ottawa: DND, 1980, 268.

Klann, Glen. "Morale Victories: How Leaders can build Positive Energy." *Leadership in Action*, Vol. 24, No. 4 (September/October 2004): 7–12.

Manning, Frederick J. "Morale, Cohesion and Esprit de Corps." In *Handbook of Military Psychology*. Edited by R. Gal and D. Mangelsdorff. New York: John Wiley & Sons, 1991, 453–470.

Marshal, S.L.A. *Men Against Fire*. Alexandria, VA: Byrrd Enterprises, Inc., 1947.

Murphy, Peter J. and Kelly M. J. Farley. "Morale, Cohesion, and Confidence in Leadership: Unit Climate Dimensions for Canadian Soldiers on Operations." In *The Human in Command: Exploring the Modern Military Experience*. Edited by Carol McCann and Ross Pigeau. New York: Kluwer Academic/Plenum Publishers, 2000, 311–331.

North Atlantic Treaty Organization (NATO) Task Group HFM 081/RTG. "A Leader's Guide to Psychological Support Across the Deployment Cycle." Draft document, 26 October 2006.

Stouffer, Samuel A. *The American Soldier: Combat and its Aftermath*. Princeton, NJ: Princeton University Press, 1949.

# 30

# MOTIVATION

by Phyllis P. Browne and Robert W. Walker

For a military leadership handbook, there are obvious "layers" in the study and application of *motivation* as a human behaviour. This chapter briefly covers three of these "layers": the "generic" motivation of people, the human motivation applied to a working life, and the role of leadership in worker, work team, or organizational group motivation.

At the foundational level, there is the "everyday," generic, human, motivation(s) of life. It is these motivations or the collective motivation, with its broad supportive theories and its observable human behaviours, that would apply to people at personal levels, family levels, team or group levels, community levels, societal levels, and even at global levels, i.e., throughout much of one's life and one's activities and interests.

Upon this foundation of everyday motivation is that primary, situation-specific, focused human motivation experienced and expressed within work groups or situations, within working organizations or as part of life in work-site institutions. For this handbook, the work-site institution of importance is a military one,[1] however, this chapter on motivation will deal with the general environments of people in everyday work life — the challenges and responsibilities, the positives and negatives of a working lifestyle that integrate with motivation, the satisfaction acquired from successes and accomplishments, the recognition and rewards that result from personal successes and professional development, a sense of partnership with colleagues, and the sharing of a unit cohesion. The Canadian Forces (CF) is presented as an institutional example.

Having identified generic motivation and work motivation, the third level for review, in line with the title of this handbook, is that of leadership and the responsibilities of leaders with respect to motivation in the workplace. To generate and ensure an effective level of motivation among team members, what must leaders address? To what must leaders attend? Perhaps member well-being and commitment? Perhaps recognition of a job well done? The CF once again is utilized as an institutional example.

Therefore, this chapter will provide relevant background information on aspects of human motivation with some relevant theories, will explore the interaction of motivation and organizational work factors, and will conclude by addressing some requisite roles and responsibilities important to leaders committed to motivating others affiliated with their organizations.

## "GENERIC" MOTIVATION DEFINED

*The Canadian Oxford Dictionary* defines the verb *motivate* as "supply a motive to; cause [a person] to act in a particular way; stimulate the interest of [a person in an activity]."[2] Canadian university psychology professors Gary Johns and Alan Saks define motivation as "the extent to which persistent effort is directed toward a goal."[3] There are four basic characteristics of this definition upon which the authors have elaborated: effort, persistence, direction, and goals. Johns and Saks contend that persistence is reflected in the application of effort to the work activity, i.e., effort is indicated by the individual's behaviour in performing the "job" to achieve the goal. Michael Maccoby, a well-known consultant on issues of leadership, strategy, and organization, contends that the demonstration of self-expression, hope, and fear, collectively, is accepted universally as one of the most important motivations to work. Individuals are generally motivated to work to satisfy an emotional necessity, directed by values and opportunity.[4]

To better understand what motivates people to engage in work, it is necessary to go beyond Johns, Saks, and Maccoby, and to identify those dynamic values that determine people's needs. Although this will be addressed more thoroughly in the section below on theories of motivation, a basic understanding of these dynamic values is pivotal.

First, the backdrop of motivational behaviour for a society or culture is the collection of the larger societal and/or national norms for behaviour. For example, Canadians generally adhere to and respect the legislated Canadian rights and freedoms for citizens, the roles and responsibilities of the individual worker or professional who supports those freedoms, and the personal and professional values of dedication, loyalty, integrity, and tenacity that working members learn and share.

## **MOTIVATION AND THE WORKPLACE: SOME BACKGROUND**

Early research in the late 19th and early 20th centuries addressed organizational productivity and the development of the individual's potential in the organization. Ideas about motivation in the workplace are rooted in the origins of Scientific Management, which was initiated in the late 19th century by "efficiency expert" Frederick Taylor.[5] The process later became known as "Taylorism," with a focus aimed at high productivity levels through the segregation of work activities into short, self-contained actions. Motivation was interpreted as the consequence — an added incentive — that resulted from efficiency in work design practices, predominantly on industrial-era production lines. (Think Henry Ford!) Motivation was seen as encouraging individuals to participate in efficient work activities in order to enhance the quality, performance, and the scope of their responsibilities and, ultimately, to influence organizational outcomes. Taylorism was about task-orientation and what individuals, in working groups or not, could achieve independently, but leader attention to workers' work circumstances and well-being collectively received limited emphasis.

Because human motivation often goes much beyond individual efficiency, including a sharing of ideas, team-building, unit cohesion and open-mindedness, a "human relations" and worker-behaviour focus remained to be addressed, beyond the earlier Taylorism/work behaviour focus. The Human Relations (HR) movement of the 1930s and 1940s addressed human behaviour in organizations, including motivation, leadership, and group dynamics.[6] The movement postulated that high morale and motivation led to high productivity, as opposed to the Taylorism approach to high productivity as leading to satisfaction, high motivation, and morale. The HR movement continues to this day. (Think Bill Gates!)

From the Taylorism and Human Relations movements came contradictory approaches, thoughts and theories on how leaders should treat, or motivate, followers. Maccoby has stated that work orientation is generally manifested in the needs of workers, followers, and subordinates to be responsible, to achieve, and to create. Motivated behaviour of the follower may serve to achieve both the goals of the organization as well as that employee's personal goals. As such, the motivation can be of an intrinsic and/or an extrinsic nature, or a combination of both.[7] This intrinsic/extrinsic dichotomy deserves some explanation.

*Intrinsic Motivation*

Intrinsic motivation is usually self-directed and often exists at a profound and emotional level. Put differently, motivation that comes from within the person is intrinsic motivation.[8] Examples of intrinsic motivation include job satisfaction, feelings of accomplishment, and performing to high levels of expectation. Choosing to participate in activities specifically for their enjoyment, becoming involved in sports or choosing to participate in volunteer work, are all activities that are typically intrinsically motivated or self-directed. When motivation is self-determined or self-initiated, it is linked to positive and important consequences and will be positively associated with future work intentions.[9] Intrinsic motivation is a personal and internal experience that can be manifested through the achievement of an external outcome.

It is important to understand that intrinsic motivation can be tampered with and can affect the outcome of the activity for which the individual had been motivated to perform. For example, a person who takes the initiative to organize social activities for the organization for which s/he is subsequently rewarded may begin to reinterpret the motivation for this self-directed behaviour. Such reinterpretation may result in a false understanding of the behaviour because the personal incentive is rewarded, and so may be perceived as extrinsically motivated.[10]

*Extrinsic Motivation*

Extrinsic motivation is often externally influenced and can be manipulated. From the organizational perspective, two of the most basic forms of extrinsic motivation are supervision and reimbursement. However, within the work environment, although extrinsic motivators may be applied or introduced by the organization, intrinsic motivators may also reinforce them, thereby creating an interaction between intrinsic and extrinsic motivators. Take, for example, a situation where a reward or recognition may be extended to an employee for having met an important deadline or for developing a specific project. The accomplishment of either of these could have been a personal challenge for the individual generated by intrinsic motivation to overcome that challenge. In such a situation, the accomplishment is realized on two levels, as an organizational outcome and as a personal achievement.

Organizations now attempt to tune into the intrinsic motivations of work members in order to motivate them. Maccoby stated in *Why Work?*, that, in such organizations, employees are "engaged by their responsibilities, by the challenges that are part of their jobs."[11] Maccoby also explains that people who work in motivational organizations see the organization's goals as meaningful and worth achieving. To this extent, according to Maccoby, the organization is connecting with both intrinsic and extrinsic motivation. Improvement in productivity can result.[12]

## **THEORIES OF MOTIVATION**

The concept of motivation has been explored by a significant number of scholars, resulting in various theoretical approaches to addressing the concept. Among the earliest, prominent and foundational works on motivation are:[13]

## Abraham Maslow's Hierarchy of Needs Theory

Maslow's theory seems to have set the foundation from which other theorists moved in different exploratory directions in theorizing the concepts of motivation. In a nutshell, Maslow provides a hierarchy of five critical levels of needs by which the individual is motivated, and identifies these sequentially as physiological (e.g., oxygen, food, and water); safety needs; the need for love, affection, and a sense of belonging (i.e., giving and receiving love and affection, and group membership or affiliation); the need for esteem; (i.e., the desire for stability, a high level of self-respect as well as respect from others); and the need for self-actualization (e.g., a person's need to be, and to accomplish that which the person feels destined to be or accomplish, e.g. becoming a writer, artist, or musician, etc.).[14] The needs hierarchy forms a pyramidal structure from the physiological base to the apex of self-actualization.

## Frederick Herzberg's Motivators and Hygiene Factors

This motivational theory is patterned on Maslow's needs hierarchy, but is characterized as a "motivators" (human) and "hygiene" (maintenance) theory. The motivators affect satisfaction and no satisfaction while the "hygiene" factors affect dissatisfaction and the absence of dissatisfaction.[15] In other words, the motivators that induce satisfaction are not the opposite of factors that induce dissatisfaction. Instead, factors that are opposite to satisfaction or job satisfaction, for instance, are those that generate no satisfaction or no job satisfaction. Likewise, factors that are opposite to dissatisfaction or job dissatisfaction are those that induce no dissatisfaction or no job dissatisfaction. To exemplify the argument, factors that respond to needs such as hunger and thirst are human motivators, while factors that induce pain-avoidance behaviour are "hygiene" factors and relate to the environment or workplace.[16] Like Maslow's needs hierarchy, the motivators cannot be utilized until all the hygiene factors or needs are satisfied.[17]

*Douglas McGregor's XY Theory*

This theory of motivation, on the other hand, identifies two types of approaches by leaders/managers in relation to organizational development and strategies of motivation of employees. The Theory X approach assumes that, generally, people are lazy, reluctant to work, and must be threatened with punishment in order to pursue organizational objectives. This theory usually generates poor results through its authoritarian style and embedded threats of punishment. The Theory Y approach provides people with the opportunity to grow and develop, and can result in better performance and results.

*David McClelland's Motivational Needs Theory*

This "Needs Theory," developed in the 1960s, determines that, over time, people's experiences will result in three different types of needs specific to impacting motivation of organizational behaviour relevant to employees and organizations. The Need for Achievement indicates a desire to perform a role better or more efficiently, to solve problems or to master a complex task; the Need for Affiliation assumes a desire to establish and maintain good relationships with others; and the Need for Power reflects the desire to control others, to be responsible for them, and to influence their behaviour. McClelland concluded that the Need for Power was the most effective for a leader's success. He further contended that it was more important for the leaders to have the ability to influence people and that ultimately, the successful leader would have a greater need for power than for achievement.

Although these theories date from the middle years of the 20th century, and have been challenged and/or corroborated in whole or in part by other theorists, they remain representative of current theories of motivation. The human dimension increasingly has become a common thread in organizational policy and a function of organizational outcome, and has gained prominence as a critical component of organizational success. Consequently, due to a better understanding of and increasing value for the complementary nature of these two factors, there is noticeable reciprocity in the attention paid to member motivation and organizational outcomes.

## **WELL-BEING, ORGANIZATIONAL EFFECTIVENESS, AND MOTIVATION**

To realize organizational effectiveness, it is important for leaders to "… unlock the talent, experience, wisdom and common sense of the many within [the] organization by making work easier, simpler, quicker, rewarding, safer …"[18] and it is important for leaders to be "… responsible for coherent and effective policies … as well as robust programs that provide … for all members, at all times, and in all places. Members' circumstances vary widely though, so the challenge for institutional leaders is to ensure that rewards are equitable (proportional to merit), accurately reflect needs, and are applied systematically.[19]

Organizations employ various techniques to motivate their employees through new or improved policies that enhance employee well-being and therefore influence positively its organizational outcomes. Johns and Saks, for example, identify the motivational aspects of Starbucks' mission statement that include "providing a great work environment, treating people with respect and dignity, a commitment to diversity, and making all employees partners."[20] In recognizing the value of members' motivation, many organizations have ensured open-mindedness in leaders for accommodating cultural and intellectual diversity in their staffs; flexible work schedules for employees, including double-income and single parents; profit-sharing programs in which employees as shareholders benefit from their own productivity; and informal work attire. Such techniques promote the inclusiveness of employees, the valuing of their opinions and ideas, recognizing their personal life circumstances, and giving voice to them in the decision-making process.

## MOTIVATION IN THE CANADIAN FORCES

Any approach to leadership must be comprehensive and signify a genuine concern and respect for people and the quality of their conditions of service.... Looking after people involves taking care of their physical, intellectual and emotional well-being ...[21]

The general perception by the public is that it takes a special, or at least a specific, type of individual to join the military. Using the CF as an example of a motivational organization, it can be seen that the CF to a large degree introduces elements of motivation as early as the recruitment and recruit training processes when potential members are made aware of incentives that give much consideration to member well-being, including opportunities for personal and professional development.

CF personnel, having the motivation to become CF members, possessing career ambitions, and generally having been socialized and trained together, will be more or somewhat motivated to engage in national and international tours of duty. Even with some level of reluctance to serve on dangerous assignments, soldiers are socialized to understand their commitments to their country and its needs. Their responsibilities for these needs and those of their colleagues are expected to supersede their own personal preferences and security.

Military organizations such as the CF develop policies and practices that recognize performance and reward it. This is based on two important purposes, to enhance or sustain performance, and to indicate to the team or group what successful performance really looks like. The reward motivators are manifested in the practice of public recognition of performance outcomes that serve two primary purposes. In the first instance, it serves as a motivator for other individuals and, in the second, it serves as personal satisfaction for the recipient.

The CF is an organization in which member well-being is specifically linked to organizational effectiveness, and is fittingly reflected in policies that embrace personal and professional development of its members. This is also a clear example of the interconnectedness or fusion of extrinsic and intrinsic motivation where both employer and employee become the direct beneficiaries. The CF is somewhat unique in that incentives/motivators designed to enrich the aptitude, abilities, and knowledge skill sets of the individual are, indirectly, also critical components of organizational effectiveness. Consequently, externally-introduced motivators interact with intrinsic needs in a collaborative interaction essential to the high priority of the organization of member well-being.

## **LEADERSHIP AND MOTIVATION**

> Leading People is about leading in all work environments, but fundamentally it is about something much more profound; it is about leading people in integrated operations, that is, domestic operations, conflict situations, and combat.... Whether aboard ship, on patrol, or in hostile airspace, or supporting these activities in myriad ways, CF leaders must be prepared to lead in times of uncertainty, ambiguity, confusion, danger ...[22]

Sound military leadership is an absolute prerequisite for motivation among unit members. Effective military leaders will inspire followers and generate confidence in the group. However, these leaders have a challenging role in keeping their troops motivated, particularly because of the new security environment in which military service takes place. This new security environment, globally, creates a circumstance wherein the motivational role that is critical to mission success needs to be assumed by and shared with numerous team members. The challenge, then, among all of the other challenges of establishing security, is to ensure a cooperative and effective team-leadership approach for motivating team leaders and followers. Being decisive and being persuasive are key characteristics of a good leader, however, the means to convey such characteristics can be, collectively, the team members. They are the conduit(s) through which the motivation to take effective action is conveyed, and therefore they contribute greatly to the probability of success.

From a patriotic perspective, leaders are responsible to impart to soldiers that the missions are important to Canada, and that the members have the support of the government and the people. Lieutenant-Colonel Banks, in relating his experience in Croatia in 1994, observed that "Canadian soldiers are willing to take risks, sometimes great risks, but they must know and understand the reason why." He further stated that "They are not Wellington's 'scum of the earth' and they cannot be flogged into battle. If they do not understand why, or do not identify with the reason, they may demonstrate reluctance ..."[23]

In military service, it is important for the leader to assure the unit members that their personal welfare is of great concern, and that "it is not just about patriotism." The unit's welfare includes everything from the basic availability of food and clean clothes, to the emotional and social needs satisfied by contact with buddies, home, and family. It is also critical for soldiers to be assured that their families are being treated with dignity and respect. The absence of any of these factors can influence individual and unit effectiveness and can become demotivators for the unit members.

Morale and cohesion play central roles in maintaining high levels of motivation. The leader has an important role in sustaining this unit morale and cohesion and, when the leader has a visible presence, is an effective communicator, and displays continuous attention to members, s/he is displaying and living that important role. A leader, when socializing with subordinates and demonstrating this concern for them, indirectly will reduce stress-related, poor-performance, outcomes.[24] The contention is that "if behaviour can be predicted and explained, it can often be controlled or managed."[25] Where behaviour is controlled or managed as a motivational tool, the objectives must be to inspire the individuals and to instill confidence in their abilities.

## **SUMMARY**

This chapter on motivation has reviewed the motivational aspects of Western society life in general and some fundamental theories of motivation that relate to such circumstances. With this foundation of generic motivation having been established, motivation then needed to be examined within the context of the working environment as a set of motivational circumstances. Additionally, this chapter explored the essential responsibilities of leaders to keep their members appropriately motivated and to maintain a collective unit motivation for those members.

In summary, the attention of large numbers of scholars over generations, individuals who work with theories of human behaviour, has resulted in the identification of the factors of most significance for the motivation of today's workers and their leaders, the organizational outcomes, and the members' commitment to the institution. Two of the major outcomes of such studies are that motivation plays a major role in organizational success, and, in order to maximize the impact of that dimension, it is critically important to address the needs of members and, as much as possible, to ensure their work satisfaction and their well-being and security. An organization's success and effectiveness have never depended more than they do today on team members' motivational dispositions. Accordingly, its leaders' attributes and abilities to generate that very strong motivation among working members are of primary significance in determining high-quality mission successes and organizational outcomes.

## NOTES

1. See the Chapter 5 on Combat Motivation in this handbook. It is written by an experienced combat officer and leader who addresses the unique challenges of military leaders to generate combat motivation when commanding military members during warfighting.
2. *The Canadian Oxford Dictionary*, the foremost authority on current Canadian English (Don Mills, ON: Oxford University Press, 2001), 947.

3. Gary Johns and Alan M. Saks, *Organizational Behaviour: Understanding and Managing Life at Work*, 6th ed. (Toronto: Pearson Prentice Hall, 2005), 134.
4. Michael Maccoby, *Why Work? Motivating the New Workforce*, 2nd ed. (Alexandria, VA: Miles River Press, 1995).
5. Johns and Saks, 15.
6. Donald D. White and David A. Bednar, *Organizational Behavior: Understanding and Managing People at Work* (Toronto: Allyn and Bacon, 1986), 9.
7. Maccoby, 43.
8. Genene Koebelin, "Motivation Techniques: Inspiring Your Co-Workers to Excellence," from Human Performance Development course, Suffolk University, Spring 1999, http://www.work911.com/performance/particles/kobmot.htm, (accessed January 2007).
9. Nancy Otis and Luc G. Pelletier, "A Motivational Model of Daily Hassles, Physical Symptoms, and Future Work Intentions among Police Officers," *Journal of Applied Social Psychology*, Vol. 35, No. 10 (2005): 2193–2214.
10. Cookie White Stephan and Walter G. Stephan, *Two Social Psychologies*, 2nd ed. (Belmont, CA: Wadsworth, 1990).
11. Maccoby, 221.
12. *Ibid.*
13. Robert G. Owens, *Organizational Behavior in Education*, 2nd ed. (Upper Saddle River, NJ: Prentice Hall, 1981).
14. *Ibid.*
15. William J. Bartley, "Motivation and Leadership in Your Business," *Journal of Arboriculture*, Vol. 3, No. 4 (1977): 65–68.
16. *Ibid.*
17. Victor H. Vroom, "Leadership and Human Behavior," *Organizational Behavior and Human Performance*, Vol. 4, No. 3 (1964): 142–175.
18. http://www.accel-team.com/productivity/addedValue_00.html (accessed April 10, 2007).
19. Canada. *Leadership in the Canadian Forces: Leading the Institution*, (Kingston: DND, 2007), 112.
21. Canada. *Leadership in the Canadian Forces: Leading People*

(Kingston: DND, 2007), 56.
22. *Leading People*, 9.
23. Lieutenant-Colonel D. Banks, "Reflections of Operational Service" in *In Harm's Way: On the Front Lines of Leadership — Sub-Unit Command on Operations*, ed. Colonel Bernd Horn (Kingston: Canadian Defence Academy Press, 2006), 7.
24. Sanela Dursun, *Human Dimensions of Military Operations: The Construct of Personnel Tempos and Its Relationship with Individual and Organizational Well-Being*. Defence Research & Development Canada, CORA TR 2006-29 (Ottawa: DND, 2006).
25. *Ibid.*, 13.

## SELECTED READINGS

Canada. *Leadership in the Canadian Forces: Leading the Institution.* Kingston: DND, 2007.

Canada. *Leadership in the Canadian Forces: Leading People.* Kingston: DND, 2007.

Capstick, Colonel Mike, Lieutenant-Colonel Kelly Farley, Lieutenant-Colonel Bill Wild (Ret'd), and Lieutenant-Commander Mike Parkes. *Canada's Soldiers: Military Ethos and Canadian Values in the 21st Century.* Ottawa: Department of National Defence, 2005.

Johns, Gary and Alan M. Saks. *Organizational Behaviour: Understanding and Managing Life at Work,* 6th ed. Toronto: Pearson Prentice Hall, 2005.

Latham, Gary P. *Work Motivation: History, Theory, Research and Practice (Foundations for Organizational Science).* Thousand Oaks, CA: Sage Publications, 2006.

Maccoby, Michael. *Why Work? Motivating the New Workforce.* Alexandria, VA: Miles River Press, 1995.

Walker, Robert W., ed. *Institutional Leadership in the Canadian Forces: Contemporary Issues.* Kingston: Canadian Defence Academy Press, 2007.

# 31
# PHYSICAL FITNESS
## by Bernd Horn

It is a timeless truth that physical fitness is vitally essential to soldier effectiveness. Overwhelmingly, when combat veterans are asked how leaders can best prepare their subordinates for battle, the response is a resounding, "Make sure they are physically fit." Similarly, Brigadier-General David Fraser, having recently returned from commanding NATO forces in Afghanistan, emphasized to the Canadian infantry association that, "Fitness is critical — soldiers will be tested and they have to keep going no matter how hard the conditions."[1] The crux of the argument is hardly surprising. After all, the fitter the person, the farther s/he can go, at a faster rate, with more weight, less food and less rest, and still think clearly and make precise, timely decisions at the end. Quite simply, physical fitness is fundamental to the military profession. It ensures alertness, stamina, and vitality. It is key to mission success.

## WHAT IS PHYSICAL FITNESS?

The British Army takes a very pragmatic approach to fitness. They define it as "the ability to respond instantly and effectively to the physical and psychological demands of operations, and the ability to maintain that response over prolonged periods."[2] The Canadian Army breaks it down into component parts:

1. Aerobic capacity — the ability of lungs, heart, blood vessels, and muscles to take in oxygen, deliver it to

the working muscles and other tissues, and have the working muscles use the oxygen to provide energy for work.

2. Muscular strength and endurance — the ability of muscles to generate forces in a particular movement.

3. Power — the ability of your muscles to generate forces at a high speed.

4. Flexibility — the ability of your muscles and joints to move through their full range.[3]

In essence, physical fitness is the capacity to perform those physical activities required of the body. The higher the level of fitness, the greater the threshold of activity an individual can undertake. In the military context, fitness is essential to completing operations, particularly in demanding, harsh environments (whether in terms of climate, terrain, or combat) and those conducted with heavy loads and at a high tempo.

## THE BENEFITS OF PHYSICAL FITNESS

The U.S. Surgeon General's 2007 report has noted that physical fitness carries significant health benefits such as reduced risk of coronary heart disease, hypertension, colon cancer, and diabetes.[4] This is significant when one considers that currently 66 percent of Americans are overweight or obese.[5] Although the military normally represents a healthy component of society, physical fitness is also vital to the effectiveness, health, and well-being of military personnel. In fact, a high level of physical fitness:

* contributes significantly to the fighting effectiveness and general health of individuals;
* develops robustness which leads to a reduction in susceptibility to injury;

* provides individuals with more energy;
* makes individuals more alert and productive;
* acts as a moderator of both sleep needs and adaptability to disrupted or deprived sleep;
* improves mood and imparts a sense of well-being;
* builds team spirit and cohesion at the sub-unit and unit levels;
* ensures health and readiness for deployments to austere, harsh, and hostile environments;
* assists individuals to cope with stress (helping them relax and feel less tense);
* assists with the maintenance of morale;
* builds self-confidence;
* assists with the development of soldiers, particularly recruits; and
* acts as a powerful vehicle for the inculcation and sustainment of the military ethos.

In addition, physical fitness provides a number of health benefits:

* increasing the metabolism of the body (i.e. muscles use more calories, the body burns more calories);
* reducing resting blood pressure and resting heart rate;
* reducing cholesterol and triglyceride levels;
* increasing high density lipoprotein levels;
* helping individuals to lose extra pounds and stay at the ideal weight; and
* motivating people to cut down or stop cigarette-smoking.[6]

## **DIET AND FITNESS**

A proper diet is essential to physical fitness. Good foods prior to exercising include vegetables, fruits, breads, pasta, potatoes, rice, soups, skinless chicken, and fish. Maintaining a healthy body weight will make it easier to maintain a fitness program. It also will boost energy and contribute to a general sense of well-being. The following will assist in maintaining proper body weight:

- eating a nutritious breakfast;
- avoiding plain sugars in the form of candies, jams, soft drinks, and rich desserts;
- cutting down on items high in fat, such as butter, margarine, and oils, fried foods, preserved meats, and junk food snacks such as chips; and
- if drinking alcohol, doing so in moderation.

In summary, with respect to diet and fitness, one should:

- eat lots of carbohydrates — pasta, grains, potatoes, fruits, and vegetables;
- get sufficient protein;
- minimize the intake of fat — oils, butter, creamy sauces, meats, etc.; and
- drink plenty of fluids throughout the day.[7]

## **GENDER AND FITNESS**

It is generally accepted that women do not have the same aerobic capacity or strength as men due in large part to differences in body size and composition. The lower aerobic capacity of women is due to a smaller muscle mass, a smaller volume of blood, and lower hemoglobin concentrations in the blood. Women generally are about two-thirds as strong as men but, relative to their size, women can gain as much or more strength than men following similar training programs. Nonetheless, despite the differences, women respond to aerobic and strength training programs in the same manner as men.[8]

## PHYSICAL FITNESS AND LEADERSHIP

As explained throughout this chapter, physical fitness is fundamental to soldiering. It impacts soldier well-being, unit morale and cohesion, operational effectiveness, and mission success. There is no better investment to be made by a commander or military leader than to ensure the physical fitness of their subordinates. However, to be effective, leaders cannot just espouse the importance of fitness, they must live their words. Physical fitness must be a lead-by-example activity. Leaders must exhibit fitness and lead fitness activities.

In fact, for junior leaders, physical fitness training is an excellent vehicle for demonstrating leadership. Normally, a newly arrived junior officer is the least experienced individual in the organization. As such, junior officers must look to their more experienced subordinates for assistance. However, physical fitness training is one area where junior officers can set the tone and provide the example of leadership. Generally, subordinates will accept the fact that a junior leader with no experience will need time to develop with regards to the many technical skills required in the military, however, a lack of physical fitness in leaders normally is not tolerated.

Aside from setting the example, leaders also have the responsibility of creating a conducive environment for physical fitness activities. They must create opportunity and time for fitness training, stress its importance, and ensure all members are participating in useful and safe fitness activities. The latter issue cannot be over-emphasized. Physical fitness training conducted improperly (e.g. running on hard surfaces in combat boots, no warm-up exercises, poor stretching techniques, push-ups with rucksacks) can have an opposite effect. In fact, it actually can lead to injury (both short term and long term). As such, leaders must ensure that physical fitness education, which includes, for instance, information on exercising and training, and diet and nutrition, is promulgated to their subordinates.

In the end, leaders play an important role in the physical fitness levels of their organizations. Leaders set the tone — they set the example. Normally, the morale, esprit de corps, and effectiveness of a unit is evident in the levels of physical fitness exhibited by its leaders and subordinates.

## SUMMARY

Physical fitness is fundamental to the profession of arms. It is at the very heart of mission success. The most intelligent, technically proficient leaders and soldiers are of little value if they are unable to perform physically on the battlefield. If they cannot function because of fatigue due to physical exhaustion, the leaders and soldiers become liabilities instead of contributing members of the team. High levels of physical fitness allow individuals to carry heavy combat loads longer and farther despite harsh climatic conditions and difficult terrain. High levels of physical fitness allow those same individuals to operate with less sleep and to cope better with stress. Equally important, high levels of physical fitness allow those individuals to endure all of the above and still think clearly and make timely decisions in conditions where lives may be at risk. In the end, physical fitness leads to mission success.

## NOTES

1. Brigadier-General David Fraser (former commander Multi-National Brigade, Regional Command Sector South, Kandahar, Afghanistan), presentation to Canadian Infantry Association, Annual General Meeting, Edmonton, 25 May 2007.
2. Ministry of Defence (UK), *Soldier Management: A Guide for Commanders* (London: Army, 2004), 109.
3. Canada, *Army Fitness Manual b-gl-382-003/pt-001* (Ottawa: Land Force Command, ND), 1.
4. Jennifer Cumming and Craig Hall, "The Relationship Between Goal Orientation and Self-Efficacy for Exercise," *Journal of Applied Social Psychology*, Vol. 34, No. 4 (2004): 747.
5. *CNN Live*, 20 April 2007. The report also noted that one hour of exercise today adds two hours to life later.
6. Canada, Public Service Health Bulletin 16 — Exercise and Well-Being

(Ottawa: Health and Welfare Canada, ND).

7. *Army Fitness Manual*, 7–8.
8. Ibid., 5.

**SELECTED READINGS**

Canada. *Army Fitness Manual, B-GL-382-003/PT-001*. Ottawa: Land Force Command, ND.

Cumming, Jennifer and Craig Hall. "The Relationship Between Goal Orientation and Self-Efficacy for Exercise." *Journal of Applied Social Psychology*, Vol. 34, No. 4 (2004): 747–763.

Goodspeed, Michael. "It's Time for a Fitness Transformation." *Canadian Military Journal*, Vol. 6, No. 1 (Spring 2005): 67–68.

Ministry of Defence (UK). *Soldier Management: A Guide for Commanders*. London: Army, 2004.

Wilson, Philip M., Wendy Rodgers, Chris Blanchard, and Joanne Gessell. "The Relationship Between Psychological Needs, Self-Determined Motivation, Exercise Attitudes and Physical Fitness." *Journal of Applied Social Psychology*, Vol. 33, No. 11 (2003): 2373–2392.

# 32

# THE PROFESSIONAL DEVELOPMENT FRAMEWORK

by Robert W. Walker

The changing nature of the global security environment necessitates an aggressive approach to continuous professional development of Canadian Forces (CF) leaders. Current and ongoing CF transformation initiatives magnify the demand for tenacity, decisiveness, and versatility in CF leader roles, reflect the increasing challenges for professionalism in leadership, and underscore the pronounced need to enhance leader capacities through effective learning.

The Canadian Forces Leadership Institute (CFLI) examined leadership and leader learning requirements through a substantial research effort using military literature and the generic leadership literature, as well as interviews with military leaders. The CFLI first generated a CF effectiveness model as the institutional backdrop against which the requisite leader elements (also referred to in military reports as metacompetencies or "domains") and their attributes were profiled. This ensemble of effective requisite leader elements then was integrated with the continuum of leader levels (junior, intermediate, advanced, senior) to generate a Professional Development Framework (PDF). Next, attention was given to the appropriate professional development — the relevant subject matter and the appropriate learning strategies and developmental methodologies — across the levels of leadership and across the requisite leader metacompetencies.

## CF EFFECTIVENESS: A MODEL OF OUTCOMES

The CF is hierarchic and bureaucratic in nature, with a profession of arms embedded within it. The integration of these two components, the organization and the profession — necessary to achieve institutional effectiveness — is accomplished through leading people, mainly at the tactical level, and leading the institution at the operational and strategic levels. As per Figure 1, the four outcomes — mission success, member well-being and commitment, internal integration, external adaptability[1] — constitute organizational effectiveness, while the conduct value of military ethos infuses these outcomes with professional standards, a professional ideology, and a professional effectiveness.

Collectively, then, Institutional or CF Effectiveness is the consequence of Organizational Effectiveness and Professional Effectiveness.

**Figure 1: Canadian Forces Effectiveness**
**Essential Outcomes &**
**Conduct Values**

- Mission Success (Primary Outcome)
- External Adaptability (Enabling Outcome)
- Military Ethos (Conduct)
- Internal Integration (Enabling Outcome)
- Member Well-being & Commitment (Enabling Outcome)

## A LEADER FRAMEWORK OF REQUISITE CAPACITIES

CF leaders must see the world in terms of its paradoxes and contradictions in order to balance the competing institutional demands represented at Figure 1. Seeing institutional dynamics this way demands a systemic perspective of global activities and a dramatic flexibility in outlook. Accordingly, leaders, with experience and advancement, require strong cognitive/thinking capacities and social/behavioural capacities, the capacities to respond to and shape change in a learning-organization setting, with technical expertise and institutional knowledge, integrated with a professional ideology that supports a mastery of the profession of arms.

The five leader metacompetencies or elements — **Expertise, Cognitive Capacities, Social Capacities, Change Capacities,** and **Professional Ideology** — collectively constitute a leader framework. Brief descriptions of the leader elements and their attributes are provided in Table 1. A total of 16 attributes required of all CF leaders are nested within the five elements. Each attribute in turn would consist of a collated grouping of competencies specific to position, level, and role. The interrelationship of the five leader elements could best be depicted as an assembly of joined puzzle pieces, i.e., a schematic that would visually represent the interconnectedness and interdependency of the leader elements that, only collectively, make effective leadership possible.

## A PROFESSIONAL DEVELOPMENT FRAMEWORK

Figure 1 provided the model for CF Effectiveness. Table 1 describes the five requisite leader elements or capacities, and their attributes, needed to generate CF effectiveness. Next, a Professional Development Framework (Figure 2), first introduced in this book's introduction, incorporates the requisite leader elements arrayed against a progressive professional development process stretching from junior to senior leader levels. This framework is a template for generating expertise, for developing effective leader capacities at the tactical, operational and strategic levels, and for inculcating an appropriate Professional Ideology.

## Table 1: A Leader Framework: Five Elements and Sixteen Attributes

| A Framework of Five Leader Elements | Sixteen Attributes (in bold) Within Five Elements Across the Leader Continuum<br>The focus, scope, and magnitude of competencies for responsibilities related to the leader attributes will vary with rank, leader level, position, etc., and usually increase with time in the CF, rank, seniority, and credibility. |
|---|---|
| Expertise | Expertise consists of **technical** (clusters, e.g., combat arms, sea trades, aircrew) and **specialist** (Military Occupation Classification) proficiencies, an understanding and development of the **military and organizational** environments, and the practice and eventual stewardship of the profession of arms, with the capacities to represent and transform the system through applications at the **strategic and institutional** levels. |
| Cognitive Capacities | Cognitive capacities consist of a problem-solving, critical, **analytic** "left-brain" competence to think and rationalize with mental discipline in order to draw strong conclusions and make good decisions; plus an innovative, strategic, conceptually **creative** "right-brain" capacity to find novel means, "outside the box" ends, and previously undiscovered solutions to issues and problems. |
| Social Capacities | Social capacities consist of a sincere and meaningful behavioural **flexibility** to be all things to all people, with authenticity, combined with **communications** skills that clarify understanding, resolve conflicts, and bridge differences. These capacities are blended with an **interpersonal** proficiency of clarity and persuasiveness, **team** relationships that generate coordination, cohesion, trust, and commitment, and **partnering** capabilities for strategic relations building. |
| Change Capacities | Change capacities involve **self**-development, with risk and achievement, to ensure self-efficacy; **group**-directed capacities to ensure unit improvement and group tranformation; and all with an understanding of the qualities of a CF-wide learning organization, the applications of a **learning organization** philosophy, and the capacity of strategic knowledge management. |
| Professional Ideology | Professional ideology consists of an acute awareness of the unique, theory-based, discretionary body of knowledge at the core of the profession, with an **internalized ethos** whose values and beliefs guide the application of that knowledge. The discretionary nature of military knowledge requires keen judgment in its use and involves **moral reasoning** in thinking and acting, shaped by the military ethos. Professional Ideology underpins a leader exemplar with **credibility/impact** who displays character, openness, assertiveness, and extroversion that ensures the necessary effect by and from the leader. |

## PROFESSIONAL IDEOLOGY

Professional Ideology occupies a privileged position in the PDF. The other four elements across the top left side of the Framework are present in most effective organizations. Only when these elements are shaped by a professional ideology (depicted in Figure 2 as outgoing concentric rings, like an old-fashioned sonar burst) do all five elements coalesce into a collective, interdependent, "Rubik's Cube" of effective leader elements needed for achieving complete institutional effectiveness.

Significantly, Professional Ideology claims a unique, discretionary, theory-based body of military knowledge authoritative in a functional and cognitive sense, and a military ethos that guides and adjudicates how that knowledge is used. The theory-based knowledge consists of the General System of War and Conflict comprising policy, strategy, operational art and tactics. The military ethos consists of three components, Beliefs and Expectations about Military Service, Fundamental Canadian Values, and the Core Military Values.[2] Professional Ideology demands doing good work over self-interest, and effectiveness over efficiency. Professional

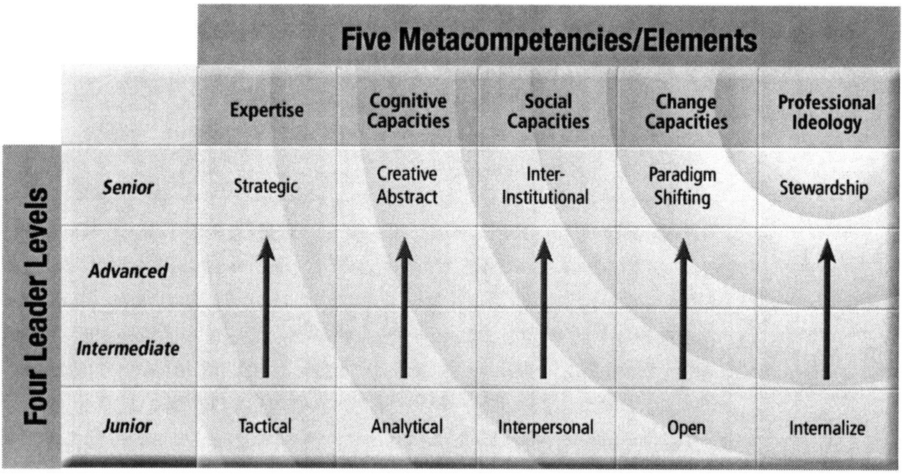

| Four Leader Levels | Five Metacompetencies/Elements | | | | |
|---|---|---|---|---|---|
| | Expertise | Cognitive Capacities | Social Capacities | Change Capacities | Professional Ideology |
| Senior | Strategic | Creative Abstract | Inter-Institutional | Paradigm Shifting | Stewardship |
| Advanced | ↑ | ↑ | ↑ | ↑ | ↑ |
| Intermediate | | | | | |
| Junior | Tactical | Analytical | Interpersonal | Open | Internalize |

**Figure 2: The Professional Development Framework**

Ideology stands in contrast to bureaucratic ideology (managerialism) and market ideology (entrepreneurialism).

A fully developed PDF would populate the cells or compartments created by the five elements cross-tabulated with the four leader levels, to represent a seamless, balanced leader development process. The leader elements, leader levels, and foci for professional development are presented in more detail at Appendix A to this chapter in a format that emphasizes that development at the more expansive leader levels — advanced and senior — are founded on development at the less expansive leader levels — junior and intermediate. Additionally, the foci reflect the transition in leadership from a predominantly leading-the-people emphasis at junior and intermediate levels to a predominantly leading-the-institution commitment at advanced and senior levels.

## **PROFESSIONAL DEVELOPMENT METHODOLOGIES FOR LEADER ELEMENTS**

With respect to professional development, the five elements traditionally have been addressed through the conventional PD pillars of experience, training, education, and self-development. There is no intention to change this PD construct, but the Canadian Defence Academy continues to review the dependence on, and effectiveness of, each pillar, and the learning methodologies within each of them. Methodologies need to be balanced between those appropriate for the more concrete elements — expertise, for example, where acquisition of fundamental skills and knowledge is emphasized predominantly through pedagogical, group-taught, teacher-centred, lecture approaches to information dissemination — and adult-learning, student-centred, self-initiated, experiential learning methodologies appropriate for the less concrete and more abstract capacities — complex cognitive, social/behavioural, change-shaping — and professional ideology.

As examples, for aspiring advanced and senior leaders focusing on institutional leadership in the complex 21st century, increased emphasis on adult-learning strategies would be most appropriate to develop leader expertise related to institutional and strategic capabilities (e.g., understanding and influencing "how Ottawa works"); Cognitive Capacities (e.g., complex organizational wisdom, creative thinking); Social Capacities (e.g., leader flexibility, conflict resolution, effective communication for partnering with international organizations); Change Capacities (e.g., learning-organization information-sharing applicable to hi-tech combat communications); and Professional Ideology (e.g., proactively becoming exemplars, custodians, stewards of the profession, using moral/ethical reasoning, balancing autonomous thinking with conformity/team membership, etc.). The need exists to explore further the potential of adult learning strategies and methodologies beyond the all-students-same-material-same-method classroom lectures and pedagogical methodologies.

## **CF APPLICATIONS**

The Professional Development Framework is relevant and applicable in all aspects of CF activities. Six current, micro through macro examples of the integration of CF initiatives and the PDF's metacompetencies are:

* *The applicability of 360° assessment feedback to individual leader development*
    Leader capacities benefit from self-awareness. Valid feedback from superiors, peers, and subordinates acquired through a well-designed and well-conducted multi-rater 360º feedback process is one means for acquiring self-awareness. The Canadian Forces College has expanded its utilization of 360° processes to achieve this.

* *Senior CF leaders' professional development needs*
    Substantial research plus interviewing of

senior officers and NCOs determined the specific subject content and effective learning circumstances required. Contrasted with valid and well-founded needs, learning initiatives needed to be convenient and succinct for the CF's institutional leaders with little time to be absent from their worksites. Solutions demanded developmental initiatives with the most beneficial subject matter and the most concise learning situations, in line with the PDF.

* *Effective professional development methodologies and learning strategies*

    The enhancement of leadership as an institutional requirement is best addressed through a variety of leader development methodologies balanced between those appropriate for the more "concrete" leader elements, for example, technical Expertise, and those more appropriate for the less concrete and more abstract capacities such as complex Cognitive and Social processes, Change shaping, and stewardship of Professional Ideology. The latter require adult-learning, student-centred, self-initiated, actively engaged learning processes.

* *Review and reform in the CF's current human resources / personnel systems*

    The PDF, as the instrument for categorizing the requisite human/leader capacities, can act as a conceptual foundation for reviewing and restructuring current HR policies and programs. The Framework is especially important as a tool for senior leaders of the institution when addressing HR programs and procedures with the intent to improve them and increase their relevance.

* *Evolving operational leader challenges in warfighting while nation-building*

    Current CF member deployments, for example, in Afghanistan, reflect the most recent and tragic complexities in leadership challenges. Recent research in Afghanistan[3] identified unique tactical, operational, and strategic leadership challenges evolving from the singular characteristics of leading soldiers in an Afghan version of the "three block war" — warfighting, conflict resolution, and nation-building through humanitarian endeavours. These challenges demand truth, duty, and valour in its individual leaders leading people, and tenacity, decisiveness, and versatility in its leader corps leading the CF institution. These challenges demand highly developed leader capacities as represented in the Professional Development Framework.

* *CF Transformation 2005[4] and its 6 CDS Principles*

    The PDF was designed to be applicable across the micro-to-macro, junior-to-senior CF circumstances of leadership. At the broad, institutional level, CF Transformation 2005 initiatives require extreme complexities in leader behaviours. The PDF identifies the requisite capacities, along with a system of professional development, for ensuring that CF professionalism and strong leadership is available to achieve the Transformation.

## SUMMARY

The Professional Development Framework, incorporating the most effective learning strategies, is a comprehensive model for expanding the depth and breadth of leader learning. It supports a shift from a pedagogical sufficiency ("pass") model of professional development to a mastery ("excel") model of performance, human resourcing, and leadership, a shift that can lead to the identification of the highest potential — professionally developed military leaders — rather than simply the military members basically suited for their next appointments.

The overriding and continuous theme for the CF is *Transformation*. The ongoing 2005 transformation must be led by a focus on the leadership and professional enhancement of members in their capacities to master the challenges of the 21st century. The Professional Development Framework is a crucial linchpin in this overall effort.

# Appendix A: Professional Development Framework

| | | EXPERTISE<br>Tactical to Strategic | COGNITIVE CAPACITIES<br>Analytical to Creative/Abstract |
|---|---|---|---|
| **LEADER LEVEL** | **SENIOR** | **Security Expertise**<br>• Scope and content moves from knowledge to expertise with accompanying expansion to a strategic understanding of the domain of security.<br>• Shift from knowledge to expertise requires ability to apply the philosophy and principles that govern the generation and employment of military capacities (knowledge + philosophy = expertise) and strategic, institutional co-existence among peer ministries, foreign defence agencies.<br>• Expertise at this stage clearly is dependent upon the complementary development in Professional Ideology, and a full understanding of the profession of arms. | **Knowledge Creation**<br>• Able to generate, organize, and manage the theory-based body of knowledge applied across the profession.<br>• This goes beyond the analytic, creative, and judgment capacities needed to adapt the profession to the external environment, and expands to include the obligation to update and extend the profession's unique body of knowledge so as to ensure that the profession is discharging all of its responsibilities to society in the most effective manner.<br>• Strong parallel to cognitive capacities at advanced academic post-graduate levels; masters the particular academic discipline but also generates new knowledge. |
| | **ADVANCED** | **Defence Knowledge**<br>• Shift from information to knowledge, incorporating a broad understanding of the CF and defence as key components of security and government functions.<br>• The shift from information to knowledge requires the additional perspective of understanding the rationale and purpose of intended actions; and the generalized outcomes that are to be achieved (information + purpose = knowledge). | **Mental Models**<br>• Uses inductive and deductive reasoning skills to create, adapt, and generalize knowledge both from one's own previous learning and experiences, and from other domains such as professional literatures.<br>• Conducts abstract reasoning and draws on appropriate professional orientation to be able to understand desired outcomes.<br>• Aware of assumptions embedded in the military way of framing issues, testing working hypotheses, operating within the academic discipline of military thinking. |
| | **INTERMEDIATE** | **Military Information**<br>• How MOC contributes to larger formation capabilities.<br>• Understanding not only what to do but the context in which this occurs (data + context = information).<br>• Examples: Effects-based operations, impact of instability and conflicts on multinational relations, international law, civil control of military. | **Theories and Concepts**<br>• Able to reason, moving from the concrete to the abstract, from procedures and rules to principles. |
| | **JUNIOR** | **Technical and Tactical Procedures**<br>• Learning standard Military Occupational Classification (MOC) and sea/land/air procedures.<br>• For initial leader roles, acquiring an overview of such standards and procedures, and small group tactics. | **Theorems, Practical Rules**<br>• Reasoning at this level is intended to identify the appropriate task procedures, using simple theorems, practical rules, or established scientific principles/laws.<br>• When cognitive capacities interact with expertise at the junior level, the two elements function in a "cookbook" approach to problem-solving and task accomplishment. There is limited capacity for innovation. |

| SOCIAL CAPACITIES<br>Interpersonal to Inter-Institutional | CHANGE CAPACITIES<br>Openess to Paradigm Shifting | PROFESSIONAL IDEOLOGY<br>Internalizing to Stewardship |
|---|---|---|
| **Strategic-Relations Building**<br>• Relates to the concept of Leading the Institution, relies on secondary and tertiary influence processes for the senior leader to communicate institutional priorities and strategic intent across organizational systems.<br>• Builds open teams such that immediate subordinates can contribute novel ideas and can critique taken-for-granted assumptions.<br>• Externally focused capacities pertain to building and maintaining strategic relations with others engaged in the broad security arena and related national/government initiatives. | **Multi-Institutional Partnering**<br>• Focus is external, on changing others' understanding of the military as a strategic political capacity; and internal on implementing internal change initiatives.<br>• In this latter regard, there is an emphasis on the initial stages of anticipating change, effectively contributing to the change, and monitoring and adjusting initiatives over the change period.<br>• Senior leader initiatives exist to transform and improve a team or multiple units, or to attempt learning-organization applications at organizational and institutional levels. | **Stewardship of the Profession**<br>• Core capacities are related to managing collective professional identity — the key issues of articulating what the profession is, what it stands for, and what it believes in.<br>• Able to engage in very abstract reasoning, exemplified at the highest stages of moral/identity development — in particular, the capacity for independent judgment of the profession's core philosophy, ideology, and principles.<br>• This capacity is integrated with acquisition of related capabilities in Cognitive and Change Capacities. |
| **Group Cohesiveness**<br>• At this level of larger or multiple units/teams/groups, is involved in aspects of leading the institution, and applies broad influence processes to ensure internal cohesion, fostering commitment and supporting subordinate leaders while also engaging in effective boundary-spanning activities, especially in joint or multinational operations. | **Group Transformation**<br>• Able to adapt and align groups or subsystems to the broadest requirements of the institution while ensuring the tactical proficiency and effective integration of individuals and small teams/sections within the larger formation. | **Cultural Alignment**<br>• Guides framing of problems and interactions with others to apply leader influence to shape or align the extant culture to be consistent with the ethos.<br>• Contains some of the most complex challenges in achieving competing institutional effectiveness objectives — mission success versus member well-being; internal synchrony and stability versus external adaptability and experimentation. |
| **Individual Persuasion**<br>• Social skills for leading people, particularly the abilities to effectively influence others "one-on-one" or small-group, using some range of influence behaviours appropriate to the characteristics of the situation, the followers, and the individual leader. | **Self-Efficacy**<br>• Capacities at this stage are focused on the individual's abilities to monitor self-efficacy, engage in self-reflection, make early commitments to self-development, and adapt one's behaviours to the social environment/context in which one is functioning. | **Self-Regulation**<br>• Conducts basic self-regulation, avoiding obvious ethical violations and not displaying behaviours that erode the reputation, image, or credibility of the profession; essentially a journeyman stage of professionalization.<br>• Abides by the principles of the Defence Ethics Program.<br>• Capable of serving as an example. |
| **Team-Oriented Followship**<br>• Aware of group norms, minimum leader-style flexibility.<br>• Moderate communication capabilities applied through baseline interpersonal skills, reflecting an awareness of basic influence factors, group diversity issues, and non-prejudicial self-behaviour. | **External Awareness**<br>• Minimal expectation in change capacities would be a generalized orientation and awareness of changes occurring external to the CF, and the CF transformational efforts, as means of signalling the importance of practising openness to externally driven change. | **Normative Compliance**<br>• Understands the concepts and practices of the profession of arms at an introductory level. At a minimum, practices military group norms, and adheres to discipline demands.<br>• As an *ab initio* professional (apprentice), looks externally (to supervisors or codes of conduct) for guidance as to the appropriate behaviours in specific circumstances. Internalizes values minimally. |

## NOTES

1. Canada, *Leadership in the Canadian Forces: Conceptual Foundations* (Kingston: DND, 2005), 19. Also available online at http://www.cda-acd.forces.gc.ca/cfli for more detail.
2. Canada, *Duty with Honour: The Profession of Arms in Canada* (Kingston: DND, 2003), 26–31. Also available online at http://www.cda-acd.forces.gc.ca/cfli.
3. Bernd Horn, "'Outside the Wire' — Some CF Leadership Challenges in Afghanistan," *Canadian Military Journal*, Vol. 7, No. 3 (Fall 2006): 8–14.
4. See the case study by Lieutenant-General M. Jeffery, *Inside Canadian Forces Transformation*, for the historical record and review of the restructuring and transformation of the CF. (Kingston: Canadian Defence Academy Press, 2008.)

## SELECTED READINGS

Canada. *Duty with Honour: The Profession of Arms in Canada*. Kingston: DND, 2003. http://www.cda-acd.forces.gc.ca/cfli.

Canada. *Leadership in the Canadian Forces: Leading People,* Kingston: DND, 2007. http://www.cda-acd.forces.gc.ca/cfli.

Canada. *Leadership in the Canadian Forces: Leading the Institution.* Kingston: DND, 2007. http://www.cda-acd.forces.gc.ca/cfli.

Canada. *Leadership in the Canadian Forces: Conceptual Foundations.* Kingston: DND, 2005. http://www.cda-acd.forces.gc.ca/cfli.

Canada. *Leadership in the Canadian Forces: Doctrine*. Kingston: DND, 2005. http://www.cda-acd.forces.gc.ca/cfli.

Horn, B., ed. *Contemporary Issues in Officership: A Canadian Perspective.* Toronto: Canadian Institute of Strategic Studies, 2000.

Horn, B. and S.J. Harris, eds. *Generalship and the Art of the Admiral: Perspectives on Canadian Senior Military Leadership.* St. Catharines: Vanwell Publishing Limited, 2001.

Hughes, R.L. and K.C. Beatty. *Becoming a Strategic Leader.* San Francisco: Jossey-Bass and the Center for Creative Leadership, 2005.

Hunt, J., G. Dodge, and L. Wong, eds. *Out-of-the-Box Leadership: Transforming the Twenty-First Century Army and Other Top-Performing Organizations.* Stamford, CT: JAI Press, 1999.

Matthews, L.J., ed. *The Future of the Army Profession.* Boston: McGraw-Hill, 2002.

Walker, R.W. *The Professional Development Framework: Generating Effectiveness in Canadian Forces Leadership,* CFLI Technical Report 2006-01. Kingston: Canadian Forces Leadership Institute, 2006. http://www.cda-acd.forces.gc.ca/cfli/.

Wenek, K.W.J. "Defining Leadership." Unpublished Paper. Kingston: Canadian Forces Leadership Institute, 2003. http://www.cda-acd.forces.gc.ca/cfli/.

———. "Defining Effective Leadership in the Canadian Forces." Unpublished Paper. Kingston: Canadian Forces Leadership Institute, 2003. http://www.cda-acd.forces.gc.ca/cfli/.

Wong, L., S. Gerras, W. Kidd, R. Pricone, R. Swengros. *Strategic Leadership Competencies.* Carlisle Barracks, PA: Strategic Studies Institute, U.S. Army War College, 2003. http://www.carlisle.army.mil/research_and_pubs/research_and_publications.shtml.

Zaccaro, S. *The Nature of Executive Leadership: A Conceptual and Empirical Analysis of Success.* Washington, DC: American Psychological Association, 2001.

# 33
# PROFESSIONAL IDEOLOGY
by Bill Bentley

As described in Chapter 28 of this handbook, The Military Professional, at the very core of any profession is the professional ideology that defines its special area of expertise and how that expertise is to be employed. This professional ideology stands in distinct opposition to the other two major ideologies extant in the socio-economic sphere of the Western capitalist market system. These are market ideology (entrepreneurialism) and bureaucratic ideology (managerialism). Professor Eliot Freidson contrasts these three ideologies in terms of how each views and organizes knowledge. In Freidson's terminology, there are three "logics" underlying the nature, acquisition, and application of knowledge in the areas of professions, the market, and bureaucracies respectively. According to Freidson, in ideal/typical professionalism, specialized workers control their own work, while in the free market, consumers are in command and in bureaucracy, managers dominate. "Each method has its own logic requiring different kinds of knowledge, organization, career, education and ideology."[1]

Military professional ideology is essential to the effective functioning of the profession of arms. It infuses and energizes the other three attributes of military professionalism — responsibility, expertise, and identity — and embeds the profession in the society that it serves.

## IDEOLOGY DEFINED

The concept of ideology, properly understood, is a powerful tool for analyzing how professionals view their role in society. The noted sociologist Talcott Parsons has argued that, in the modern world, concern with the expression of moral commitments and with their application to practical problems, social and otherwise, has to a considerable degree become differentiated in the function of ideology.[2] Thus, for example, the ideology of professionalism asserts above all else devotion to the use of disciplined knowledge for the public good.

Ideology reflects a specific system of ideas, or a conception of the world, that is implicitly manifest in law, in economic activity, and many other manifestations of individual and collective life. But it is more than a conception of the world as a system of ideas, it also represents a capacity to inspire concrete attitudes and motivate action. To be recognized as such, an ideology must be capable of organizing humans; it must be able to translate itself into specific orientations for action. To this extent ideology is socially pervasive, in other words, "the source of determined social actions."[3]

The American sociologist Daniel Bell uses the concept of ideology in this way in his book, *The Cultural Contradictions of Capitalism*, explaining that it is in the character of an ideology not only to reflect or justify an underlying reality, but, once launched, to take on a life of its own. "A truly powerful ideology opens up a new vision of life to the imagination, but once formulated it remains part of the moral repertoire to be drawn upon."[4]

The concept of ideology, therefore, is systematic — beliefs about one topic are related to beliefs about another, different, topic; normative — to a large degree it contains beliefs about how the world ought to be; and, programmatic — it guides or incites concrete action.

## PROFESSIONAL IDEOLOGY DEFINED

Systematic, normative, and programmatic, professional ideology claims both a specialized, theory-based knowledge that is authoritative in both a functional and cognitive sense and a commitment to a transcendental value that guides and adjudicates the way that knowledge is employed. It is authoritative functionally because the knowledge in question is the only knowledge that can get the job done. It is cognitively authoritative because it is ontologically grounded, theoretically based, and can only be fully accessed intellectually. The commitment at issue is represented in the ideology by the professional's occupational ethic, or, in the case of the military profession, the military ethos. The ideology of professionalism, furthermore, argues that expertise properly warrants special influence in certain affairs because it is based on sustained systematic thought, investigation, and experience, and, in the case of individuals, accumulated experience performing work for which they have had long and appropriate training and education.

## THE KNOWLEDGE COMPONENT: THE GENERAL SYSTEM OF WAR AND CONFLICT

The systematic, theory-based knowledge at the core of the military profession is the General System of War and Conflict comprising policy, strategy, operational art, and tactics. Figure 1 represents this body of knowledge.

In this model, war is defined as the continuation of policy with the admixture of other means. As the Prussian military theorist Carl von Clausewitz noted, warfare has its own grammar but not its own logic. In war the logic is supplied by policy. Strategy, operational art, and tactics constitute warfare, the grammar of war.[5]

Strategy is defined as the art of distributing and applying military means, or the threat of such action, to fulfil the ends of policy.[6] At the interface of policy and strategy lies the domain of civil-military relations and the zone in which all of the instruments of national power are aggregated and coordinated to generate a nation's overall national security strategy.

There are two fundamental types of strategy in the General System of War and Conflict model. If the political objective is unlimited or unconditional, then the strategy of annihilation is appropriate. Decisive

# THE MILITARY LEADERSHIP HANDBOOK

**Figure 1: The General System of War and Conflict**

military victory at the strategic level is the goal and only final victory counts. Both First and Second World Wars are good historical examples. If, however, the political goal is more limited, the bipolar strategy is the appropriate choice. Here, the strategist acts on both the battle pole and the non-battle pole, either simultaneously or sequentially. The Korean War and the Gulf War are good examples.

The operational level of war may be defined as the connecting link between strategy and tactics. It is at this level that campaign plans are created and executed. Operational art is the theory and practice of preparing and conducting operations (aggregated as campaigns) in order to connect tactical means to strategic ends.

The tactical level of warfare is the realm of actual, direct combat. Here tactical manoeuvre; that is, fire and movement, is employed to physically achieve tasks assigned by the operational level commander.

In the profession of arms, this core body of knowledge is augmented by a wide range of skills and other knowledge from a wide variety of disciplines (i.e. history, political science, the natural sciences, etc.). This supplementary knowledge is distributed and differentiated throughout the profession by rank and function.

## THE ETHOS COMPONENT: THE CANADIAN MILITARY ETHOS

Military ethos is defined as "the spirit that animates the profession of arms and underpins operational effectiveness". It acts as the centre of gravity of the military profession and establishes an ethical framework for the professional conduct of military operations.

In the Canadian context the military ethos is intended to:

- establish the trust that must exist between the Canadian Forces and Canadian society;
- guide the development of military leaders who must exemplify the military ethos in their everyday actions;
- create and shape the desired military culture of the Canadian Forces;
- establish the basis for personnel policy and doctrine;
- enable professional self-regulation within the Canadian Forces; and
- assist in identifying and resolving ethical challenges.

The Canadian military ethos is made up of three fundamental components: Beliefs and Expectations about Military Service; Canadian Values; and Core Canadian Military Values.

Military professionals must internalize these:

*Beliefs and Expectations*

- Accepting Unlimited Liability — all members accept and understand that they are subject to being lawfully ordered into harm's way under conditions

that could lead to the loss of their lives.
* Fighting Spirit — this requires that all members of the Canadian Forces be focused on and committed to the primacy of operations.
* Discipline — discipline helps build the cohesion that enables individuals and units to achieve objectives that could not be attained by military skills alone and allows compliance with the interests and goals of the military institution while instilling shared values and common standards. Discipline among professionals is primarily self-discipline.
* Teamwork — this builds cohesion, while individual talent and the skills of team members enhance versatility and flexibility in the execution of tasks.

*Canadian Values:*

As a people Canadians recognize a number of fundamental values that the nation aspires to reflect. Canadians believe that such values can be woven into the fabric of society. Canadian values are expressed first and foremost in founding legislation such as the Constitution Act 1982 and the Charter of Rights and Freedoms. Other key values that affect all Canadians are anchored in a number of foundational legislation and articulated in their preambles.

*Core Military Values:*

* Duty — obliges members to adhere to the rule of law while displaying dedication, initiative, and discipline in the execution of tasks.
* Loyalty — entails personal allegiance to Canada and faithfulness to comrades; it must be reciprocal and based on mutual trust.

* Integrity — unconditional and steadfast commitment to a principled approach to meeting obligations. It calls for honesty, the avoidance of deception, and adherence to high ethical standards.
* Courage — both physical and moral. Courage allows a person to disregard the cost of an action in terms of physical difficulty, risk, advancement, or popularity.

The values, beliefs, and expectations reflected in the Canadian military ethos are essential to operational effectiveness, but they also serve a more profound purpose. They constitute a style and manner of conducting military operations that earn for soldiers, sailors, and Air Force personnel that highly regarded military quality — honour.

## THE COMPETING IDEOLOGIES: MARKET AND BUREAUCRACY

Contrary to the professional's claim that only specialists who can do the work are able to evaluate and control it, both the ideology of the market and bureaucratic ideology claim a general kind of knowledge, superior to specialized expertise, that can direct and evaluate it.

Market ideology claims that ordinary human qualities informed by everyday knowledge and skills and fuelled by self-interest enable the individual, properly motivated, to learn whatever is necessary to make all economic or political decisions. Denying the professional expertise any unique status, market ideology falls back on its own special kind of preparation for positions of leadership — an advanced but general form of education that they believe equips them to direct or lead specialists as well as ordinary citizens in the pursuit of profit.

The noted sociologist C.W. Mills warned of the dangers to professionalism posed by bureaucracy as early as 1956, arguing that bureaucracy was becoming such a dominant force in modern society that professions were increasingly "being sucked into administrative machines,"[7] where knowledge is standardized and routinized into the bureaucratic apparatus and professionals become mere managers. The bureaucratic ideology claims the authority to command, organize, guide, and supervise the activities of professionals. It denies authority to professional expertise by claiming a form of knowledge that is superior to specialization because it can organize it rationally and efficiently. Those who espouse this view of knowledge from within the ranks of bureaucracy or management could be called "elite generalists."

## **STEWARDSHIP OF THE PROFESSION**

Ensuring that professional standards are established and strengthened is certainly one of the most important roles for institutional leaders in the Canadian Forces (CF). Continuous execution of this responsibility is essential to enhancing operational effectiveness and is the primary means of protecting the profession from the insidious encroachments of both market ideology and especially bureaucratic ideology. Stewardship must focus on both components of professional ideology — the maintenance and growth of the core body of knowledge at the heart of the profession and the promulgation of the military ethos to align military culture with its dictates. Given the inclusive nature of the profession of arms in Canada (see Chapter 28 on The Military Professional), stewardship is the responsibility of both officers and senior Non-Commissioned Members (NCMs) serving in the regular and reserve force. Stewardship is, therefore, formally defined as the special obligation of officers and non-commissioned members who, by virtue of their rank or appointment, are directly concerned with ensuring that the profession of arms in Canada fulfils its organizational and professional responsibilities to the Canadian Forces and Canada. This includes the use of their power and influence to ensure the continued development of the profession, its culture, and its future leaders to meet the expectations of Canadians.

## SUMMARY

The military profession is always under a degree of pressure that threatens to erode its basis tenets and characteristics. These threats include those posed by the competing ideologies discussed above, as well as societal changes that may be anathema to true professionalism. In addition, new roles and responsibilities in the uncertain security environment of the 21st century must be accommodated. This requires new forms of expertise while retaining a clear sense of identity as members of the profession of arms. It is the role of professional ideology, properly nurtured, to bind the profession firmly and to ensure its future vitality.

## NOTES

1.  Eliot Freidson, *Professionalism* (Chicago: University of Chicago Press, 2001), 6.
2.  Talcott Parsons, "The Professions," *Encyclopaedia of the Social Sciences*, 545.
3.  J. Larrain, *The Concept of Ideology*, (Athens, GA: University of Georgia Press, 1979), 80.
4.  Daniel Bell, *The Cultural Contradictions of Capitalism* (London: Heineman, 1976), 60.
5.  Carl von Clausewitz, *On War*, Michael Howard and Peter Paret, eds. (Princeton: Princeton University Press, 1976), 146. See also General Sir Rupert Smith, *The Utility of Force* (London: Allen Lane, 2005).
6.  Canada, *Leadership in the Canadian Forces: Leading the Institution* (Kingston: DND, 2007), 42.
7.  C.W. Mills. *White Collar* (New York: Oxford University Press, 1956), 112.

## SELECTED READINGS

Bengt, Abrahamsson. *Military Professionalism and Political Power.* London: Sage Publications, 1972.

Bentley, Bill. *Professional Ideology and the Profession of Arms in Canada.* Toronto: Canadian Institute of Strategic Studies, 2005.

Freidson, Eliot. *Professionalism.* Chicago: University of Chicago Press, 2001.

Horn, Bernd and Stephen J. Harris, eds. *Generalship and the Art of the Admiral.* St. Catharines, ON: Vanwell Press, 2001.

Larrain, J. *The Concept of Ideology.* Athens, GA: University of Georgia Press, 1979.

Smith, General Sir Rupert. *The Utility of Force: The Art of War in the Modern World.* London: Allen Lane, 2005.

# 34

# SELF-DEVELOPMENT

by Brent Beardsley

Self-development is a critical requirement in all professions. As such, self-development is considered an individual responsibility to learn and to stay current with the professional body of knowledge and skills above and beyond that which is provided in the overall formal professional development system. Within the military profession, self-development is recognized as one of the four pillars of the professional development system along with education, training, and experience.[1]

Self-development differs from the other three pillars or regimes of professional development in that it is based solely on the self-motivation of each individual member of the profession. Increasingly, in all professions, due to the rapid and massive expansion of the professional body of knowledge and skills, self-development is being recognized as growing in importance in overall professional development.

In the military profession, subject to service requirements, the institution will encourage and support every member of the profession for training, education, experience, and other self-development opportunities over and above those offered by the institution.[2] Self-development can only be effectively and optimally applied if each and every member of the profession is expected to, and is given the opportunities to take an active role in the planning of their personal self-development program over the period of their entire career. For a personal professional self-development program to be successful, it must be deliberate, continuous, sequential, and progressive throughout the career of the individual.

## WHAT IS SELF-DEVELOPMENT?

Self-development is defined in the military profession as "professional development over and above the formal requirements of the profession but related in some way to career requirements."[3] Self-development may take the form of education, training, or experience above that which is formally provided by the profession in the overall professional development system. Some self-development opportunities will be supported by the profession in the form of professional journals, professional development seminars, suggested professional reading lists, or financial and/or time support for academic upgrading. Other self-development opportunities will be an individual responsibility, like watching television in a second language to enhance second-language comprehension, personal and professional reading, or pursuing personal sports or fitness activities. In either case, the key characteristic of self-development is that it is an individual responsibility to conduct professional development that will enhance career progression over and above that which is formally provided by the profession. Individuals get out of their self-development programs what they put into those programs.

Self-development is a lifelong learning process. In the military profession, it begins upon enrollment and is expected to be conducted until retirement. While the military profession must be a learning organization,[4] the will and actual conduct of the self-development portion of that learning is individually planned and conducted. It must be clearly understood that self-development is an individual responsibility that is increasingly being viewed in all professions as a critical component of an overall professional development system and, in theory, may become the primary mode of professional development in the future.

Throughout a career, the military professional must take advantage of various postings, employment positions, activities, or any other opportunity that presents itself in which to conduct self-development. The increasing access to distance learning from remote locations provides an excellent opportunity and example of how essential support for self-development can be obtained.

## TYPES OF SELF-DEVELOPMENT

For the overall professional military development system to be effective, each member of the profession must first accept responsibility for career-long involvement and commitment to all aspects of their development. The self-development pillar or regime of the overall professional development system will consist of formal or informal components in the form of education, training, or experience opportunities.

Formal opportunities like part-time academic upgrading (education), part-time second language enhancement (training), or volunteering for a specific task to gain a new and potentially valuable experience (experiential opportunity) are structured activities that may be supported or even provided by the profession on a voluntary basis to provide career-enhancement opportunities.

Informal opportunities like personal reading and research (education), watching television in a second language to enhance comprehension skills (training), or taking advantage of an unforeseen opportunity like public speaking in a second language (experiential), are not structured activities and may not be formally supported or provided by the profession. However, these can be valuable activities in self-development. The most essential skill that every professional must master is reading, both in speed and comprehension abilities.

The best approach to self-development consists of an individually developed program, based on personal factors (e.g., family responsibilities, professional responsibilities, availability) and personal choices (e.g. interests, occupational specialty) that combines both formal and informal components to achieve the career goals determined by the individual professional for their personal career aspirations and requirements.

## THE PROFESSIONAL THEORY-BASED BODY OF KNOWLEDGE

Every profession has a discretionary, theory-based, professional body of knowledge, as a part of its professional ideology, which must be mastered systematically as the individual advances in their profession.[5] In the military profession, this core body of knowledge supports the lawful and ordered application of military force in support of government policy.[6] The essential components of that body of knowledge will be provided formally by the CF Professional Development System through education, training, and experience. However, above and beyond that which is formally provided, self-development should contain a focus on enhancing one's knowledge and understanding of the General System of War and Conflict and the essential subjects that contribute to that system. The military professional must enhance their understanding of the levels of war and conflict, namely the tactical, operational, strategic, and policy levels. They must understand the types of war or conflict/conflict resolution, namely total and limited, the supporting functions of command, management, and leadership in both theory and in practice, and the theory and practice of civil-military relations. In addition, military history, awareness of technology, military professionalism, ideology and ethos, human behaviour, and an understanding of culture, communications skills, and ethics are all key subjects that need to be addressed in a professional self-development program.[7]

Upon entry into a profession, the professional body of knowledge will be a critical component of the early and essential socialization process into the profession. As a professional advances in rank, responsibilities, and appointments, the study of the professional body of knowledge will be much more substantive and intellectually challenging.

In addition to the study of the military professional body of knowledge, some members of the profession are dual professionals (e.g. doctors, dentists, lawyers) and will also have to include the study of the professional body of knowledge related to their other profession. (See Chapter 28 on The Military Profession and Chapter 32 on the CF Professional Development Framework.[8])

## THE COMPONENTS OF SELF-DEVELOPMENT

There are five key components of an effective self-development program:

## SELF-DEVELOPMENT

1. Every military professional must live a life that is balanced between their personal, family, professional, leisure, and self-development activities. Over-emphasis on any one or some of the above at the expense of others potentially will have an overall damaging impact on the individual and can ultimately prevent the individual from achieving his/her potential career achievements. It is very important to live a balanced life through the skilled use of time management that provides sufficient time for all of life's priorities.

2. As a basics requirement in a self-development program, every military professional must ensure their personal health, physical, and psychological well-being are continually maintained and even enhanced in order to ensure they are fit to serve. Attention to diet, a continuous physical fitness program, avoidance of threats to health (e.g. refraining from smoking, over-consumption of alcohol, abuse of prescription or non-prescription drugs), obtaining sufficient rest, avoiding threats to occupational health and safety hazards and stress reduction must be a core feature of a self-development program. (See Chapter 31 regarding Physical Fitness).

3. The military professional must ensure sufficient personal time for personal recreation, relaxation, and time dedicated to family and friends. Healthy personal time, time for personal relationships, and time to fulfill personal obligations are important components in a self-development program and are critical in ensuring member well-being.

4. The military professional must ensure sufficient time for the requirements of their employment, which will vary depending on the nature of their job (e.g. garrison employment or operational tours).

5. The military professional must then ensure that there is sufficient time for a personal self-development program after completing a self-development assessment.[9]

## **THE SELF-DEVELOPMENT ASSESSMENT**

A self-development assessment asks a series of personal questions. The answers to these questions, balanced against the time available after weighing the components of self development, will largely determine the self-development program of the individual military professional:

1. What do I need/want to achieve in my life and career?
2. What do I know or am able to do at my current level and at the next level?
3. What do I need to know or be able to do at my next or future levels?
4. What is the difference between what I know or am able to do and what I need to know or be able to do to achieve my personal goals?
5. Within this difference, what are my priorities?
6. How can I acquire the knowledge and skills through formal or informal educational training and experience opportunities?
7. Who can help me acquire the knowledge and skills I require?

# SELF-DEVELOPMENT

8. What else, if anything, do I need in order to prepare a personal self-development or learning plan within the priorities and time available?[10]

The self-development assessment will be an ongoing process throughout one's career. The self-development program will be a lifelong process that in some cases will go beyond retirement. Priorities will change with age, status, responsibilities, rank, etc., which is to be accepted, and the program must adapt to changes in priorities. Time available for self-development will change and the self-development program must be flexible. An individual needs to ensure that their superiors in the military are aware of the goals of their personal self-development program and are kept aware of any formal support they require to conduct the program. As well, individuals must be inquisitive and aware of other sources of support for their self development program.

## **LEADERSHIP ISSUES AND SELF-DEVELOPMENT**

Every leader in the military profession must, like every other member of the profession, develop and conduct a self-development program. This is a classic case where leaders must lead by setting the example for their followers in the development and conduct of a self-development program as briefly described in this chapter. In addition, leaders must also create a climate within their teams where followers are encouraged and, where possible, are assisted in developing and conducting their personal self-development program.

## **SUMMARY**

The self-development pillar or regime is the personal component of the formal professional development system. It is a personal responsibility and each professional has an obligation to conduct a personal lifelong self-development program. However, it is a program that will largely be determined by the individual. The results and rewards of the program will be determined largely by the time and effort the individuals personally are prepared and able to put into the conduct of their self-development.

It is clearly impossible for any or every member of the profession of arms to master every subject in the professional body of knowledge or in the body of knowledge applicable to their occupation specialty. Health must be enhanced, personal responsibilities must be satisfied, time available must be determined, a balanced life must be ensured, formal work requirements must be met, and personal goals, interests, and aspirations must be ensured. It is important to understand that self-development is a lifelong and career-long process that may be both formal and informal in nature and may be supported or not by the profession. The key to success is to start early, maintain the program, adapt and change it as required, or as priorities change and develop over the long term in the pursuit of the personal program. Realism and self-discipline will be tested, but the rewards, both personal and professional, will be achieved by the continuous pursuit of a personal self-development program.

## NOTES

1. Canada, *Canadian Officership in the 21st Century (Officership 2020)* (Ottawa: DND, 2001), 1–5; Canada, *The Canadian Forces Non-Commissioned Member in the 21st Century (NCM Corps 2020)* (Ottawa: DND, 2003), 1–5.
2. *Ibid.* In both documents, on page 7, self-development is identified as a key initiative in career-long intellectual development, which is the second strategic objective of both plans.
3. Robert W. Walker, *The Professional Development Framework: Generating Effectiveness in Canadian Forces Leadership*, CFLI Technical Report 2006-01 (Kingston: Canadian Forces Leadership Institute, 2006), 31–35.
4. Peter M. Senge, *The Fifth Discipline: The Art and Practice of the Learning Organization* (New York: Doubleday, 1990), 3–5, 14. This book is a seminal work on the learning organization and these

# SELF-DEVELOPMENT

references describe and define the learning organization. The roles and responsibilities of the individuals in the learning organization are examined throughout this publication but are covered in detail in Chapter 9.
5. L.W. Bentley, *Professional Ideology and the Profession of Arms in Canada* (Toronto: Canadian Institute of Strategic Studies, 2005), 5–17. The professional body of knowledge is examined in great detail in this book as a key component of professional ideology.
6. *Officership 2020*, 7; *NCM Corps 2020*, 10.
7. Canada, *Duty with Honour: The Profession of Arms in Canada* (Kingston: DND, 2003), 50–52.
8. Bentley, 17–50. The importance of the professional body of knowledge, as a key component within the professional ideology, is defined and examined, including its relationship to other professions within the military (e.g. doctors, lawyers).
9. Stephen Covey, *First Things First* (New York: Fireside, 1994). This popular book provides a number of approaches and ideas related to self-development within overall life prioritization. The five key components in this paragraph are adapted from this book.
10. Stephen Covey, *The 7 Habits of Highly Effective People* (New York: Fireside, 1990). This highly popular book provides a number of powerful lessons for personal growth. The summary in this paragraph is adapted from that book.

## SELECTED READINGS

Bentley, L.W. *Professional Ideology and the Profession of Arms in Canada*. Toronto: The Canadian Institute of Strategic Studies, 2005.

Canada. *Canadian Officership in the 21st Century (Officership 2020)*. Ottawa: DND, 2001.

Canada. *The Canadian Forces Non-Commissioned Member in the 21st Century (NCM Corps 2020)*. Ottawa: DND, 2003.

Canada. *Duty with Honour: The Profession of Arms in Canada.* Kingston: DND, 2003.

Canada. *Leadership in the Canadian Forces: Leading People.* Kingston: DND, 2007.

Canada. *Leadership in the Canadian Forces: Leading the Institution.* Kingston: DND, 2007.

Covey, Stephen R. *The 7 Habits of Highly Effective People.* New York: Fireside, 1990.

——. *First Things First.* New York: Fireside, 1994.

Senge, Peter M. *The Fifth Discipline: The Art and Practice of the Learning Organization.* New York: Doubleday, 1990.

Walker, Robert W. *The Professional Development Framework: Generating Effectiveness in Canadian Forces Leadership.* CFLI Technical Report 2006-01. Kingston: Canadian Forces Leadership Institute, 2006.

# 35
# STEREOTYPES
by Allister MacIntyre

In Chapter 1 in this handbook on Attitudes, stereotypes were acknowledged as being associated with the cognitive component of attitudes.[1] In most respects this is an accurate depiction, but it does not paint the entire picture. This chapter will more completely flesh out the stereotype concept and explain why leaders need to appreciate the manner in which stereotypes function. When we consider the increasingly diverse military climate within which we work as a subset of Canadian society, it is easy to understand why this topic will undoubtedly become progressively more important.

It is essential to keep in mind that no two people will hold the identical attitude about a target group, object, or place because life experiences are not necessarily identical.[2] Similarly, the cognitive component of our attitudes will vary from person to person depending upon our experiences or values. Nevertheless, there will still be some shared aspects, and these aspects will be discussed in greater detail during the exploration of stereotypes. Nevertheless, the question of where stereotypes fit into this equation remains to be answered.

The tripartite model of attitudes was presented in Chapter 1. Within this conceptual model, attitudes are comprised of affective, cognitive, and behavioural components. Although leaders can use their position to control behaviour, real attitude change must be initiated within the cognitive facet of an attitude. Affective changes will follow and behavioural intentions will become internalized. The cognitive component contains everything we know, or believe to be true, about an attitude object. This type of mental representation is known as a schema, and the aspects of our schema that tend to be shared by others within our group (i.e., there is a tendency for others to believe these things to be true as well) comprises the stereotype for a given out-group.

## **STEREOTYPE DEFINED**

Indeed, the word *stereotype* is widely used by lay people and researchers alike. A journalist, Walter Lippmann, introduced the term itself in 1922.[3] Lippmann was fascinated by the fact that different people could observe the same event in dramatically different ways. He adopted the printing industry's term "stereotype" (a metal plate for making duplicates) to capture the notion that our preconceptions influence how we perceive events. Although the term *schema* is much older, with roots that can be traced back to an ancient Greek word[4] for "form," "shape," or "figure," its use as an everyday word is not as common.

Because we are unable to know every other member of society on an individual basis, the attitudes formed toward others are often based on the stereotypes (e.g., social, ethnic, religious, racial, gender, or cultural) held for the group in which we categorize individuals. A stereotype is a set of widely shared broad generalizations about the characteristics of a group of people. We classify others on the basis of their membership in a group and automatically assume that they possess a number of traits that are associated with that group.[5] Stereotypes do not necessarily result in prejudice, or pejorative attitudes. In fact, they can reflect positive traits about people just as easily as negative ones. For example, as a result of commonly held stereotypes, it may be assumed that an Asian person will be more reserved, or a black person will be a good dancer. Neither of these stereotypes have negative connotations. Steven Penrod has even suggested that stereotypes are "essential for organizing the multiplicity of experiences we have from infancy through adulthood."[6] However, Penrod also acknowledges the negative side to stereotypes. He adds that, "… when our stereotypes are so rigidly maintained that no new information can modify them, they no longer contribute to our understanding of the world, but in fact insulate us from reality."[7]

Despite the fact that everyone seems to have their own appreciation of what a stereotype is, there is a continuing debate amongst theorists with respect to the nature and defining characteristics of stereotypes. This may be, at least in part, a consequence of the fact that stereotype research has been conducted in many diverse fields including clinical, experimental, and social psychology, as well as sociology.[8] For many years, stereotypes were viewed as being "bad" and they have even been treated from time to time as "bizarre or pathological phenomena."[9] In keeping with this image, stereotypes have been described as: unjustified generalizations made about an ethnic group, rigid impressions that conform very little to facts, exaggerated beliefs, inaccurate and irrational overgeneralizations, and clusters of preconceived notions.[10]

Current attitudes regarding stereotypes are less harsh. There is virtually total agreement that stereotypes are cognitive constructs, and generally widespread consensus that "an ethnic stereotype is a set of beliefs about the personal attributes of the members of a particular social category."[11] As stated more recently by social psychologists Don Taylor and Fathali Moghaddam, the current "… trend is to view stereotyping as a basic cognitive process that is neither desirable nor undesirable in and of itself."[12] We have even progressed to a point where it is accepted that stereotypes are universal constructs that are used by everyone as an aid in processing information about our social environment and stereotypes are both inevitable and usually quite functional for effective social interaction.[13]

Stereotypes could be considered bad if our adherence to them was so rigid that we were unable to differentiate between individual members of a social group. However, according to South African psychologist John Duckitt, there "is no evidence indicating that stereotyped beliefs are any more rigid, over generalized, or incorrect than any other widely held category-based generalizations."[14] Furthermore, social psychologists Jussim, McCauley, and Lee assert that although "people often perceive differences among groups, we are not aware of a single study identifying a single person who believed that all members of a social group had a particular stereotype attribute."[15] Although beyond the scope of this chapter, the numerous ways through which stereotyping can be viewed as a useful construct have been identified elsewhere.[16]

## CATEGORIZATION

The underlying process that is considered to be responsible for stereotyping is categorization. When we encounter someone new, we tend to search for salient features that may provide some clues with respect to how this person might be categorized. Some of the more obvious indicators are gender, skin colour, articles of clothing, and language. Once we succeed in categorizing someone, our stereotypes for whatever group(s) we have selected become activated. As stated by social psychologist Patricia Devine, the "the target's group membership activates, or primes, the stereotype in the perceiver's memory [which makes] other traits or attributes associated with the stereotype highly accessible for future processing."[17] In this manner, stereotypes offer faster access to meaningful associations and they facilitate subsequent inferences.[18]

In the Canadian Forces, the categorization process can be extremely rapid because of the uniforms worn by serving members. When CF members walk into a room filled with other CF members, they are able to quickly pigeonhole all of the people in the room. One broad category in use is officers versus non-commissioned members. Another almost equally broad categorization stems from Army, Navy, and Air Force membership. We can readily discern experience (using ribbons and medals, including a rough estimate of years of service) and occupation (using hat badges and other accoutrements). All of these categorizations carry with them an associated stereotype and, if we delve even deeper, we will discover that two male Army infantry officers can generate distinct stereotypes simply because they belong to two different regiments. These stereotypes, in turn, will influence how we behave in the presence of these colleagues, superiors, and subordinates.

Russell Fazio, an Ohio State professor who specializes in the study of attitudes, states that when one encounters "a target person who is categorizable in multiple ways, people do not necessarily "see" the same person. Yet, how the target is categorized will determine what stereotypes, attitudes, and expectations are activated and, ultimately, influence judgment and behavior."[19] This is true to a degree, we can certainly view the same person and have divergent categories come to mind. However, it is also possible that, even with the use of identical categories, and the activation of the same stereotype, our attitudes will differ. In fact, it is almost self-evident that the attitudes that we hold for a particular group will influence or shape how we interpret the consensual stereotype that we hold for the group. In other words, the knowledge and beliefs aspects of the stereotype may be consensual but, dependent upon our independently held related attitude, the final judgment of the attitude can differ. Consequently, how people within a given target group may be judged can range from stingy to frugal, arrogant to confident, or indecisive to prudent. The essence of this notion is captured in the following passage:

> Many of the terms in a stereotype seem to have the same denotation, but vary in connotation. Thus, the Japanese who were considered intelligent before World War II became sly during the war. The Jews are now intelligent, but were formerly thought to be shrewd. Changes in a group's stereotype seems to move along an evaluative dimension, while the denotative aspect remains similar. This shift along the evaluative dimension provides the possibility of predicting the content of a group's stereotype in the event of a shift in attitude.[20]

## **SUMMARY**

In summary, stereotypes and attitudes are closely linked concepts, with stereotypes being associated with the cognitive component of an attitude. While a stereotype is simply a mental representation that we have for a particular group of people, attitudes are more emotion-laden, are described as being either positive or negative, and are often the consequence of stereotypes. Even though stereotypes may be convenient, they tend to lack accuracy because, as generalizations about people, they are unscientific and hence unreliable.[21] According to psychologists James

Calhoun and Joan Acocella, stereotypes are basically false because they exaggerate "the differences between groups (for example, Irishmen are drunks; Jews are sober), and they take no account of the millions of individual differences within groups."[22] In any case, whether true or false, stereotypes do help to account for behaviour and should not be ignored.

Stereotyping exists because we have a drive for unity and consistency in our understanding of reality. We fill in the assumed content, and structure information around a central theme. Here, the central theme is the person's membership in a specific group. The stereotypes formed are largely the result of socialization, (such as children engaging in role-playing), and these stereotypes are reinforced by our contemporaries, by school and employment systems, and by the media. As a result of persistent stereotypes, we have a tendency to ascribe characteristics to someone simply because we learn that he/she is a member of a specific group (whether ethnic, religious, cultural, or otherwise). Although it may be pleasing to think otherwise, by definition, it would be extremely rare to find a fully functioning individual who does not engage in stereotypical thinking and behaviour. The only exceptions that come to mind are young children who are still lacking in their cognitive development, or others who are suffering from some mental impairment or disability. Furthermore, people can only have negative (or positive) attitudes toward a generic group as a result of stereotypes formed for that group.

Stereotypes are subject to change. However, as with attitudes, this is neither a common nor easily rendered occurrence. A key impairment to changing stereotypes is the fact that the stereotypes themselves act as "cognitive filters" through which "we select what information to use, what to ignore, and how to interpret it."[23] As a result of this filtering, disconfirming information about stereotypes is assimilated less easily than neutral or confirming information. Amir's contact hypothesis can be invoked as evidence for change but, once again, the circumstances required for a measurable change to take place are difficult to attain. Furthermore, Amir himself suggests that our theoretical understanding of what contact involves as a potential agent of change, and what the actual underlying processes are, is very limited.[24] Additionally, psychologists Myron Rothbart and Oliver John state that the effects of inter-group contact on stereotypic beliefs is dependent "upon (1) the potential susceptibility of those beliefs to disconfirming information and the degree to which the contact setting "allows" for disconfirming events, and (2) the degree to which disconfirming events are generalized from specific group members to the group as a whole."[25] In other words, members of an out-group who are observed behaving in other than stereotypical fashion may still be considered as exceptions to the rule. On the more negative side, as already presented, these exceptions may even be viewed more disfavourably because they do not adhere to the stereotype.[26]

Additionally, a common source for information upon which we can form stereotypes is the news media (as well as other media such as television and movies). News stories tend to resort to sensationalism, thereby leading to erroneous information and faulty generalizations. It is difficult to fault this argument. Newspaper editors and the directors of television news shows are in the business of attracting readers and viewers respectively. If a story can be made more exciting by appealing to stereotypical views, then the stereotypical presentation of out-groups will continue, and this in turn will continue to reinforce the perceptions already held by viewers.

The message for leaders is they must first acknowledge that stereotypes will not only exist, they will influence attitudes. These attitudes, in turn, will have an impact on behavioural intentions. Particular emphasis must be placed on dispelling negative stereotypes that are associated with specific target groups. This is critical as we work in increasingly diverse environments and carry out military missions throughout the far corners of the world. Changing stereotypes is not easy, but no task worth doing ever is. Leaders must strive to alter beliefs, instill new knowledge, and behave in a manner that is worthy of emulation.

## NOTES

1. The cognitive component of an attitude is actually referred to as a schema. It is when specific characteristics of an attitude object are

shared by other members of our reference group that stereotypes come into being.
2. S.P. Marshall, *Schemas in Problem Solving* (New York: Cambridge University Press, 1995).
3. D.L. Hamilton, S.J.Stroessner, and D.M. Driscoll, "Social Cognition and the Study of Stereotyping," in *Social Cognition: Impact on Social Psychology*, eds. P.G. Devine, D.L. Hamilton, and T.M. Ostrom (New York: Academic Press, 1994).
4. Marshall, *Schemas in Problem Solving*, 1–7.
5. J.F. Calhoun and J.R. Acocella, *Psychology of Adjustment and Human Relationships* (New York: Random House, 1978).
6. S. Penrod, *Social Psychology* (Englewood Cliffs, NJ: Prentice Hall, 1983), 394.
7. *Ibid.*, 394.
8. R.D. Ashmore and F.K. Del Boca, "Conceptual Approaches to Stereotypes and Stereotyping," in *Cognitive Processes in Stereotyping and Intergroup Behavior*, ed. D.L. Hamilton (Hillsdale, NJ: Erlbaum, 1981).
9. *Ibid.*, 10.
10. See D.M. Taylor and F.M. Moghaddam, *Theories of Intergroup Relations: International Social Psychological Perspectives*, 2nd ed. (New York: Praeger, 1994).
11. Ashmore and Del Boca, "Conceptual Approaches to Stereotypes and Stereotyping," 13.
12. Taylor and Moghaddam, *Theories of Intergroup Relations: International Social Psychological Perspectives*, 161.
13. J. Duckitt, *The Social Psychology of Prejudice* (Westport, CT: Praeger, 1992).
14. *Ibid.*, 16.
15. L.J. Jussim, C.R. McMauley, and Y. Lee, "Why Study Stereotype Accuracy and Inaccuracy?" in *Stereotype Accuracy: Toward Appreciating Group Differences*, eds. Y. Lee, L.J. Jussim, and C.R. McCauley (Washington, DC: American Psychological Association, 1995), 7.
16. For a comprehensive review, see S.T. Fiske, "Stereotyping, Prejudice,

and Discrimination," in *The Handbook of Social Psychology, Volume II*, 4th ed., eds. D.T. Gilbert, S.T. Fiske, and G. Lindzey (New York: McGraw-Hill, 1998).
17. P.G. Devine, "Stereotypes and Prejudice: Their Automatic and Controlled Components," *Journal of Personality and Social Psychology*, Vol. 56, No. 7 (1989): 5–18.
18. S.M. Anderson, R.L. Klatzky, and J. Murray, "Traits and Social Stereotypes: Efficiency Differences in Social Information Processing," *Journal of Personality and Social Psychology* 59 (1990): 192–201.
19. R.H. Fazio, "Further Evidence Regarding the Multiple Category Problem: The Roles of Attitude Accessibility and Hierarchical Control," in *Stereotype Activation: Advances in Social Cognition, Volume XI*, ed. R.S. Wyer, Jr. (Mahwah, NJ: Erlbaum, 1998), 107.
20. N.R. Cauthern, I.E. Robinson, and H.H. Kraus, "Stereotypes: A Review of the Literature 1926–1968," *Journal of Social Psychology* Vol. 84, No. 119 (1971): 103–125.
21. J.W. Vander Zanden, *Social Psychology* (New York: Random House, 1977).
22. Calhoun and Acocella, *Psychology of Adjustment and Human Relationships*, 240.
23. J.E. Alcock, D.D. Carment, and S.W. Sadava, *A Textbook of Social Psychology* (Toronto: Prentice Hall, 1988), 198.
24. M. Rothbart and O.P. John, Social Categorization and Behavioral Episodes: A Cognitive Analysis of the Effects of Intergroup Contact. *Journal of Social Issues* 41 (1985) 81–104.
25. *Ibid.*, 81.
26. See also, M.D. Storms, "Attitudes Toward Homosexuality and Femininity in Men," *Journal of Homosexuality* 3 (1978): 257–263.

**SELECTED READINGS**

Ashmore, R.D. and F.K. Del Boca. "Conceptual Approaches to Stereotypes and Stereotyping." In *Cognitive Processes in Stereotyping and Intergroup Behavior.* Edited by D.L. Hamilton. Hillsdale, NJ: Erlbaum, 1981.

Duckitt, J. *The Social Psychology of Prejudice.* Westport, CT: Praeger, 1992.

Fiske, S.T. "Stereotyping, Prejudice, and Discrimination." In *The Handbook of Social Psychology, Volume II,* 4th ed. Edited by D.T. Gilbert, S.T. Fiske, and G. Lindzey. New York: McGraw-Hill, 1998.

Hamilton, D.L., S.J. Stroessner, and D.M. Driscoll. "Social Cognition and the Study of Stereotyping." In *Social Cognition: Impact on Social Psychology.* Edited by P.G. Devine, D.L. Hamilton, and T.M. Ostrom. New York: Academic Press, 1994.

Jussim, L.J., C.R. McMauley, and Y. Lee. "Why Study Stereotype Accuracy and Inaccuracy?" In *Stereotype Accuracy: Toward Appreciating Group Differences.* Edited by Y. Lee, L.J. Jussim, and C.R. McCauley. Washington, DC: American Psychological Association, 1995.

Marshall, S.P. *Schemas in Problem Solving.* New York: Cambridge University Press, 1995.

Taylor, D.M. and F.M. Moghaddam. *Theories of Intergroup Relations: International Social Psychological Perspectives,* 2nd ed. New York: Praeger, 1994.

# 36
# STRESS AND COPING
by Allister MacIntyre and Colin Bridges

Life cannot exist without stress. At one time or another, everyone has suffered the consequence of too much to do and too little time; the anxieties associated with life's uncertainties; conflicting demands; the lack of control at work, school or home; never enough money, and the anguish of coping with a loved one's illness or death. But why do some people handle stress better than others? Why is one person able to calmly step in front of a crowd to speak publicly, while others cringe in horror at the mere thought? How is it possible for an airborne soldier to relish the opportunity to leap out of an airplane, yet shrink away from cruising in a submarine? An effective leader will understand the nature of stress, be aware of the consequences, and appreciate that coping mechanisms can be developed to help people deal with stress.

This chapter will enable leaders to recognize their own stress reactions and be aware of the symptoms of stress in their followers. The content will also provide examples of cognitive appraisal strategies and other stress-management techniques for the consideration of leaders. Leaders should continuously manage the levels of strain within subordinates by not being a source of work stress through their own behaviours, by buffering stress through the provision of social support, and by doing what they can to increase resistance to stressors while enhancing wellness.[1] The likelihood of reaching any of those three goals can be increased by actively improving leadership skills, providing stress management and stress inoculation training, and developing transformational leaders.[2]

Engineers can calculate with precision the strain created by stress distributions on different materials of various shapes, densities, and compositions.³ Their findings will be robust in terms of repeat examinations because the reactions obtained in inert matter will be consistent across trials. This is not the case for stress in humans. Stress reactions will vary tremendously and we cannot even predict with any certainty if a particular event or stimuli will be stressful for specific individuals. Since the days of French physiologist Claude Bernard,⁴ biologists have consistently observed variability (upon the introduction of controlled stimuli) in the adaptive changes of individuals to maintain their steady state. In other words, as biological creatures, our responses are anything but predictable. This characteristic will be addressed more completely, but it is important to understand at this point that it is due to the subjective nature of what constitutes a threat, or benefit, to each person that the same stressor can be *perceived* differently.

## WHY SHOULD LEADERS BE CONCERNED ABOUT WORK STRESS?

First and foremost, leaders should always be interested in the well-being of their subordinates. After all, leaders cannot exist without followers, and mental health can be every bit as devastating as physical health. Leaders who fail to let subordinates know that their physical and emotional well-being is important will have difficulty instilling any sense of loyalty and intrinsic motivation toward work. But, the costs to an organization can be overwhelming. Almost 20 years ago, Fletcher reported that 60 percent of the absences from work could be attributed to work stress and, in the United Kingdom, 100,000,000 work days were lost yearly because of occupational stress.⁵ Quillian-Wolever and Wolever cite California Department of Mental Health statistics illustrating that stress associated disorders could account for 60–90 percent of visits to medical facilities.⁶ Finally, Kelloway and Francis refer to research findings that demonstrate that stress-related business losses in the United States rose from $60 billion per year to a staggering $150 billion per year in just an eight-year period (1982–1990).⁷ This is, without a doubt, a leadership issue of monumental concern.

## **WHAT IS OCCUPATIONAL STRESS?**

A universally accepted definition of stress continues to be elusive. Over a quarter of a century ago, Hans Selye, one of the pioneers of stress research, indicated that even though almost four decades of laboratory work had been conducted on the physiological mechanisms of the adaptation to stress, "... few people have taken the trouble to find out what stress really is."[8] To some extent this remains unchanged today. For example, one recent approach states that, "Stress is a condition or feeling experienced when a person perceives that demands exceed the personal and social resources the individual is able to mobilize."[9] Implicit within this definition is the assumption that all stress is bad. While this meshes nicely with a layperson's usage of the term, it does not paint the complete picture. There is no doubt that such adverse experiences are negative; and this sort of stress is known as *distress*. However, stress can also be positive, it can be stimulating and, when levels are not intolerable, it can actually boost performance.

Experiences of intense joy (euphoria), for example, winning the Stanley Cup (if that remained an unattained life goal), can be stressful. Watching the birth of one's child certainly generates some stressful moments. Butterflies in the stomach before speaking to an audience provide a signal that one is experiencing stress, with the senses and awareness becoming enhanced. These positive responses to stress are known as *eustress*. Eustress also encompasses the joy of self-expression. Organizational leaders typically attain their positions of power and climb the corporate ladder through the experience of success within "stressful" promotion competitions. Having endured this sort of competitive process, these leaders should be aware that positive growth opportunities can be gained by matching tasks to the interest and skill levels of their subordinates or capabilities. They can also help to manage stress levels by ensuring the cohesiveness of sections.

Psychologist Stephan Motowidlo and his colleagues argued that stress should be viewed as, "an intervening variable with antecedent causes and behavioural consequences."[10] Psychologists Pratt and Barling expanded upon this notion and provided some clarity by differentiating between the concepts of stressor — stress — strain.[11] Stressors refer to the objective characteristics of the stimuli introduced into the environment. Stress is the subjective interpretation of those stressors. Strain is an individual's response resulting from the subjective experience of this stress (this can be along the physical, emotional, behavioural, and/or psychological dimensions). In simple terms, a stressor is anything (e.g., a person, place, thing, activity) that produces a stress response. The stress itself is this very subjective perception of the stressor. The strain is the consequence (e.g., physiological, affective, behavioural, psychological) of the stressor. The process is linear in that a stressor will lead to stress and the stress will generate strain.

With this sequential relationship in mind, it is easy to grasp some of the answers to the questions posed at the beginning of this chapter. The critical aspect of this process takes place at the subjective level of an individual's perception. If a person does not perceive jumping out of an airplane as a stressful event, then it is not. It really is as simple as that. In fact, if the airborne soldier finds this aspect thrilling, then the sensation is actually more in line with eustress rather than distress. Yet, despite the fact that the term eustress has been in existence for decades (attributed to Hans Selye), as recently as 2003, health psychologists Nelson and Simmons pointed out that it is still largely misunderstood as an aspect of stress.[12] They argued that eustress and distress could be viewed as two aspects of the stress response, one positive and one negative. From their perspective, a stressor can simultaneously generate positive and negative responses and people can only fully appreciate this response by taking into consideration both aspects. Nelson and Simmons liken eustress and distress to hot and cold water respectively. Pure cold water (distress) can be very uncomfortable, but the addition of some hot water (eustress) can make the experience bearable. Using a bathtub analogy, they reflect that, "Few individuals take a totally cold bath (distress) or a totally hot bath (eustress),"[13] and they refer to the increase in likelihood of an individual being displaced from their "comfort zone" during prolonged experience of stress toward either end of that continuum. This cold/hot approach has intuitive appeal because individuals can readily appreciate that even distressful events can have positive aspects. Speaking in front of an audience can be petrifying while simultaneously offering rewards in the form of boosts to esteem and confidence. As a more extreme example, consider the loss of a loved one who has been battling a terminal illness. No one can deny the anguish associated with the loss, but neither can one escape the sense of relief that comes with the realization that the pain and suffering has mercifully come to an end.

## **TYPES OF STRESS**

The differentiation between eustress and distress is just one way to categorize the stress experience. When differentiating in this manner, a person is classifying the stress reaction itself as being either positive or negative. But, it is equally vital that he or she also gain an appreciation for the nature of the stressors themselves. It has been said that "… stressors may vary along several dimensions: frequency of occurrence, intensity, duration, and predictability (time of onset)."[14] Stressors can also be categorized by specific types. The three types of stressors, those objective stimuli that are viewed as stress-provoking, are typically classified as daily stressors, acute stressors, or chronic stressors.[15] In the wake of events like the destruction of New York's World Trade Center, an additional category, referred to as catastrophic stressors, has been added to the list.[16]

Daily stressors (referred to in some studies as *hassles*) are the irritating, frustrating, distressing demands that sometimes characterize everyday transactions with one's environment. Being late for an appointment, missing a bus, or being cut off in traffic are examples of daily stressors. Acute stressors are normally severe (but dependant upon coping skills), are of short duration, have a more specific onset (it is easy to determine the precise commencement of the stressor), are frequently equated with life events, and their intensity decreases rapidly over time. This would include car accidents, an injury, or the death of a loved one. Chronic stressors are demanding on an ongoing and relatively unchanging basis, their onset is not clearly identifiable, and their intensity remains fairly constant over time. An oppressive work atmosphere, job insecurity, or ongoing family strife would be examples of chronic stress. Finally, catastrophic stressors have many characteristics in common with acute stressors but tend to be on a larger scale and may involve many people or even the whole world. Catastrophic stressors are not limited to terrorist acts. A catastrophic stressor may include natural disasters (e.g., Hurricane Katrina in New Orleans) and non-terrorist acts of violence against humanity (e.g., the 2007 massacre at Virginia Polytechnic Institute and State University ["Virginia Tech"]).

## **WHAT CAUSES DISTRESS AND STRAIN?**

A case has already been made for stressors to often have both positive and negative aspects. So why does distress take place and how does strain fit into the equation? When the negative aspects of a stressor overwhelm the positive aspects, distress will play the more prominent role in the process. Furthermore, the effects of stress are known to be cumulative, so each incident does not contribute to an effect in isolation. A series of hassles can bring a person to a breaking point, and someone who is already suffering from a form of chronic or acute stress will have far less tolerance for what others would have viewed as a minor incident. Thus, someone snaps and responds with hostility when even a casual criticism is offered with constructive intentions by a co-worker. In fact, stressors have been said to have a lingering effect, "… such that each episode leaves behind a residue that may add up across stressful exposures."[17]

The costs of stress and strain for organizations have already been noted. Distress and strain are the primary culprits and, on a more human scale, the impact on individuals who suffer distress can be traumatic on many levels. Although low levels of distress can actually enhance arousal and increase effectiveness, there is an optimum point for performance beyond which functioning will drop. This performance/arousal function is defined by the Yerkes-Dodson law and is visualized as a curve-shaped inverted "U" with the optimal plateau resulting when just the right range of arousal is experienced.[18] Leaders need to understand that low levels of stress can actually enhance performance, but they also need to be aware when their subordinates have crossed the boundary from performance facilitation to impairment.

## **SIGNS AND SYMPTOMS OF STRAIN**

Occupational stressors, in combination with the other stressors of life, will eventually lead to strain. Leaders need to be alert and watch for the signs and symptoms of strains as they become apparent in themselves and in their followers. But, what are some of the signs and symptoms? Researchers have shown that work stress can be reliably linked to outcomes such as mental and physical health, mood, depression, anxiety, job dissatisfaction, propensity to quit, spousal abuse, and burnout.[19] This should not be viewed as an exhaustive list, as the possible work stress outcomes are endless. Taking into consideration Hans Selye's General Adaptation Syndrome,[20] which outlines the reaction to extreme stress for both humans and animals, one is no further ahead in terms of being able to identify the occurrence of strain. This syndrome includes the fight-or-flight Alarm Reaction (whereby hormones are secreted to mobilize the body's defenses), the stage of resistance (the body adapts to the stressor), and the stage of exhaustion (since the body has a finite capacity to adapt). The real dilemma for leaders is knowing how to look for signs that these strains are taking place.

Stress outcomes, or strains, are usually clustered into three primary categories: psychological, physiological, and behavioural.[21] Psychological strains would include negative attitudes, feelings of worthlessness, moodiness, depression, nervousness, and forgetfulness. Physiological strains are characterized by hypertension, insomnia, stomach cramps, diarrhea or constipation, fatigue, headaches, neck or back pain, and general poor health. Finally, behavioural indicators of strain include aspects like withdrawal, active protesting, excessive smoking or drinking, absenteeism, argumentativeness, and exploiting people. Other theorists include categories like cognitive responses (e.g., attention span, judgment, decision-making) and emotional responses (e.g., anger, impatience, anxiety). Whatever classification system is adopted, it is easy to appreciate that many of the outcomes are not easy to observe directly. However, even psychological and physiological outcomes will normally be accompanied by a behavioural response of some nature.

Once leaders have become aware of their own distress reaction, they may be able to consider what specific stressor, cumulative stressors, or latent internal characteristics may have triggered those reactions. They can also anticipate similar reactions in their subordinates and be on the watch for the warning signs. Armed with this knowledge, leaders will be able to consider strategies that may be able to minimize the intensity of future strain when exposure to similar stressors is anticipated. By taking this proactive approach, they will be able to identify the common characteristics in distressing episodes, and recognize the strain in people across multiple incidents.

## MANAGING STRESS

Selye's General Adaptation Syndrome may be applicable to both animals and people, but the additional consideration of human thoughts and feelings has led to the cognitive-transactional theory of stress.[22] The cognitive-transactional theory contends that stress is an ongoing process of constantly assessing demands and calculating capabilities to resist. This leads to a reciprocal relationship that people have with their environment through which they continuously evaluate the sources and types of impacts that may be coming their way.[23] It has been argued that, "Coping resources available to an individual can halt or reverse the cascade at any point. Such resources include skills, talents, knowledge, and support networks, all of which should be included in an effective stress management program."[24] This means that, unlike animals, humans can take an active role in anticipating and coping with stress.

The management of stress can follow a three-tiered model of prevention.[25] Primary interventions in the workplace focus on eliminating or reducing the source of the stressors. Secondary interventions are directed at helping employees to identify the sources of stress and developing strategies to reduce the impact. Finally, tertiary interventions are designed to help employees who have not been able to cope effectively and are now suffering from strain.[26]

## **PRIMARY INTERVENTIONS**

Leaders need to pay particular attention to workplace factors that contribute to stress and strain. Some factors will be unavoidable, but whenever conditions can be altered for the better, effective leadership can play a critical role. Job conditions that may lead to distress include the design of tasks, management style, interpersonal relationships, work roles, career concerns, and environmental concerns. Leaders also need to be aware of work role factors like role conflict (carrying out one task conflicts with the completion of another), ambiguity (employees are unsure of their responsibilities or lack clear direction), and overload (too much to do in the time available that can be managed successfully). According to Nelson and Simmons, "An example of an organizational resource that may be important for generating both manageability and hope is information. Role ambiguity has been shown to have a strong negative impact on hope."[27] These authors also emphasize that supervisors who are open and supportive can help to increase satisfaction, reduce the role ambiguity, and create a foundation that will generate a feeling of hope.

Leaders need to create a healthy work environment. This is only possible with an ongoing focus on aspects like task characteristics and job design, distribution of workload, realistic deadlines and expectations, understanding the strengths and weaknesses of subordinates, instilling a sense of teamwork, clear direction, fair treatment of employees, showing appreciation, ergonomic changes where possible, and control over negative environmental conditions.

## **SECONDARY INTERVENTIONS**

An exploration of all the possible stress-management techniques that can be brought to bear is beyond the scope of this chapter. Furthermore, there is no one single technique that will be effective for all employees and all types of stress. But, leaders do need to be aware that a variety of stress-management methods are available and they must be supportive by encouraging subordinates to participate in programs that may meet their needs. Leaders and their subordinates might be able to pursue some of these stress-management techniques: use of imagery to cue tranquil setting, meditation, yoga, listening to the sounds of nature, progressive muscle relaxation, massage, tai chi, exercise, walking, and hobbies.[28] The likelihood of the successful use of any of their favourite techniques will be constrained by the setting and anticipated time pressures.

## **TERTIARY INTERVENTIONS**

By definition, tertiary interventions will only take place after strain has become overwhelming and negative consequences have taken place. Leaders need to take advantage of workplace programs (e.g., employee assistance programs, alcohol and drug treatment programs) to ensure that a valuable asset, the person suffering strain, is not lost to the organization. Leaders also need to encourage employees to seek help from professional sources and do everything possible to protect their subordinates from additional stressors. Finally, it is a leadership responsibility to create a welcoming atmosphere when the person returns to the workplace following treatment. The Canadian Forces has taken a lead role in this regard through their recognition of the impact of Post Traumatic Stress Disorder (PTSD) and the long-term support of those suffering from PTSD.

## **SUMMARY**

The consequences of work stress can be devastating for individuals and organizations. However, with effective leadership and workplace support, the consequences do not have to be dire. Leaders need to be aware that stress will exist, individuals will respond to stressors differently, and one size does not fit all when it comes to interventions and treatment.

## NOTES

1. Kelloway, E.K. and L. Francis, *Stress: Definitions, Interventions, and the Role of Leaders* (Kingston: Canadian Forces Leadership Institute, Research Report #CR02-0619, 2003).
2. *Ibid.*
3. B.H. Gottlieb, *Coping with Chronic Stress* (New York: Plenum Press, 1997), 49–52.
4. C. Bernard, *Leçons sur les phénomènes de la vie commune aux animaux et aux végètaux,* Vol 2. (Paris: Ballière, 1879).
5. B. Fletcher, "The Epidemiology of Occupational Stress," in *Causes, Coping and Consequences of Stress at Work,* eds. C.L. Cooper and R. Payne (New York: John Wiley & Sons, 1988).
6. R.E. Quillian-Wolever and M.E. Wolever, "Stress Management at Work," in *Handbook of Occupational Health Psychology,* eds. J.C. Quick and L.E. Tetrick (Washington, DC: American Psychological Association, 2003).
7. Kelloway and Francis, *Stress: Definitions, Interventions, and the Role of Leaders.*
8. H. Selye, *Stress Without Distress* (Toronto: McClelland and Stewart, 1974), 25.
9. R. Schwarzer and S. Taubert, "Tenacious Goal Pursuits and Striving Toward Personal Growth: Proactive Coping," in *Beyond Coping: Meeting Goals, Visions and Challenges,* ed. E. Frydenburg (London: Oxford University Press, 2002), 21.
10. S.J. Motowidlo, J.S. Packard, and M.R. Manning, "Occupational Stress: Its Causes and Consequences for Job Performance," *Journal of Applied Psychology* 71 (1986): 618–629.
11. L. Pratt and J. Barling, "Differentiating Between Daily Events, Acute, and Chronic Stressors: A Framework and its Implications," in *Occupational Stress: Issues and Developments in Research,* eds. J.J. Hurrell, L.R. Murphy, S.L. Sauter, and C.L. Cooper (London: Taylor and Francis, 1988).
12. D.L. Nelson and B.L. Simmons, "Health Psychology and Work

Stress: A More Positive Approach," in *Handbook of Occupational Health Psychology*, eds. J.C. Quick, and L.E. Tetrick (Baltimore, MD: United Book Press, 2003).
13. *Ibid.*, 101.
14. Kelloway and Francis, 2.
15. Pratt and Barling.
16. Kelloway and Francis.
17. J.E. Singer and L.M. Davidson, "Specificity and Stress Research," in *Stress and Coping: An Anthology*, 3rd ed., eds. A. Monat and R.S. Lazarus (New York: Columbia University Press, 1991), 38.
18. Quillian-Wolever and Wolever.
19. A.T. MacIntyre, "Some Effects of Daily Work Role Stress on Same Day and Next Day Emotional Exhaustion" (unpublished master's thesis, Queen's University, 1989).
20. H. Selye, *The Stress of Life* (New York: McGraw Hill, 1956), 38.
21. T.A. Beehr, *Psychological Stress in the Workplace* (New York: Routledge, 1995).
22. R. Schwarzer and S.T. Steffen, "Tenacious Goal Pursuits and Striving Toward Personal Growth: Proactive Coping," in *Beyond Coping: Meeting Goals, Visions and Challenges*, ed. E. Frydenburg (London: Oxford University Press, 2002), 20.
23. S. Folkman, R.S. Lazarus, C. Dunkel-Schetter, A. DeLongis, and R.J. Gruen, "Dynamics of a Stressful Encounter: Cognitive Appraisal, Coping, and Encounter Outcomes," *Journal of Personality and Social Psychology*, Vol. 50, No. 5 (1986): 992–1003.
24. Quillian-Wolever and Wolever, 361.
25. C.G. Hepburn, C.A. Loughlin, and J. Barling, "Coping with Chronic Work Stress," in *Coping With Chronic Stress*, ed. B.H. Gottlieb (New York: Plenum Press, 1997).
26. *Ibid.*
27. Nelson and Simmons, 114.
28. Quick, J.C., Tetrick, *Handbook of Occupational Health Psychology* (Baltimore: United Book Press, 2003), 365–371.

**SELECTED READINGS**

Beehr, T.A. *Psychological Stress in the Workplace.* New York: Routledge, 1995.

Cooper, C.L. and R. Payne, eds. *Causes, Coping and Consequences of Stress at Work.* New York: John Wiley & Sons, 1988.

Gottlieb, B.H. *Coping with Chronic Stress.* New York: Plenum Press, 1997.

Heart and Stroke Foundation of Canada/Canadian Mental Health Association. *Coping With Stress.* Toronto: Risk Factor Series, 1997.

Monat, A., and R.S. Lazarus, eds. *Stress and Coping: An Anthology,* 3rd ed. New York: Columbia University Press, 1991.

Quick, J.C., and L.E. Tetrick. *Handbook of Occupational Health Psychology.* Baltimore: United Book Press, 2003.

# 37

# TEAMS

by Robert W. Walker, Allister MacIntyre, and Bill Bentley

Leadership does not exist in a vacuum. By definition, leaders influence others, usually a number of followers, but also peers or superiors, all of these people, in various circumstances. Accordingly, leaders are destined to carry out their leadership activities and influence through others, predominantly, presumably, within a group or team context. To be effective, therefore, leaders must possess a good appreciation of the team, group or unit characteristics, its group dynamics, its structures and functions. ("Team" will be the collective noun or adjective, i.e., teams or team leadership, used in this chapter.)

But, what is a team? What does a team represent to a leader? Can a leader create the "right" team? How does a leader ensure that the team's individuals collectively can be a team with strong influence? What member characteristics are most meaningful, and most relevant, to the success of team objectives?

To a leader, a team may represent a cluster of people with the potential to serve as a set of intertwined conduits through which leader influence can be exercised in order for goals and missions to be achieved. In his review of leadership in organizations, renowned leadership researcher Gary Yukl concluded that leader behaviours can influence team effectiveness but, most likely, he hypothesized, in conjunction with the leader's impact on several intervening variables or through those intertwined conduits.[1] For example, leaders could influence the efforts of subordinates by clarifying their roles and developing their abilities, organizing the structure of work, encouraging cooperation and teamwork, providing necessary resources, and handling any required external negotiations. There are perspectives other than Yukl's, of course, as reflected in innumerable publications on leadership, many of them, regrettably, only indirectly referring to team leadership. As this 21st century commences, it is important and timely to determine what is known and not known about team leadership, particularly that which is relevant to military teams?

## **DEFINING TEAM LEADERSHIP**

The questions, the definition, and the complexities of the phenomenon of team leadership reflect the challenges and the major issues facing any team leader. As a first step, a "team" definition is needed. An appropriate working definition, provided by distinguished behavioural scientist and prolific author Eduardo Salas and others in their 1992 book on team performance, is:

> A distinguishable set of two or more people who interact, dynamically, interdependently, and adaptively toward a common and valued goal/objective/mission, who have been assigned specific roles or functions to perform and who have a limited life-span of membership.[2]

This definition emphasizes that team members typically possess unique responsibilities and can make critical contributions to the collective in a milieu of high interdependence. The leader's attention must be focused upon the structure and strength of the relationship of the collection of individuals who, potentially, will have great impact on the success of the goals, objectives, and missions through supporting and magnifying the influence of the leader. These members exert influence in all directions on other team members, i.e., the leader is not the only team member exerting an influence. Coordination and synchronization among members' contributions are required to achieve team goals. Identification of individual member strengths, full information exchange, and constant monitoring are needed to successfully integrate team responses and outcomes.

Usually, in studies of human influence phenomena, a useful definition is but the first step, and a substantial literature, in this case on team leadership, usually would be the next reference point. That would serve as a foundation for future study and action. However, this sequential progress is undermined by the fact that much of the previous leadership research and many of the leadership theories addressing influence are not directly applicable to team leadership. As leadership guru Stephen J. Zaccaro and colleagues noted as recently as 2001, "Despite the ubiquity of leadership influences on organizational team performance and the large literatures on leadership and team/group dynamics, we know surprisingly little about how leaders create and handle effective teams."[3]

Zaccaro et al. observed that, despite this substantial leadership research and a separate broad literature on group/team dynamics, previous studies tended to focus on leader influence on collections of subordinates *individually*, but without attending to how leadership fosters the integration of subordinates' actions, i.e., how leaders promote team processes necessary for successful outcomes.[4] These earlier studies and theories tended to focus on dyadic exchanges, on influential relationships of two persons, one a leader, and one usually a follower, although sometimes a peer or superior. For these earlier studies, leadership influences were assessed against individuals' performance expectations and subordinates' individual outcomes, not against leadership effects on collective team performance. However, within a team, there exist numerous dyads, triads, and multiple-member relationships fuelled by the shared or distributed leadership among team members with responsibilities that generate multiple relationships of varying influence.

This intertwined-conduits factor, in itself, would discourage theorists and researchers from choosing to do team leadership research. The circumstance, therefore, for fledgling or emergent leaders, military or otherwise, is one of major concern and hesitation, for these leaders are left to figure out and understand their teams extemporaneously, unfortunately bereft of directly applicable research, reference points, and sound knowledge.

## IDENTIFYING TEAM LEADER FUNCTIONS AND PROCESSES

Fortunately, some research initiatives have existed and do exist. Longtime leadership researcher E.A. Fleishman and his colleagues, including Stephen Zaccaro, approached leadership in teams from the perspective of "functional social problem-solving" performed through four requisite leadership functions.[5] Their work was widely accepted as it reflected those leader responsibilities to team members necessary for enhancing the probabilities of success. They were to:

* search out and structure the requisite information important to team members;
* use, coordinate, and communicate that information to team members for problem-solving in the service of team goal attainment;
* acquire, develop, and motivate team members; and
* acquire and utilize material resources in support of and necessary for team actions.

Importantly, these requisite leader functions then needed to be partnered with a number of leader-directed team processes in order to contribute to team effectiveness:

* cognitive processes like information sharing and processing, and participative debriefs;
* motivational processes integrated with team cohesion and a sense of collective efficacy;
* affective (or team emotional "tone") processes like social imitation and interpersonal liking; and
* coordination processes involving orientation functions, matching capabilities to roles, and team timing exercises.

A second, more recent review by university professor C. Shawn Burke and colleagues, an article that is part of a 2006 special issue of the journal *The Leadership Quarterly*, was dedicated to team leadership research. Burke et al. put forward team leadership as poised for major advances both in terms of science (design, measurement, methods) and practice (multi-level issues, cross-level effects, internal/external factors).[6] This article identified the lack of integration of leader behaviours and team performance outcomes. Their major analysis of dozens of research initiatives resulted in a framework being created that depicted some strong relationships among certain team leaders' task-oriented and/or people-oriented behaviours, and the team outcomes and team successes. Potential team leaders' mechanisms (consideration, empowerment, transformational behaviours) could facilitate team learning for broader team development. Importantly, the authors recommended that the focus in future become an examination of team leadership theories and not theories of leadership applied to teams.

Such an expansive analysis of leader behaviours and team outcomes highlight two points, first, the importance to leaders of understanding team dynamics, as well as the individual team member characteristics that lead to teams working together successfully to achieve common goals and, second, the dearth of straightforward findings of benefit and use to team leaders attempting to enhance team performance and effectiveness.

To accomplish such team tasks, to succeed in the mission, team leaders must ensure that a number of team members, to work together successfully, are able to:

* be engaged fully in distributed-leadership circumstances;
* make decisions while maintaining motivation;
* coordinate their activities independently and collectively;
* function together, as well as function together with team leaders;
* maintain their primary focus, team results; and
* see leadership as part of the team structure, not a separate entity.

By engaging in these behaviours that affect other team members, the potential for team members to accomplish team objectives will be positively influenced. And, when the focus is on military leadership, frequently exercised in dangerous, even life-threatening, circumstances, such team dynamics are of unimaginable importance.

## **LEADERS IN EFFECTIVE MILITARY TEAMS**

The recent publication *Leadership in the Canadian Forces: Doctrine*, states that generic leadership (not effective military leadership) can be defined as "… directly or indirectly influencing others, by means of formal authority or personal attributes, to act in accordance with one's intent or a shared purpose."[7] This definition is intended to be value-neutral (making no allusions to good or bad leadership, good or bad results) and is not inconsistent with other generic leadership views.

However, when value-neutral is replaced by value-laden leadership, when results of leadership are gauged for their effectiveness or degrees of success, such a revised definition for leadership becomes even more important to reflect the interests of a military institution. The 2005 doctrinal manuals, *Leadership in the Canadian Forces: Doctrine* and *Leadership in the Canadian Forces: Conceptual Foundations*, followed by the two 2007 applied publications, *Leadership in the Canadian Forces: Leading People* and *Leadership in the Canadian Forces: Leading the Institution*, specifically address the subject of military team leadership.[8] The applicable definition for the CF is: "Effective Canadian Forces leadership is directing, motivating and enabling others to accomplish the mission professionally and ethically, while developing or improving capabilities that contribute to mission success."[9]

Unfortunately, even the military leadership literature of the latter half of the 20th century tended to focus on research of immediate (i.e., dyadic, one-on-one, leader-follower) leadership, with little regard given to the different levels of leadership within a team, nor the multiple sets of relationships within teams. Additionally, most teams, military and otherwise, have some sort of leadership hierarchy, which adds another dimension to the multiple sets of team member relationships. A sports team, for example, might extend from team owner to team manager, to coach, to assistant coach, to team captain, to players, and perhaps even to support staff like equipment managers and "water boys." Different levels of leadership will co-exist in a group, and the behaviours at the different levels will have different precursors and different consequences. This is yet one more variable for the attention of team leaders in military settings.

It has become almost a tenet of military doctrine that teamwork is key to the success of any mission. As articulated in *Leading People*, a military team can be described as a group whose members share a common sense of purpose, are committed to the success of the team over their own individual successes, and share a high mutual trust and accountability.[10] Teamwork enables a group to respond rapidly and more easily to the complex, dynamic, and life-threatening situations that are becoming the norm in the modern military world. Further, being part of a team generally results in higher motivation for the more educated CF members of today, who are thereby offered the opportunity to have a greater influence on the organization to which they belong, an organization that is addressing the new global security environment.

The following attributes characterize an effective team, and a central element of military team building is the enhancement of these attributes within the team:[11]

- understanding of, mutual agreement on, and identification with, the primary task;
- open communication among all team members;
- appropriate and complementary member skills;
- mutual support and well-developed group skills (so that any differences can be managed effectively), thus creating a comfortable atmosphere about the team;
- an agreed means of decision-making that allows all team members to support the decision ultimately reached;

* appropriate exercise and distribution of leadership; and
* mutual trust among all team members.

## **TEAM LEADER ATTRIBUTES**

The credibility attributed to and the impact "allowed from" a team leader, functioning in a climate of trust between the leader and those led, are positively related to such qualities as conscientiousness, fair play, and cooperation. Whether credibility is founded mainly on a leader's demonstrated competence, a leader's obvious care and consideration for others, or the perceptions of a leader's character (integrity, dependability, consistency, loyalty, openness, and fairness), the evidence supporting this common understanding is compelling and robust. It follows that an important part of the leader's job is to build and maintain credibility such that healthy trust relationships with followers, peers, and superiors can result, and leader impact will occur.

*Trust*

Trust is the foundation upon which teams are built. The greater the bonds of trust the stronger will be the team's esprit de corps and effectiveness. Effective military teams depend on this atmosphere of mutual trust — trust in peers and subordinates and, above all, trust in the leader. This trust is positively related to individual and group performance, to persistence in the face of adversity, to the ability to withstand stress, to job satisfaction, and to commitment to continued service. Leaders build trust by:[12]

* demonstrating high levels of proficiency and professional competence in the performance of core functions, and taking advantage of opportunities to enhance their professional expertise and competence;

* exercising good judgment in decisions that affect others, and not exposing people to unnecessary risks;
* showing trust and confidence in their team members by giving them additional authority and involving them in decisions where circumstances allow;
* demonstrating concern for the well-being of their team members, representing their interests, and ensuring they are supported and taken care of by the organization;
* showing consideration and respect for others, being polite, friendly and approachable, treating members fairly (without favour or discrimination) and being consistent;
* being professional in bearing and conduct;
* maintaining high standards and honest and open communications;
* leading by example, sharing risks and hardships, and refusing to accept or take special privileges; walking the talk; and
* keeping their word and being counted upon to honour their obligations.

Building trust is an arduous, time-consuming process. Once gained, it must be jealously guarded, for it is easy to lose and difficult to regain. Following are some guidelines for leaders to sustain the trust of their followers:[13]

* Demonstrate that you are not self-serving. If team members see the leader using his or her position, using the members themselves as tools, or using the organization to meet the leader's personal goals or career ambitions, the leader's credibility will be undermined. Careerism is easily recognized and is not acceptable.

* Be a team player. Leaders support their team when they are under pressure. This does not mean that leaders do not acknowledge deficiencies and shortcomings. However, trusted leaders stand up for their people in an appropriate manner and in accordance with the military ethos.
* Practise openness. Keep people informed. Mistrust comes as much from what members do know as from what they do not know. Explain decisions and disclose all relevant information.
* Be fair. Before making decisions, examine how others will perceive things. Give credit where it is due, and be objective and impartial in evaluations.
* Be transparent. Share your thoughts with your team. Ensure the members are fully informed of all matters that concern them. A cold and unfeeling leader does not generate trust. One can be strong and human at the same time.
* Show consistency in the basic values that guide your decision-making. Mistrust comes from not knowing what to expect. Let your values and beliefs consistently guide your decisions and actions.
* Maintain confidences. People will inherently provide details and confidences about their personal lives. There is no surer way to lose an individual's trust than to betray this confidence.

*Cohesiveness and Teamwork*

Trust, however, is only one element of a strong, effective team; cohesiveness and teamwork are also needed. Effective leaders understand that, although trust fuels cohesiveness and teamwork, leaders must also generate these traits in their team members. Leaders can do this by applying the guidelines below:[14]

* A clear purpose. Make sure that the team has clearly articulated joint goals.
* Participation. Ensure that team members share information, ideas, and knowledge and contribute to task completion.
* Civilized disagreement. Controlled disagreement (for example, a difference of opinions) is a natural occurrence and can have a positive impact on team dynamics. It can also have a very negative impact if it is not properly managed. Effective teams manage the disagreement and resolve the conflict. In the dialogue designed to resolve disagreement the leader allows everyone to have his or her say.
* Open communications. Promote members' communicating openly and with trust.
* Active listening. Ensure that team members listen to each other, both to understand information and to help resolve interpersonal conflict.
* Stifling rumours. Rumours create confusion and can undermine the mission. They inevitably have bad effects on morale. Rumours are dealt with by providing maximum, honest information in an atmosphere that encourages discussion and questions. Effective leaders maximize transparency in their actions and decisions.
* Climate. Set a comfortable atmosphere in which to work, an atmosphere that is free of derision and where each member feels valued.
* A say in decision-making. Allow members to express disagreement and to voice opinions in a respectful manner to superiors and the team. There needs to be a clear understanding that although solutions and ideas put forward by team members may not be

adopted, nonetheless, their contributions are both valued and welcome.
* Sharing experiences over time. Collective experiences allow team members to learn each other's strengths and weaknesses and to maximize their combined efforts.
* Sharing hardships over time. Members learn to rely on comrades, trust their judgment, and bond with them.
* Delegating leadership. Distribute and delegate responsibility whenever circumstances permit in order to achieve greater buy-in and commitment commensurate with skills and experience in the team.
* Embracing diversity. Identify and integrate the unique contribution of each member.
* Understanding and following policies and procedures.
* Keeping superiors informed.
* Being a good team member and follower, as appropriate.

*Commitment and Support*

Finally, military leaders are reminded that the accomplishment of mission success is only possible when leaders have team members who are committed, engaged, and supportive. Leaders and team members are equally important to the achievement of the team's vision and established tasks. While they remain true to the military ethos, the role of team members is to assist leaders in becoming more effective. Team members make things happen, and in the absence of the leader, exemplary members can step in as the situation demands. By doing so, they are able to enhance the continuing development of their own leadership skills, knowledge, and attitudes.

Team members can be relied upon to complete routine and assigned tasks, maintain co-operative working relationships, share in various leadership functions, and support the development of present and future leaders. Effective members support the team leader to fully realize his or her strengths and to compensate for his or her weaknesses. When necessary, followers should provide constructive criticism to assist the leader. Competent team members are not afraid to enter into open, honest, and frank (but respectful) discussions with their leader. Team members will be most effective when they are committed and supportive thusly:[15]

* Know what is expected. Team members must know their duties and responsibilities, their performance and behaviour standards, and their range of authority.
* Establish and maintain contact with the leader. This will minimize inconsistent or unclear messages.
* Take initiative and keep the leader informed. Team members take the initiative to solve problems that block the achievement of mission success.
* Provide accurate information and feedback. Members provide their leader with clear, accurate, and timely information that may affect the quality of his or her decisions, whether it is good or bad news.
* Support change. Inherently, most change initiatives are met with resistance. Team members must provide support and encouragement to the leader.
* Support the team. The member must be an effective team player.
* Provide alternative ideas and options. Assist the leader in identifying weaknesses or deficiencies in a plan or decision. Provide timely ideas and options that assist the leader in his or her planning. Never make this a personal attack or critique of the leader's skills or ability.

* Support decisions. After providing input, and once a decision by the leader has been made, fully and loyally implement the decision. Do not undermine the decision.
* Demonstrate appropriate recognition. Acknowledge the contributions of team members and leaders.
* Challenge orders or direction when necessary. There are only two instances when a team member challenges orders: first, when it is a manifestly unlawful command; second, when the member believes the order is clearly unethical. In the latter case, team members must understand that they will be held fully accountable for their decisions.

*Shaping Change*

Another challenge for team leaders, whether at junior or senior institutional levels of leadership, whether military or non-military, is to initiate, shape, and oversee change in the team or an organization surrounding a team. The challenge is to select the right members, hopefully with some of them from within the team, some possibly as subject matter experts external to the team, i.e., to compose the "right" membership in accordance with the purpose of change. Harvard Business School's John Kotter, author of *Leading Change*, has identified four key characteristics seen to be essential to effectively shape and change a team or coalition:[16]

* Position Power — through sufficient key players, main line leaders, to ensure progress, but to be seen as representative of the larger team;
* Expertise — for provision of various points of view relevant to the change at hand;
* Credibility — with enough people with strong reputations, who are trusted, responsible for firm

pronouncements that will be accepted; and
* Leadership — by inclusion of proven leaders familiar with the process of implementing change.

## SUMMARY

This chapter constitutes an introduction to team leadership, effective military teams, team leader functions and processes, and effective team leader attributes. The aspects that have had the greatest impact on understanding team leadership have been addressed: how well team leaders are aware of the team dynamics supportive of mission success; the attributes and strengths of and among the team members; and the interactions that take place among the leader, the team members, and the task or mission requirements.

Leadership abilities in support of team dynamics can be professionally developed, enhanced, and nurtured, as espoused in CF doctrine and manuals. Key elements for team leader effectiveness have been presented in this chapter and, as the CF manual *Conceptual Foundations* advocates, the integrative leadership model for the CF incorporates team leadership as an essential aspect.[17] This CF model was heavily based on Gary Yukl's Multiple-Linkage Model, as Yukl's is one of the few models with a systems approach and demonstrated linkages among various sub-groups of variables.[18] The four major sub-groups of variables are: leader characteristics and behaviour; individual and team variables; situational variables; and outcomes. This comprehensive model incorporates, among other factors — personal and positional power influences, leader competence and attributes, leadership influences, situational constraints and environmental aspects, and task features — the very important factor of team dynamics. The main entity still lacking in all of this research, applications, and model-building is a broad and substantial research base required to more fully understand team leadership!

## NOTES

1. G. Yukl, *Leadership in Organizations,* 5th ed. (Englewood Cliffs, NJ: Prentice Hall, 2002), 12–13.
2. E. Salas, T.L. Dickinson, S. Converse, and S.I. Tannenbaum, "Towards an Understanding of Team Performance and Training," in *Teams: Their Training and Performance,* eds. R.W. Swezey and E. Salas (Norwood, NJ: Ablex Publishing, 1992), 3–29.
3. Stephen J. Zaccaro, Andrea L Rittman, and Michelle A Marks, "Team Leadership," *The Leadership Quarterly,* Vol. 12, No. 4 (2001): 451–483.
4. *Ibid.,* 452–453.
5. E.A. Fleishman, M.D. Mumford, S.J. Zaccaro, K.Y. Levin, A.L. Korotkin, and M.B. Hein, "Taxonomic Efforts in the Description of Leader Behavior: A Synthesis and Functional Interpretation," *Leadership Quarterly,* Vol. 2, No. 4 (1991): 245–287.
6. C. Shawn Burke, Kevin C. Stagl, Cameron Klein, Gerald F. Goodwin, Eduardo Salas, and Stanley M. Halpin, "What Types of Leadership Behaviors are Functional in Teams? A Meta-Analysis," *The Leadership Quarterly* 17 (2006): 288–307.
7. Canada. *Leadership in the Canadian Forces: Doctrine* (Kingston: DND, 2005), 3.
8. *Doctrine,* 2005; Canada. *Leadership in the Canadian Forces: Conceptual Foundations* (Kingston: DND, 2005); Canada. *Leadership in the Canadian Forces: Leading People* (Kingston: DND, 2007); Canada. *Leadership in the Canadian Forces: Leading the Institution* (Kingston: DND, 2007).
9. *Conceptual Foundations,* 2005, 30.
10. *Leading People,* 2007, 76.
11. *Ibid.,* 68.
12. *Ibid.,* 70–71.
13. *Ibid.,* 71–72.
14. *Ibid.,* 74–76.
15. *Ibid.,* 76–77.
16. John Kotter, Leading Change (Boston: Harvard Business School Press, 1996), 57–58.

17. *Conceptual Foundations*, 121.
18. *Ibid.*, 120.

**SELECTED READINGS**

Burke, C. Shawn, Kevin C. Stagl, Cameron Klein, Gerald F. Goodwin, Eduardo Salas, and Stanley M. Halpin. "What Types of Leadership Behaviors are Functional in Teams? A Meta-Analysis." *The Leadership Quarterly* 17 (2006): 288–307. (This is one of seven articles in this special journal edition that addresses Team Leadership.)

Canada. *Leadership in the Canadian Forces: Leading People* (Kingston: DND, 2007).

Hughes, Richard L. and Katherine C. Beatty. *Becoming a Strategic Leader*. San Francisco: Jossey-Bass and the Center for Creative Leadership, 2005.

Katzenbach, J.R. *Teams at the Top: Unleashing the Potential of Both Teams and Individual Leaders*. Boston: Harvard Business School Press, 1998.

Kotter, John P. *Leading Change*. Boston: Harvard Business School Press, 1996.

Nadler, D. A. "Leading Executive Teams." In *Executive Teams*. Edited by D. A. Nadler, J. L. Spencer and Associates. San Francisco: Jossey-Bass, 1998.

Zaccaro, Stephen J., Andrea L. Rittman, and Michelle A. Marks. "Team Leadership." *The Leadership Quarterly*, Vol. 12, No. 4 (2001): 451–483.

# 38
# THEORIES
by Emily Spencer

Effective leadership may be dependent on the leader, the follower, the situation, or any combination of these factors. This chapter has been organized according to the main leadership aspects addressed in each model or theory. The purpose of this chapter is to familiarize the reader with the basic concepts underlying a number of models and theories. It does need to be emphasized that this chapter's content is representative of the mainstream leadership research conducted through a number of decades, but it does not, cannot, represent the full scope and magnitude of all research, theories, etc.

## TRAIT THEORY

Pioneer leadership researchers were confident that personality traits essential for leadership effectiveness could be identified through empirical research. Physical characteristics, aspects of personality and aptitudes, were areas that were most often studied during early research on leadership traits.[1] Although different researchers identified a variety of leadership traits and characteristics, it is generally thought that there are five major leadership traits: intelligence, self-confidence, determination, integrity, and sociability.[2] Possession of these leadership traits was believed to be the essential component for exhibiting leadership behaviour.

## COGNITIVE RESOURCES THEORY

The cognitive resources theory is a situational model that deals with the cognitive abilities of leaders. This theory explores the conditions under which cognitive resources, such as intelligence and experience, relate to leadership effectiveness. It is hypothesized that group performance is a construct of a complex interaction between two leader traits, intelligence and experience, one type of leader behaviour, directive leadership, and two aspects of the leadership situation, interpersonal stress, and the nature of the task.[3]

The first proposition of cognitive resources theory is that leader ability contributes to group performance when the leader is directive and followers require guidance. The second proposition is that perceived stress influences the relation between intelligence and decision quality. The third, and final, proposition is that perceived stress moderates the relation between leader experience and performance.[4]

## LEADERSHIP SKILLS MODEL

This model is similar to the trait theory, but instead of focusing on leader traits, skills possessed by the leader are considered to be the most relevant component for effective leadership. Like the trait theory, the leadership skills model is concerned with leadership behaviour. The skills model, however, recognizes that leadership behaviour cannot be removed from its social context.[5] Additionally, the skills model suggests that leadership potential is developed through experience.[6]

The leadership skills model emphasizes that leadership is a social phenomenon.[7] Knowledge is the central leadership skill that is referred to in the model.[8] Knowledge in social perceptiveness[9] is considered essential for effective leadership. The leader is expected to use his/her knowledge to facilitate problem-solving even, and perhaps especially, when faced with obstacles such as a lack of resources or time constraints.

## TRANSACTIONAL LEADERSHIP

Transactional leadership is based on an exchange between leaders and followers. It is effective because it is in the best interest of followers to do what the leader wants.[10] There are four types of behaviours that are associated with transactional leadership: contingent reward, active management by exception, passive management by exception, and laissez-faire leadership.

Contingent reward behaviour includes the clarification of what is expected of followers in order to receive rewards.[11] Rewards, such as money and time off, are used as incentives to motivate followers to perform. Management by exception refers to leadership that utilizes corrective criticism, negative feedback, and negative reinforcement.[12] It can either be active or passive. A leader employing the active form of management by exception is always on the lookout for problems and takes corrective actions immediately following a minor mistake or rule violation by a follower. A leader using the passive form does not monitor followers as closely, and only reacts to problems once they have occurred.[13] Laissez-faire leadership is descriptive of a leader who acts indifferently to followers and who is not concerned with the mission. This type of leader abdicates all leadership roles and responsibilities. Laissez-faire is often considered a non-leadership factor.[14]

## PSYCHODYNAMIC APPROACH TO LEADERSHIP

The psychodynamic approach to leadership is based on the premise that an individual's first experience with leadership begins on the day that he/she is born. This happens through exposure to parents as leaders. Developed by psychoanalyst Sigmund Freud and his disciple Carl Jung, the theory puts forth a variety of concepts that surround leadership in a family setting.[15]

## PARTICIPATIVE LEADERSHIP

Participative leadership is interactive and allows followers some influence over the leader's decisions. It may occur in many ways. The following four decision procedures are generally regarded as distinct and meaningful and can be ordered along a continuum beginning with non-participative autocratic decisions, and raging to the highly participative action of delegation.

Autocratic decisions are when the leader makes a decision alone. Consultation is when the leader asks followers for their opinions and ideas, and then makes the decision alone after seriously considering their suggestions and concerns. Joint decisions are when the leader meets with followers to discuss the problem and to formulate a decision; the leader has no more influence over the final decision that any other participant. Finally, delegation is when the leader gives authority to an individual or group to make a decision.[16]

## **LEADER MEMBER EXCHANGE THEORY**

Leader-member exchange (LMX) theory describes the role-making process between a leader and an individual follower. It describes how leaders develop specific relationships over time with different followers. The basic premise behind the theory is that leaders and followers mutually define the follower's role and, as they do so, leaders develop a separate exchange relationship with each individual follower. The exchange relationship usually takes one of two forms. According to the theory, most leaders establish a special exchange relationship with a small number of trusted followers who function as assistants, lieutenants, or advisors. These followers are then classified as being in the "in-group." In the exchange relationship with the remaining followers, who are thought of as being in the "out-group," there is relatively little mutual influence.[17] The Key to the LMX theory is that there is a dyadic relationship between the leader and each individual follower.

## **SOCIAL EXCHANGE THEORY**

There are many different theories based on social exchanges. Most forms of social interactions are based on an exchange of benefits or favours, either material, psychological, or both.[18] Often it is through a variety of social exchanges that an individual emerges as the leader of a group. A simple demonstration of competence and loyalty to the group may shape the expectations others form about the leadership role that an individual should play in the group. His/her influence over group decisions is then compared to that of other group members. Furthermore, an individual who has demonstrated good judgment accumulates "idiosyncratic credit" which allows him/her more latitude than other group members to deviate from nonessential group norms. If a leader steers the group in an innovative way that turns out to be successful, his/her expertise is confirmed, and the leader is awarded more influence and status by the group. If the leader's proposal turns out to be a failure, however, the group is likely to rethink the terms of the exchange relationship.[19]

## **LPC CONTINGENCY MODEL**

Fred Fiedler's Contingency Model of leadership is a situational theory that emphasizes the relationship between leader characteristics and the situation.[20] The model describes how the situation moderates the relationship between leadership effectiveness and a trait measure called the least preferred co-worker (LPC) score.[21] The LPC score is determined by asking a leader to rate, on a set of bipolar adjectives, the one person with whom he/she worked least well. The score offers a description of the leader's emotional reaction to an individual who represents an obstacle to goal attainment.[22] A leader who is generally critical in rating the least preferred co-worker will obtain a low LPC score. Fiedler describes this type of leader as "task oriented." Contrarily, a leader who is strongly motivated to have close interpersonal relationships with other people will receive a high LPC score. This type of leader is described by Fiedler as "relationship oriented." Fiedler characterizes leaders who receive a medium LPC score as "socioindependent."[23]

According to Fiedler, the relationship between a leader's LPC score and leadership effectiveness depends on a complex situational variable called "situational favourability" (also called "situational control"). Fiedler defines favourability as the extent to which the situation gives a leader control over followers.[24] Three aspects of the situation are considered to mediate this relationship: the quality of leader-member relations, the leader's position power, and the task structure.[25]

## HERSEY AND BLANCHARD'S SITUATIONAL LEADERSHIP MODEL

Paul Hersey and Ken Blanchard proposed a contingency theory of leadership that prescribes the use of a different pattern of leadership behaviour depending on the "maturity" of an individual follower.[26] Maturity includes two related components. The first, "job maturity," refers to a follower's task-relevant skills and technical knowledge. This component is directly reflective of the follower's ability to perform a task. The second, "psychological maturity," is indicative of the follower's self-confidence and self-respect. This aspect influences a follower's willingness to perform a task.[27]

Given follower readiness and leader behaviour, Hersey and Blanchard propose four possible leadership styles, each one relating to a particular combination of follower and leader behaviour. The first is "telling," which represents a directive style of leader behaviour. The leader simply tells followers what to do. It is recommended for followers who have both low job maturity and low psychological maturity. The second style is "selling," which represents a style of leader behaviour that is both directive and supportive. Here the leader convinces followers of the importance and necessity of task accomplishment. It is recommended for followers who have low job maturity and high psychological maturity. The third is "participating," which represents a supportive style of leader behaviour and involves the leader and follower interacting to determine the proper course of action for a given situation. It is recommended for followers who have high job maturity and low psychological maturity. The fourth is "delegating," which is characterized by a style of leader behaviour that is lacking in both supportiveness and directiveness.[28] Delegation is recommended for followers who have both high job maturity and high psychological maturity.

## PATH-GOAL THEORY OF LEADERSHIP

The path-goal theory of leadership was developed to explain how leadership behaviour can influence the satisfaction and performance of an individual follower. It emphasizes the relationship between the leader's style and both the characteristics of the follower and the situation.[29] The theory proposes that a leader's behaviour is motivating or satisfying to the follower if the behaviour increases the attractiveness of goals while, simultaneously, increasing follower confidence in achieving them.[30] The leader is, therefore, very active in guiding, motivating, and rewarding followers in their work. In short, the leader steers the follower down a path to his/her goals by selecting behaviours that are best suited to an individual follower's needs and the situation. In doing so, the leader also navigates the follower around obstacles that lie on the path toward goal achievement.[31]

## MULTIPLE LINKAGE MODEL OF LEADERSHIP

The multiple-linkage model builds upon earlier models of leadership and group effectiveness. It was developed by Gary Yukl, and proposes that the overall impact of specific leader behaviours on group performance is complex and is composed of four sets of variables.[32] The four variables are: managerial behaviours, intervening variables, criterion variables, and situational variables.

Two of these four variables are influential in determining leader effectiveness. One factor, intervening variables, refers to the immediate effects that the leader's behaviour has on followers' job performance. Intervening variables comprise follower effort, follower ability, role clarity, organization of the work, cooperation and teamwork, resources and support, and external coordination.[33] The second set of factors that moderate the leader's impact on group performance are situational characteristics.[34] Two situational variables that influence follower effort are the formal reward system and the intrinsically motivating properties of the work itself.[35] In this model, intervening variables may be directly affected by situational characteristics and situational variables may directly affect intervening variables.[36] In the short term, the job of the leader is to correct deficiencies arising in the intervening variables. In the long term, the job of the leader is to improve situational factors.[37]

## **LEADERSHIP SUBSTITUTE THEORY**

Leadership substitute theory is a leadership model that identifies situational aspects that reduce or eliminate the need for a leader. The theory identifies two types of situational variables that affect the need for a leader: substitutes and neutralizers. Substitutes act instead of a leader and make leader behaviour redundant. Neutralizers prevent a leader from acting effectively.[38]

## **VROOM AND YETTON'S NORMATIVE DECISION MODEL**

V.H. Vroom and P.W. Yetton developed a leadership model that specifies which decision procedures should be most effective in each of several specific situations. They stipulate that the overall effectiveness of a decision depends on two intervening variables: decision quality and decision acceptance by followers.[39] Decision quality is the objective aspects of the decision that affect group performance regardless of any effects mediated by decision acceptance. Decision acceptance refers to the degree of follower commitment in implementing a decision effectively.[40]

## **CHARISMATIC LEADERSHIP**

*Charisma* is a Greek word that means "divinely inspired gift."[41] The term has since been used to describe a leader whose followers think that s/he is endowed with exceptional qualities. Such a leader is thought to yield influence over his/her followers because of these special powers instead of needing to use traditional or formal forms of authority.[42]

Charisma is believed to be attributed to leaders who advocate a unique vision, yet one that lies within the range of acceptability by followers. It is thought more likely to be attributed to leaders who act in unconventional ways to achieve their vision. Furthermore, leaders who make self-sacrifice, take personal risk, and incur high costs to achieve their vision are more likely to be viewed as charismatic.[43]

## **TRANSFORMATIONAL LEADERSHIP**

Transformational leadership is a process that changes and transforms individuals. It is often associated with ethics and involves long-term goals.[44] It is not thought to involve an exchange between leader and follower such as exists for transactional leadership. Instead, transformational leadership focuses on the process by which the leader engages with followers, and together create a connection that raises each of them to higher levels of motivation and morality.[45] A transformational leader must be attentive to follower needs and motivation, and tries to help followers reach their full potential.[46]

According to B.M. Bass, one of the leading theorists on transformational leadership, the leader transforms and motivates followers by making them more aware of the importance of task outcomes, inducing them to transcend their own self-interest for the sake of the organization or team, and activating their higher order needs.[47] It is hypothesized that follower motivation and performance are enhanced more by transformational leadership than by transactional leadership.[48]

## **SERVANT LEADERSHIP**

Robert K. Greenleaf is credited with having developed the servant leadership model. This model is based on the idea of the servant as a leader,[49] or, more conceptually, of a leader having the duty to serve his/her followers. Servant leadership was created as an attempt to link previous paradoxes concerning leadership. Task accomplishment is a focus, yet it is also recognized that leaders should be aware of the social implications associated with task accomplishment. Leadership effectiveness is another concern but global efficiency, the concern for long-range human and environmental welfare, is equally weighted in Greenleaf's model.[50]

The servant leadership model goes a step beyond the transformational leadership models. Servant leadership stresses ethical practice, whereas only certain transformational theorists suggested that ethical behaviour is a necessary component of transformational leadership.[51]

## SUMMARY

This chapter provides an introduction to leadership models and theories. Although it is possible to organize leadership models in a variety of ways, conceptualizing the leader-follower-situation triadic relationship is important; it is a reminder that leadership does not occur in a vacuum. It is also crucial to understand this relationship when exploring leadership in the Canadian Forces (CF).[52] The CF is a prime example of an organization in which leaders, followers, and the situation all contribute to the outcome of a mission.

## NOTES

1. Aspects of personality and aptitudes were both measured by psychological tests. G. Yukl, *Leadership in Organizations*, 4th ed. (Upper Saddle River, NJ: Prentice Hall, 1998), 235.
2. P.G. Northouse, *Leadership: Theory and Practice* (Thousand Oaks: Sage Publications, 1997), 17.
3. Yukl, *Leadership in Organizations*, 286.
4. *Ibid.*, 287–288.
5. M.D. Mumford, S.J. Zaccaro, F.D. Harding, T.O. Jacobs, and E.A. Fleishman, "Leadership Skills for a Changing World: Solving

Complex Social Problems," *Leadership Quarterly* 11 (2000): 26.
6. *Ibid.*, 24
7. *Ibid.*, 24.
8. Knowledge is considered to reflect a collection of key facts and principles pertaining to the characteristics of objects lying in a certain domain. It is not considered simply an accumulation of information. *Ibid.*, 20.
9. Social perceptiveness refers to the ability to have insight into the needs, goals, demands, and problems of followers and the social setting in which they are situated. In such a way, leadership skills model can be considered to involve a three-way interaction between the leader, follower, and the situation. The emphasis is, however, on the interaction between the leader and the situation and followers are often though to be part of the situation. *Ibid.*, 19.
10. Northouse, *Leadership: Theory and Practice*, 137.
11. Yukl, *Leadership in Organizations*, 326.
12. Northouse, *Leadership: Theory and Practice*, 138.
13. *Ibid.*, 138.
14. Northouse, *Leadership: Theory and Practice*, 139; and Yukl, *Leadership in Organizations*, 326.
15. Northouse, *Leadership: Theory and Practice*, 184.
16. Yukl, *Leadership in Organizations*, 123.
17. *Ibid.*, 150.
18. *Ibid.*, 189.
19. *Ibid.*, 189.
20. Howell, *Understanding Behaviors for Effective Leadership*, 38.
21. Yukl, *Leadership in Organizations*, 283.
22. J.P. Howell and D.L. Costley, *Understanding Behaviors for Effective Leadership* (Upper Saddle River, NJ: Prentice Hall, Inc., 2001), 38.
23. *Ibid.*, 38.
24. P.B. Smith and M.F. Peterson, *Leadership, Organizations and Culture* (London: Sage Publications, 1988), 17–18.
25. Howell and Costley, *Understanding Behaviors for Effective Leadership*, 38; and Yukl, *Leadership in Organizations*, 283.

26. Yukl, *Leadership in Organizations*, 270.
27. *Ibid.*, 270.
28. Howell and Costley, *Understanding Behaviors for Effective Leadership*, 41.
29. Northouse, *Leadership: Theory and Practice*, 88.
30. Howell, *Understanding Behaviors for Effective Leadership*, 42.
31. Smith and Peterson, *Leadership, Organizations and Culture*, 21.
32. Howell and Costley propose only two sets of factors whereas Yukl proposes four. Howell and Costley, *Understanding Behaviors for Effective Leadership*, 46 and; Yukl, *Leadership in Organizations*, 276.
33. Yukl, *Leadership in Organizations*, 276.
34. Howell and Costley, *Understanding Behaviors for Effective Leadership*, 46.
35. Yukl, *Leadership in Organizations*, 276.
36. *Ibid.*, 276.
37. Howell and Costley, *Understanding Behaviors for Effective Leadership*, 47.
38. Yukl, *Leadership in Organizations*, 273.
39. *Ibid.*, 127.
40. *Ibid.*, 128.
41. *Ibid.*, 298.
42. *Ibid.*
43. *Ibid.*, 302.
44. Northouse, *Leadership: Theory and Practice*, 130.
45. *Ibid.*, 131.
46. B.M. Bass and B.J. Avolio, *Improving Organizational Effectiveness Through Transformational Leadership* (Thousand Oaks: Sage Publications, 1994), 14.
47. *Ibid.*, 14–15.
48. Yukl, *Leadership in Organizations*, 325.
49. R.K. Greenleaf, foreword to *The Power of Servant Leadership* (San Francisco: Berret-Koehler Publishers, Inc., 1998), x.
50. Greenleaf, foreword to *The Power of Servant Leadership*, x.
51. For Burns, transformational leadership had to be ethical, however,

for Bass, leadership had no ethical requirement. Yukl, *Leadership in Organizations*, 327.

52. See the suite of CF leadership manuals given in the Selected Readings list.

**SELECTED READINGS**

Bass, B.M. and B.J. Avolio. *Improving Organizational Effectiveness Through Transformational Leadership*. Thousand Oaks: Sage Publications, 1994.

Canada. *Leadership in the Canadian Forces: Leading the Institution*. Kingston: DND, 2007.

———. *Leadership in the Canadian Forces: Leading People*. Kingston: DND, 2007.

———. *Leadership in the Canadian Forces: Conceptual Foundations*. Kingston: DND, 2005.

———. *Leadership in the Canadian Forces: Doctrine*. Kingston: DND, 2005.

Hammond, J.W. "First Things First: Improving Canadian Military Leadership." *Canadian Defence Quarterly* 27 (1998): 6–11.

Howell, J.P. and D.L. Costley. *Understanding Behaviors for Effective Leadership*. Upper Saddle River, NJ: Prentice Hall, 2001.

Northouse, P.G. *Leadership: Theory and Practice*. Thousand Oaks: Sage Publications, 1997.

Smith, P.B. and M.F. Peterson. *Leadership, Organizations and Culture*. London: Sage Publications, 1988.

Taylor, R.L. and W.E. Rosenbach. *Military Leadership: in Pursuit of Excellence*. Colorado: Westview Press, 2000.

Yukl, G. *Leadership in Organizations*, 4th ed. NJ: Prentice Hall, 1998.

# 39

# TRUST

by Jeffrey Stouffer, Barbara Adams,
Jessica Sartori, and Megan Thompson

*A leader capable of inspiring trust is especially valuable in bringing about collaboration among mutually suspicious elements in the constituency. The trust the contending groups have for such a leader can hold them together until they begin to trust one another.*[1]

— John Gardner

The importance of trust has long been recognized. Confucius (551–479 BC), as an example, explained that in order to satisfy one's people, a leader needed to provide them with armament, ample food, and trust. Should an emperor have to eliminate any of the three, he should start with armament, and then food, because "if the people don't trust you, you have nothing to stand on." The teachings of Confucius continue to resonate today as Patrick Lencioni, author of the *New York Times* bestseller *The Five Dysfunctions of a Team* asserts, "No quality or characteristic is more important than trust."[2] In fact, today, trust is considered "a good that markets and organizations can't get enough of."[3] That being said, the reality is that the word "trust" is liberally sprinkled in both the research and Canadian Forces (CF) literature with limited explanation about how to foster trust and/or the manner in which it operates. As Diego Gambetta, a professor of sociology, notes that "scholars tend to mention trust in passing, to allude to it as a fundamental ingredient or lubricant, an unavoidable dimension of social interaction, only to move on to deal with less intractable matters."[4] To address this shortcoming, the CF has recently devoted increasing attention to the importance of trust in several key leadership publications and is increasingly supporting an ambitious program of research exploring trust in military operations undertaken by Defence Research and Development Canada (DRDC — Toronto).[5] These efforts highlight the growing recognition of the value of trust and the need for leaders to understand the undeniable impact that trust plays on individual, team, and ultimately mission success. Thus, the intent of this chapter is to bring together what is known about trust to strengthen CF leadership development.

## **THE POWER OF TRUST**

Trust is part of an organization's social capital or psychological and behavioural contract between members of an organization and "represents an important human dimension of military effectiveness."[6] If properly nurtured and harnessed, trust will contribute to valued organizational outcomes. If ignored or mismanaged, trust can undermine both individual, team, and organizational effectiveness. Based on the research literature, researchers from DRDC Toronto argue that trust in leaders and in teams has several critical impacts.[7] For example, trust:

- \* enhances morale, cohesion, and fosters a positive team climate;
- \* promotes more open communication processes and has a positive effect on the amount of information sent and shared with superiors;
- \* reduces adversarial tactics and conflict.[8] For example, when trust between members is low, task conflict is perceived as resulting from relationship conflicts, whereas when trust is high between members, task conflict is interpreted as just that and not a relationship conflict;[9]
- \* lowers defensive monitoring of others (actions

to ensure that expectations are being fulfilled). Defensive monitoring can be time-consuming and, as a result, hinder task performance. This is not to say that leaders should avoid monitoring, rather, that a prudent approach to monitoring be adopted. As Roderick Kramer asserts, "Trust the shuffler, but cut the deck anyway;"[10]

* helps to reduce the requirement for organizational and unit controls, which allows increased focus on task completion. Some organizational controls, however, may reduce trust. For example, research has shown that defensive monitoring techniques such as technologies to monitor employees to help ensure compliance and deter inappropriate behaviour may in fact have the opposite effect and may serve to undermine trust;[11]

* has a positive effect on team performance.[12] It has been shown that "trust in leadership allows team members to suspend their questions, doubts, and personal motives and instead throw themselves into working toward team goals;"[13]

* is related to a variety of organizational citizenship and work behaviours like conscientiousness, sportsmanship, civic virtue, courtesy, and helping others before oneself;

* promotes job satisfaction and organizational commitment as well as decreases intentions to quit;[14] and

* induces problem solving that is productive and creative. Under conditions of low trust, problem-solving is degenerative and ineffective.[15] "If people are preoccupied with protecting their backs or actively engaged in self-preservation behaviors, creativity will be one of the first casualties."[16]

While the above factors highlight the significance of trust within organizations, it appears that trust will become increasingly critical for the CF. As an example, Alan Okros, professor of psychology, contends:

> ... that the CF's role in the new security domain is that the building and maintenance of trust relationships has now expanded from being the means by which missions were accomplished — ensuring trust among military members, between the military and society ... to now becoming the ends to be achieved. Trust building is the new mission objective and has to be approached in very different ways than maintaining stalemates or defeating an opposing force on the battlefield.[17]

This suggests that in future operations, CF leaders increasingly will be required to show competency in building and maintaining trust at multiple levels (e.g., interpersonal, organizational, and with external agents).

## **TRUST DEFINED**

Although there is no agreement on a single definition of trust in the research literature, a widely accepted view is that "trust is a psychological state comprising the intention to accept vulnerability based upon positive expectations of the intentions or behaviour of another."[18] As such, trust is commonly seen as a psychological state involving positive beliefs and expectations about other people. Trust, however, can also promote different choices or behaviours that put these beliefs and expectations into action. Trusting another person, for example, may bring about more risk-taking when working with this person because of one's belief in the person's skills and abilities. While this definition focuses largely on individual interactions, other researchers highlight the importance of trust within groups. As an example, Francis Fukuyama, a renowned trust expert, views trust as "the expectation that arises within a community of regular, honest, and cooperative behaviour, based on commonly shared norms, on the part of other members of that community."[19] In this sense, it is incumbent on leaders to understand that trust is not limited to interactions between individuals and must be understood in terms of the general community.

## **TYPES AND NECESSARY CONDITIONS**

*Person and Category-Based Trust*

Although many types of trust are reported in the research literature, person and category-based trust have garnished considerable attention and appear most relevant to military environments. People are likely most familiar with person-based trust. This form of trust requires obtaining the necessary knowledge and experience about another person prior to making the decision to trust. Person-based trust develops over time and through exposure to situations that permit one to understand the motives, values, and beliefs of another person that enables one to increasingly predict the behaviour of this person. Over time, this information helps to alleviate the risk inherent in working with this person.

Many situations, however, do not provide sufficient time for person-based trust to emerge. Category-based trust can develop or exist without personal contact and in the absence of shared values and experiences. This form of trust is based on indirect information about another person and can be provided by the categories to which a person belongs.[20] Knowing that another person occupies a specific role, for example, may influence trust expectations about them because membership in this role can be seen to provide information about this person. In this sense, expectations regarding another's trustworthiness can arise from: rank; regimental affiliation and traditions; information about another's completed training or previous roles occupied; and established organizational expectations and rules that ensure role compliance. Similarly, indirect information about another person's trustworthiness can also influence trust. For example, third parties can communicate information regarding leader trustworthiness; thus, one's reputation can serve to amplify or diminish trust. Thus, with category-based trust, indirect knowledge of a person can circumvent the requirement for personal knowledge.

*Necessary Conditions*

As a concept, trust is most critical in situations that involve risk, vulnerability, and the need to be interdependent with other people.[21] Trust is only meaningful when there are both potential costs (of having one's trust betrayed), as well as the potential benefits that come from trusting another person (e.g., getting accurate information that one could not otherwise get). As such, the need to trust other people is strongest when something critical is at stake, when there is both the possibility that one's trust will be betrayed (e.g., a teammate will fail to perform reliably) as well as a cost to failing to trust (e.g., one has no hope of performing a task alone). Arguably, each of these conditions is inherent and readily observable in military operations. Therefore, it is critical that military professionals have an expert knowledge of the factors that are important in the development of trust.

*Factors That Promote Trust*

The importance of trust for leaders is undeniable. To fully appreciate and utilize trust to one's advantage, leaders must be cognizant of the factors associated with the development of trust. Factors found to influence judgments of trustworthiness are typically divided into three categories and include; qualities of the trustor, qualities of the trustee (the person being trusted), and qualities of the interaction.[22]

People differ in their propensity to trust and typically fall on a continuum from being unwilling to trust, to exhibiting blind trust. Research has demonstrated that one's predisposition to trust is initially evoked in new situations but, as time progresses, this tendency or propensity to trust becomes less significant as relationships develop.[23]

Previous trust experiences/histories (successes and violations of trust) also influence the willingness to resist or cultivate future trust relationships (i.e., successive positive trust experiences reward trusting behaviour).

Qualities of the trustee commonly argued to influence person-based trust include competence, integrity, benevolence, and predictability.[24] Competence is the belief or perception that the trustee has the ability or talent to complete a task or specific job (i.e., trust in another's abilities). Integrity involves consistency between a trustee's words and actions, and includes being fair, seeking subordinate input, and adhering to honest and open communication. Integrity "flows from values-based ethical behavior."[25] Benevolence means showing a genuine concern for the trustor independent of self-interest. In describing the dynamics between leaders and their personnel, clinical psychologist Henry Cloud argues that "trust is built when leaders have a genuine interest in knowing them, knowing about them, and having what we know matter."[26] Predictability simply refers to doing what one is expected to do (i.e., consistency in behaviour). Although competence, integrity, benevolence, and predictability are distinct characteristics, trust can develop with varying degrees of each of these factors.

The way in which leaders and their personnel interact will also influence the building trust. Commonly noted influences include communication styles, similarity, and shared goals.[27] Open communication styles that facilitate the sharing of information as well as provide evidence of one's values, beliefs, and attitudes will help both parties find common ground, match similarities, and determine the extent to which goals are shared. As one sees others as being similar, the assumption is that they will behave in a similar manner, thus, enabling a more accurate prediction of subsequent behaviour.

With respect to trust in groups, when one is accepted into a group, the tendency is to see oneself as more similar to the group and dissimilar to those outside the group. Trust within the group is understood and a degree of mistrust exists when members of one group interact with members of another group. This dynamic can be effective in small military teams but, if mismanaged, can lead to group dysfunction.

## **HOW CAN LEADERS PROMOTE TRUST?**

Military leaders at all levels are in the obvious and unique position to act as the catalysts in promoting, developing, and maintaining trust. As U.S. Army Lieutenant-General (ret'd) Walter F. Ulmer asserts, "Competent military leaders develop trust."[28] In other words, "trust does not come with a pay-check, it has to be earned."[29] Leader actions, being highly visible, set the tone for either the development or deterioration of trust. This makes it even more critical that leaders understand how their actions can facilitate or debilitate the development of trust. Without doubt, knowledge of the factors that contribute to the development of trust can effectively be put into action with moderate effort and the application of existing espoused CF leadership practices. For example, several means through which leaders can build trust are:[30]

* demonstrating direct and tangible evidence of high levels of proficiency and professional competence in the performance of core functions and taking advantage of opportunities to enhance their professional expertise and competence;

* exercising good judgment in decisions that affect others and not exposing people to unnecessary risks, including ensuring that taskings are consistent with individual or team abilities;

* showing trust and confidence in followers by giving them additional authority and involving them in decisions where circumstances allow. Research has shown that utilizing transformational leadership techniques[31] and allowing participation in the decision-making process and providing feedback can lead to increased trust;[32]

- demonstrating concern for the well-being of followers, representing their interests, and ensuring they are supported and taken care of by the organization. Trusted leaders devote time to learning about their personnel and what is important to those personnel;

- showing consideration and respect for others, being polite, friendly, approachable, as well as treating followers fairly and consistently (without favour or discrimination);

- being professional in bearing and conduct. Trusted leaders set high standards for themselves and demonstrate accountability. They also exemplify the best values and practices of the CF;

- maintaining honest and open communications. Trusted leaders communicate their intent, ensure it is understood, and strive to achieve transparency in communications to mitigate the debilitating effects that gossip and rumour can evoke;

- leading by example, sharing risks and hardships, and refusing to accept or take special privileges;

- keeping their word and being counted upon to honour their obligations. Trusted leaders do what they say they are going to do. This includes applying rules in a fair and consistent manner. Research has shown that procedural justice, in terms of how organizations behave (e.g., performance appraisals, award professional development opportunities etc.), is a significant predictor of trust;[33] and

* promoting commonalities that people share (values, goals, attitudes) and stressing tolerance for dissimilarities (i.e., understanding and leveraging factors associated with the development of both person and category-based trust).

Of marked interest is the striking similarity of trust-building behaviours and current CF leadership principles. Understanding and practicing these principles will set the stage for the development of trust in any organization. But what happens when trust is violated?

## **RESPONSE TO TRUST VIOLATIONS**

Regardless of adherence to the above espoused actions and the best intentions of leaders, trust violations, whether real or perceived, accidental or intentional (premeditated or opportunistic), undoubtedly will occur and represent a formidable leadership challenge. Regardless of the type of violation, most people would likely agree, as Roderick Kramer claims, "Trust is easier to destroy than create."[34] Similarly, the Dutch saying, "Trust arrives on foot but leaves on horseback,"[35] adeptly captures the slow growth but potentially rapid decline of trust when violations occur. As an explanation, researchers have suggested that actions that undermine trust are more visible and carry more weight in judgment than do trust-building events.[36] As such, leaders must be aware of actions or behaviours that can result in betrayal or trust violations.

Violations of trust are often predicated on: rule violations; honour violations (shirking responsibilities, breaking promises, lying, stealing ideas, and secret/confidence disclosures); abusive authority; and damaging another person's identity through public criticism, false or unfair accusations, and insults.[37] The breaking of a psychological contract or acting in ways inconsistent to stated or shared values can also lead to a trust violation and can reduce the social capital in organizations. Although most would agree that such actions are obvious and clearly represent betrayal, less obvious to leaders are the subsequent repertoire of responses exhibited by those violated and the often resulting effects on morale, cohesion, performance, and mission success.

Research confirms that there is no uniform behavioural response to a breach of trust. Trust violations can lead to changes in beliefs and expectations as well as promote specific behaviours. Breaches can evoke powerful "feelings of anger, hurt, fear, and frustration."[38] Behaviourally, some people will act to get even with or punish the betrayer (i.e., seek revenge or resort to violence), others will dissolve the relationship, some will attempt to renegotiate or restore the relationship, and others will simply ignore the betrayal.[39] Other research has also shown that trust violations may promote revenge fantasies, result in private confrontation and/or feuding, increase social withdrawal (work less, quit, withhold help and support)[40] and reduce employee satisfaction.[41] Regardless of the behavioural response, trust violations often lead to the reassessment of beliefs, feelings, and expectations about the trust violator.

## **REPAIRING TRUST**

Although organizations establish rules to safeguard against untrustworthy behaviours, violations (accidental or deliberate) will occur, requiring leaders to take action to repair the damage. Repairing trust can take considerable effort for both the victim and the violator and requires that each party commit to the repair process. Trust theorists have attempted to define the steps required to repair trust violations that normally include both the violator and victim recognize and acknowledge that a violation occurred, determine what caused the violation, admit that the event was destructive, and be willing to accept responsibility for the violation.[42] The next step requires that the relationship balance be restored through forgiveness or through specific actions that satisfy the victim (e.g., compensation). Repairing trust is not an easy task and takes considerable time. Even in the presence of promises, apologies, and consistent subsequent trustworthy behaviour, trust may never be fully recovered when untrustworthy behaviours are combined with deception.[43] The leader's role is to facilitate this process where feasible and appropriate and to also be aware of external referral services such as those offered through alternative dispute resolution.

## RECOMMENDATIONS FOR CREATING AND SUSTAINING TRUST

Without argument, trust is an essential organizational commodity that must be developed and sustained at every possible opportunity. As Robert Levering, co-author of *The 100 Best Companies to Work for in America* asserts, "A great place to work is one where you trust the people you work for."[44] The good news is that establishing trust is possible and for the most part, readily achievable through the application of known CF leadership practices. Research conducted by DRDC Toronto offers several guidelines to support that development and maintenance of trust:[45]

* build trust as early as possible;
* promote professional development (for leaders and subordinates);
* ensure new members of teams have opportunities to display their competence and trustworthiness and assist them in identifying with the values and goals of the team;
* utilize open and transparent communications;
* identify trust as a central value and critical team objective;
* create a strong and positive value base;
* because trust takes time to develop, minimize turnover as much as possible in military teams). "The contemporary operating environment often throws soldiers into situations where they must quickly establish working relationships with complete strangers";[46]
* demonstrate strong and positive leadership and maintain direct and personal contact with subordinates;

* promote a strong common identity (i.e., member of the CF);
* understand the antecedents of trust;
* ensure trust-eroding behaviour is eliminated (i.e., information hoarding, backstabbing, harassment, etc); and
* exercise vigilance to recognize warning signs of low or deteriorating trust (e.g., defensive monitoring, poor communications).

In addition, leader awareness of behaviours or actions that can result in trust violations and adopting practices to reduce the likelihood of violations represents a prudent strategy. Equally important, when violations occur, leaders must act immediately through direct intervention or through the assistance of organizational referral services. The development of mediation and conflict resolution skills should also enable leaders to better approach and resolve violations of trust.

## **SUMMARY**

Trust is critical to military effectiveness. It is a concept that demands rigorous leader attention, as it is closely associated with a variety of necessary and desired organizational outcomes. Leaders must work to earn the trust of followers, and acknowledge and understand the leader's important role in building and maintaining trust. That trust is fragile, is not easily restored once violated, and requires that leaders exercise vigilance and persistence in the monitoring and nurturing of trust. Trust begins with leadership.

## NOTES

1. John W. Gardner, *On Leadership* (New York: The Free Press, 1990), 33.
2. Patrick Lencioni, *Overcoming the Five Dysfunctions of a Team* (San Francisco: Jossey-Bass Publishers, 2005), 13.
3. Tom L. Roberts, Paul D. Sweeney, Dean McFarlin, and Paul H. Cheney, "Assessing Trust Among IS Personnel: A View of General Trust, Trust of Management and Inter-Organizational Trust," *Proceedings of the 37th Hawaii International Conference on System Sciences*, 2004, 2.
4. Diego Gambetta, "Can We Trust Trust?" in *Trust: Making and Breaking Cooperative Relationships,* ed. D. Gambetta (New York: Basil Blackwell, 1998), 213–237.
5. This program of research has been undertaken by DRDC Toronto (Dr. Megan Thompson, Carol McCann) in conjunction with researchers from Humansystems Incorporated in Guelph, Ontario (Dr. Barbara Adams, Dr. Jessica Sartori).
6. Canada. *Leadership in the Canadian Forces: Conceptual Foundations* (Kingston: DND, 2005), 74.
7. Barbara Adams, David Bryant, and Robert Webb, *Trust in Teams Literature Review*, Department of National Defence, 2001, DCIEM No. CR-2001-042.
8. Lynne C. D'Amico, "Examining Determinants of Managerial Trust: Evidence from a Laboratory Experiment," paper presented at the National Public Management Research Conference, Washington, DC, October 10, 2003, 3.
9. T. Simons, and R. Peterson, "Task Conflict and Relationship Conflict in Top Management Teams: The Pivotal Role of Intragroup Trust," *Journal of Applied Psychology*, Vol. 85, No. 1 (2000): 102–111.
10. Roderick Kramer, "Trust and Distrust in Organizations: Emerging Perspectives, Enduring Questions," *Annual Review of Psychology*, Vol. 50 (1999): 65.
11. *Ibid.*, 591.
12. Kurt Dirks, "Trust in Leadership and Team Performance: Evidence

from NCAA Basketball," *Journal of Applied Psychology*, Vol. 85 (2000): 1008; Kurt Dirks and Donald Ferrin, "Trust in Leadership: Meta-Analytic Findings and Implications for Research and Practice," *Journal of Applied Psychology*, Vol. 87 (2002): 618.

13. Kurt Dirks, 1009.
14. Barbara Adams, *Trust Development in Small Teams*, Defence Research and Development Canada — Toronto, DRDC No. CR-2003-016, 2003, 22.
15. Wayne R. Boss, "Trust and Managerial Problem Solving," *Group and Organizational Studies*, Vol. 3, No. 3 (1978): 331–342.
16. Manfred Kets de Vries, quoted in Rushworth Kidder, *Trust: A Primer for Global Ethics: Research Report*, Institute for Global Ethics, http://www.globalethics.org, 2.
17. Alan Okros, "Management of Trust: The Key to Military Success," Royal Military College of Canada, unpublished essay (2006): 88.
18. Denise M. Rousseau, Sim B. Sitkin, Ronald S. Burt, and Colin Camerer. "Not So Different After All: A Cross-Discipline View of Trust," *Academy of Management Review*, Vol. 23, No. 3 (1998): 394.
19. Francis Fukuyama, *Trust: The Social Virtues and the Creation of Prosperity* (New York: A Free Press Paperbacks Book, 1995), 26.
20. Roderick Kramer, 577.
21. Adams, Bryant, and Webb, 16.
22. *Ibid.*, 26.
23. Lynne C. D'Amico, "Examining Determinants of Managerial Trust: Evidence from a Laboratory Experiment," paper presented at the National Public Management Research Conference, Washington, DC, October 10, 2003, 2.
24. Adams, Bryant, and Webb, 28.
25. Christopher R. Kemp, *Trust — The Key to Leadership in Network Centric Environments* (thesis, U.S. Army War College, 2003), 16.
26. Henry Cloud, *Integrity* (New York: Collins, 2006), 56.
27. Adams, Bryant, and Webb, 30–31.
28. Lieutenant-General (ret'd) Walter F. Ulmer, quoted in *Leadership in the Canadian Forces: Conceptual Foundations* (Kingston: DND, 2005), vii.

29. M. Annison, and D. Wilford, *Trust Matters: New Directions in Health Care Leadership* (San Francisco: Jossey-Bass Publishers, 1998), 34.
30. Canada. *Leadership in the Canadian Forces: Leading People* (Kingston: DND, 2007), 70–71.
31. D. Jung and B. Avolio, "Opening the Black Box: An Experimental Investigation of the Mediating Effects of Trust and Value Congruence on Transformational and Transactional Leadership," *Journal of Organizational Behaviour*, Vol. 21 (2000): 949–964.
32. Ronald C. Nythan, "Changing the Paradigm," *American Review of Public Administration* 30 (2000), 87.
33. Julia Connell, Natalie Ferres, and Tony Travaglione, "Engendering Trust in Manager-Subordinate Relationships: Predictors and Outcomes," *Personnel Review*, Vol. 32 (2003): 569.
34. Roderick Kramer, 593.
35. Irene van der Kloet, *A Soldierly Perspective on Trust* (thesis, Tilburg University, 2005), 47.
36. Roderick Kramer, 577.
37. Robert J. Bies and Thomas M. Tripp, "Beyond Distrust: 'Getting Even' and the Need for Revenge," in *Trust in Organizations*, eds. R. Kramer and T. Tyler (Thousand Oaks: CA: Sage, 1996), 249.
38. Roy J. Lewicki and Barbara B. Bunker, "Developing and Maintaining Trust in Work Relationships," in *Trust in Organizations*, eds. R. Kramer and T. Tyler (Thousand Oaks: CA: Sage, 1996), 125.
39. Robert J. Bies and Thomas M. Tripp, 247; Jonathan J. Koehler, "Betrayal Aversion: When Agents of Protection Become Agents of Harm," *Organizational Behavior and Human Decision Processes*, Vol. 90, (2003): 245; Roy J. Lewicki and Barbara B. Bunker, 114–139; Jonathan J. Koehler, 245.
40. Robert J. Bies and Thomas M. Tripp, 255.
41. Elizabeth W. Morrison and Sandra L. Robinson, "When Employees Feel Betrayed: A Model of How Psychological Contract Violation Develops," *Academy of Management Review*, Vol. 22 (1997): 227.
42. Roy J. Lewicki and Barbara B. Bunker, 128–135; Scott Williams, "Building and Repairing Trust," *LeaderLetter* 2005 (Dayton, OH: Department of Management, College of Business, Wright State

University): 6, http://www.wright.edu/~scott.williams/LeaderLetter/trust.htm.
43. Maurice Schweitzer, John Hershey, and Eric Bradlow, "Promises and Lies: Restoring Violated Trust," *Organizational Behavior and Human Decision Processes*, Vol. 101 (2006): 15.
44. Robert Levering and Milton Moskowitz, *The 100 Best Companies to Work for in America* (New York: Currency/Doubleday, 1993).
45. Adams, Bryant, and Webb, 85–86.
46. Colonel Christopher R. Paparone, "The Nature of Soldierly Trust," *Military Review* (November–December 2002), 45.

## SELECTED READINGS

Adams, B.D. and R.D.G. Webb. *Trust Development in Small Teams* (DRDC No. CR-2003-016). Toronto: Defence Research and Development Canada, 2003.

Adams, B.D., D.J. Bryant, and R.D.G. Webb. *Trust in Teams Literature Review*. (DCIEM Report No. CR-2001-04). Toronto: Defence and Civil Institute of Environmental Medicine, 2001.

Canada. *Duty with Honour: The Profession of Arms in Canada*. Kingston: DND, 2003.

Cloud, Henry. *Integrity*. New York: HarperCollins, 2006.

Dirks, K.T. and D.L. Ferrin. "Trust in Leadership: Meta-Analytic Findings and Implications for Research and Practice." *Journal of Applied Psychology* 87 (2002): 611–628.

Fukuyama, Francis. *Trust*. New York: Free Press Paperbacks, 1995.

van der Kloet, Irene. *A Soldierly Perspective on Trust*. Tilburg, Netherlands: Tilburg University, 2005.

Kramer, R. "Trust and Distrust in Organizations: Emerging Perspectives, Enduring Questions." *Annual Review of Psychology* 50 (1999): 569–598.

Kramer, R and T. Tyler. *Trust in Organizations: Frontiers of Theory and Research*. Thousand Oaks, CA: Sage Publications, 1996.

Lencioni, Patrick. *Overcoming the Five Dysfunctions of a Team: A Field Guide*. San Francisco: Jossey-Bass Publishers, 2005.

Rousseau, D., S. Sitkin, R. Burt, and C. Camerer. "Not So Different After All: A Cross-Discipline View of Trust." *Academy of Management Review*, Vol. 23, No. 3 (1998): 393–404.

Stouffer, Jeffrey M. and Craig Leslie Mantle, eds. *In Harm's Way, Leveraging Trust: A Force Multiplier for Today*. Kingston: Canadian Defence Academy Press, 2008.

# SUMMARY

In a democratic society, the military represents the force of last resort. Its practitioners are the government's managers of violence. Whether employed domestically, or externally to conduct operations in support of the national interest as is more often the case, military personnel are consistently challenged by the complexity, ambiguity, and volatility of the environment in which they operate, particularly in times of conflict or war. As such, military leaders are required to possess and demonstrate the highest levels of ability and expertise. Entrusted with the nation's most precious resource, namely its sons and daughters, military leaders have an innate obligation to master the knowledge, skills, and attributes required in the profession of arms.

Within this reality, no aspect is more important than a sound understanding of military leadership. As stated in the introduction to this handbook, the complex requisite leadership that serves any organization with an embedded profession is an enigmatic concept that possesses different faces, is often confused with myriad terms and methodologies, and is practised by many, whether recognized or not, on a daily basis. For military leaders, this complexity is normally even more profound. The context in which they operate is one of chaos, danger, unlimited liability, and fighting spirit. In Canada, Canadian Forces (CF) leaders must also balance a number of competing institutional realities. They must address the outcomes of mission success, member well-being and commitment, internal integration of the CF's organizational components, and external integration with the global and international backdrop, all enacted within a military ethos. Clearly, the burden placed on the military leader is substantial. But, this is understandable. After all, the cost of failure is simply unacceptable.

Within this context, the Canadian Defence Academy (CDA) Press and the Canadian Forces Leadership Institute (CFLI) have recognized the need for a concise and comprehensive handbook that identifies and explains the concepts, components, and ideas related to effective military leadership. This book, *The Military Leadership Handbook*, attempts to respond to that very unique need. It is designed and structured to contribute to and increase leader effectiveness through the identification of current leader challenges, a more substantial understanding of a professional ideology in the profession of arms, and a greater comprehension of significant leadership issues. Use of this handbook should enhance the requisite leader effectiveness in military institutions and assist with mission success. However, the knowledge contained within these pages should also be of significant assistance to anyone, civilian or otherwise, wishing to attain a greater understanding of military leadership, as well as anyone studying or practising leadership in general. As the cliché states, "Knowledge is power!" and this handbook is designed to empower all who study it.

# CONTRIBUTORS

**Dr. Barbara Adams** received a Ph.D. in social psychology from the University of Waterloo in 1999. She has been working with Humansystems since 2000, and is the lead researcher in DRDC, Toronto's research program exploring trust in military contexts. Dr. Adams has undertaken several reviews of the trust literature, has developed and validated a scale to measure trust in teams, and has studied trust in military teams in both the laboratory and field environments. Other work has included moral and ethical decision-making, and human issues in network-enabled operations.

**Major Brent Beardsley** has been with the Canadian Forces Leadership Institute since 2002. He is an infantry officer with 29 years of service with The Royal Canadian Regiment, including four tours of regimental duty, a wide range of staff appointments on extra-regimental employment, and service on operations. He holds a pre-arts diploma from Sir George Williams University, a bachelor of arts degree in history from Concordia University, a post-graduate diploma in education from McGill University, and a masters of applied science in management from the Royal Military College of Canada (RMC). Major Beardsley currently is completing his second master's degree in war studies at RMC, focusing on genocide and humanitarian interventions.

**Dr. Bill Bentley** is a retired lieutenant-colonel with 35 years of experience in the Canadian Infantry. He served in both United Nations and NATO appointments, and was the Canadian Exchange Instructor at the U.S.

Army Staff College and the School for Advanced Military Studies. He currently is an associate professor at the Canadian Forces Leadership Institute. He is the author of *Professional Ideology and the Profession of Arms in Canada*. In 2006, Dr. Bentley received the Meritorious Service Medal for his contributions over 10 years to the reform of the CF Professional Development System.

**Colin Bridges** was born and raised in Australia. He has completed degrees in education and mathematics/psychology. His 11 years of service with the Australian Army Psychology Corps included the provision of psychological support services in East Timor and Bougainville. In 2002, he was voluntarily discharged from the Australian Army to immigrate to Canada with his Canadian wife. At present he is employed as a research assistant/statistical analyst at Royal Military College, and also works occasionally as a high school mathematics teacher.

**Dr. Phyllis Browne** is a graduate of Concordia and McGill Universities in Montreal and holds a Ph.D. degree in sociology. Her areas of specialization are education, social and economic change, gender issues, and labour markets. Dr. Browne was employed until 2007 as a defence scientist in the Department of National Defence, assigned to the Canadian Forces Leadership Institute of the Canadian Defence Academy, after which she was posted to Ottawa for continued employment in sociological research in national defence.

**Dr. Danielle Charbonneau** is an associate professor at the Royal Military College of Canada. She obtained her Ph.D. in clinical psychology from Queen's University in 1994. She has been working in the Military Psychology and Leadership Department since 1996, where she has been teaching courses in counselling and persuasion, and has served as acting department head for 2007 to 2008.

**Major Bradley Coates**, an air navigator, joined the Canadian Forces in 1986. Throughout his career, he has served in various operational, training, and staff positions. Since 2004, Major Coates has been the coordinator of the Canadian Forces Base Borden and Region Dispute Resolution Centre. He holds a B.A. in history from Bishop's University, an MBA from the University of Ottawa, a master of defence studies from the Royal Military College of Canada, and a graduate certificate in conflict resolution from Carleton University.

**Lieutenant-Commander (Retired) Karen D. Davis** is a senior defence scientist at the Canadian Forces Leadership Institute, holds a master of arts in sociology from McGill University and is a Ph.D. candidate at the Royal Military College of Canada. She has conducted research in the Canadian Forces for over 15 years on a range of issues including gender, leadership, culture, and cultural intelligence.

**Commander Robert S. Edwards** graduated from the Royal Military College of Canada in 1974 and subsequently received a Master's Degree in war studies. He served a full career in the Canadian Forces Navy, including the commissioning and command of HMCS *St. John's*, a Canadian patrol frigate. He has completed a number of appointments in CF professional development, including an appointment as deputy commandant at Royal Roads Military College. Commander Edwards was appointed to the Canadian delegation to NATO on defence matters from 2000 to 2004. From 2007 to 2008, Commander Edwards was the acting director of the Canadian Forces Leadership Institute of the Canadian Defence Academy.

**Dr. Rhonda Gibson**, R Psych, is a registered psychologist and trauma therapist. She has been affiliated with the Canadian Forces for over 20 years, focusing in the area of assessment, treatment, and prevention of deployment stress and its effects on military members and their families. Dr. Gibson is the author of the *Canadian Deployment Impact Scale* (1997), a self-report inventory designed to measure post-traumatic stress disorder in Canadian UN peacekeepers.

**Colonel Bernd Horn**, Ph.D., currently is the deputy commander of Canadian Special Operations Forces Command. Prior to this, from 2004 to 2007, he was the director of the Canadian Forces Leadership Institute. He is an experienced infantry officer with command appointments at the unit and sub-unit level. Colonel Horn was the commanding officer of 1 Royal Canadian Regiment/RCR (2001–2003); the officer commanding 3 Commando, the Canadian Airborne Regiment (1993–1995); and the officer commanding "B" Company, 1 RCR (1992–1993). He holds an M.A. and Ph.D. in war studies from the Royal Military College of Canada and is an adjunct professor of history at RMC.

**Dr. Daniel Lagacé-Roy** is a researcher in military ethics and leadership at the Canadian Forces Leadership Institute and an assistant professor at the Canadian Defence Academy. He currently is teaching *Military Professionalism and Ethics* at the Royal Military College of Canada. His recent edited works (2006) include *Ethics in the Canadian Forces: Making Tough Choices* (Workbook and Instructor Manual) and the *Mentoring Handbook*. Dr. Lagacé-Roy served in the Canadian Forces from 1987 to 1995 (Regular) and from 1998 to 2001 (reserves). He received his Ph.D. from L'Université de Montréal (Quebec).

**Dr. Allister MacIntyre** is an associate professor of psychology at the Royal Military College of Canada. He served the final five years of his 31-year career with the Canadian Forces as the deputy director of the Canadian Forces Leadership Institute. He has worked as a researcher in Canada and Australia, and served a three-year term as the chair of the Psychology in the Military Section of the Canadian Psychological Association. He is an adjunct professor at Carleton University and an associate with the Centre for Studies in Leadership at the University of Guelph.

**Chief Petty Officer Second Class Paul Pellerin**, (CPO2), CD, born in Grand'Mère, Québec, enrolled in the Canadian Forces in 1981. He served in HMCS *Saskatchewan* and HMCS *Terra-Nova* from 1982 to 1986. CPO2 Pellerin then opted to remuster to Intelligence. He served in a number of locations, including Valcartier, Kingston, Bagotville, and Ottawa, and he completed deployments to Cyprus, Visoko, and Sarajevo. He also served in the Canadian Embassy in Moscow. Recently, he served as the staff sergeant major at the CF School of Military Intelligence and, as of 2007, CPO2 Pellerin is serving at the Canadian Forces Leadership Institute in Kingston.

**Dr. Jessica Sartori** received a Ph.D. in applied social psychology from the University of Windsor in 2003. She has worked at Humansystems since 2004. She has been active in the trust research program undertaken for Defence Research and Development Canada (DRDC) in Toronto, as well as moral and ethical decision-making, and in research exploring communication in infantry teams. Dr. Sartori has also been involved in cognitive work analysis of work functions aboard naval ships with DRDC Atlantic.

**Lieutenant-Commander George Shorey** was with the Professional Development Section of the Directorate of Army Training — Land Force Doctrine and Training System until 2007, after which he retired from the Canadian Forces. He was a personnel selection officer with former service in the Navy as a Maritime Surface (MARS) officer. He holds a master of arts degree in psychology and has prior deployments in Cambodia and Afghanistan.

**Robert D. Sipes**, CD, served 25 years with the Canadian Armed Forces as a medical technician, including tours overseas in Gulf War 1991 and Croatia 1994. For 17 years he has been involved in Critical Incident Stress Management, trained by both the National Organization for Victim Assistance and the International Critical Incident Stress Foundation. He has been a volunteer team leader and advocate trainer with Edmonton Police Victim Services. Robert Sipes is a recipient of the Governor General's Caring Canadian and Alberta Centennial Medal for his work in Victim Services.

**Dr. Emily Spencer** has a Ph.D. in war studies from the Royal Military College of Canada. She currently works as an assistant professor with the history department at the University of Northern British Columbia. She is concurrently employed as a research officer with the Canadian

Forces Leadership Institute, where she focuses on the applicability of cultural intelligence to current Canadian Forces (CF) deployments and counter-insurgency.

**Lieutenant-Colonel Jeffrey Stouffer** enrolled in the Canadian Forces in 1988 as a personnel selection officer following completion of a master's degree in psychology from the University of Manitoba. He has since served in a variety of staff and research positions, including a one-year secondment with the Royal Canadian Mounted Police. He is currently the director of the Canadian Forces Leadership Institute.

**Dr. Megan M. Thompson** holds a Ph.D. in social psychology from the University of Waterloo. She is a defence scientist with the Collaborative Performance and Training Section at Defence Research and Development Canada (DRDC) in Toronto, and is the project manager of a research program for the CF that investigates trust violations and repair.

**Dr. Robert W. Walker**, a graduate of the Royal Military College of Canada and Queen's University, served for 21 years in the Royal Canadian Navy and the Canadian Forces — including an assistant professorship at RMC. He next served in the Royal Canadian Mounted Police for 17 years as a research psychologist and civilian member of the force. Since 2002, Dr. Walker has been an associate professor at the Canadian Forces Leadership Institute. Recently, he contributed to and edited *Institutional Leadership in the Canadian Forces: Contemporary Issues*, a 2007 publication of Canadian Defence Academy Press.

**Justin Wright** is a graduate of St. Thomas University and the University of New Brunswick in Fredericton. He holds a B.A. and a master's degree in sociology, specializing in addictions studies and social theory. He became a defence scientist in the Department of National Defence in December 2006, with his first assignment being at the Canadian Forces Leadership Institute of the Canadian Defence Academy in Kingston, Ontario.

# INDEX

Abbott, Andrew, 395
Achievement, 58, 64, 349–50, 418, 420, 439, 465
Acocella, Joan, 476
Adler, Ronald B., 65, 359
Adversaries, 32, 33
Affiliation, 66, 209, 419, 420, 529
Afghanistan, 81, 189, 194, 196, 249, 320, 429, 434, 444
Air Force, 50, 202, 204, 236, 302–03, 307, 314, 397, 457, 474
Ajzen, Icek, 25
Alcohol, 61, 279, 290, 294, 328, 330, 432, 465, 490
Ambivalence, 37, 39
Amir, Yehuda, 20–21, 476
Ang, Soon, 188–89
Antecedents, 327, 537
Anthropology, 202, 266
Argyris, Chris, 137
Aristotle, 160, 266
Army, 50, 63, 65, 98, 172, 178, 181, 191, 202, 204, 208, 221, 236, 328, 397, 403, 429, 474, 546
Arredondo, Lani, 118
Attitude Formation, 19
Attitude(s), 15–27, 32, 51, 58, 75, 78, 81, 83, 94, 107, 111, 115, 124, 142, 146, 187, 189–94, 201, 205–06, 229, 234, 281, 293, 312, 324, 329, 339, 342, 348, 358, 387, 402–03, 452, 471–77, 479, 487, 505, 531, 534
Audience, 38, 116, 118–21, 126, 192, 195, 339, 483, 485
Auerbach, Alan, 117
Australian Defence Force, 50, 173

Banks, D., 424
Barling, Julian, 483

Bartlett, F.C., 289
Bass, Bernard, 79, 161, 293
Battle Fatigue, 92
Beatty, Katherine C., 170–71
Begin, Menachem, 62
Behaviour(s), 15–16, 19, 23–26, 36, 49, 51, 57, 63, 75–76, 83, 94, 98, 111, 115, 122, 131, 141, 143–44, 147, 154, 158–61, 163–64, 173, 177–78, 184, 188, 195, 202–03, 205, 207, 229–30, 234–36, 254, 266, 271, 273, 295–96, 340–42, 347–48, 350, 356, 358–59, 361, 366–67, 372–73, 382, 405, 414–20, 424–25, 447, 464, 471, 476, 506, 508, 511–13, 516–18, 520, 527–29, 531, 535, 537, 540
Beliefs, 16–19, 26, 32, 49, 51, 118, 126, 130, 142, 187, 189–91, 193–95, 201, 208–09, 213, 229, 241, 246, 249–50, 253, 262–63, 313, 326, 346, 372, 397, 439–40, 452, 455, 457, 473, 475, 477, 503, 528–29, 531, 535
Bell, Daniel, 452
Ben-Ari, Rachel, 20
Benevolence, 531
Benhamadi, Bey, 251–52
Bennett, Jennifer J., 179–81
Bentley, Bill, 370–71, 375
Benton, Douglas, 409
Bernard, Claude, 482
Betrayal, 534–35
Biological Determinism, 310–11
Blanchard, Ken, 516
Boundaries, 52, 188, 191, 201, 205–06, 209, 217, 285
Bradley, Peter, 63
Brainwashing, 18
Bureaucracy, 393–94, 451, 457–58
Burke, C. Shawn, 497–98

Cacioppo, John, 115
Calhoun, James, 476
Canadian Charter of Rights and Freedoms, 246, 254, 456
Canadian Constitution, 233, 246, 456
Canadian Defence Academy, 9–10, 170, 441, 544
Canadian Forces Effectiveness Framework/Model, 12–13
Canadian Forces Leadership Institute, 9, 436, 544
Canadian Human Rights Commission, 252, 255
Center for Advanced Defense Studies, 190
Center for Advanced Operational Culture Learning, 188
Center for Creative Leadership, 50, 179, 181
Chang, Jui-shan, 245
Change, 13, 17–21, 23–26, 31–44, 47, 52–53, 94–95, 109, 111, 116, 120, 141, 169, 171–72, 181, 191, 195, 200, 204–06, 208, 216, 255, 276, 302–03, 309, 318, 322, 341–42, 355–63, 409, 438–39, 441–43, 447, 467–68, 471, 476, 506–08; Stages of, 42
Channel, 116–17, 119, 125–26, 229
Chaos, 31, 40, 75, 82, 110, 231, 294, 543
Character, 48–53, 55, 81, 92, 104, 142, 228, 230, 236, 278, 378, 439, 452, 501; Components of, 53; Definition of, 51, 55
Charisma, 340, 518–19
Charismatic Leadership, 518
Chief of Defence Staff (CDS), 171, 386, 397; CDS Principles, 444
Ciulla, Joanne B., 271
Clausewitz, Carl von, 372, 453
Climate, 21, 58, 107, 177–78, 203–04, 208–09, 248–49, 254, 271, 273, 315, 347, 372–73, 407–09, 430, 467, 471, 501, 504, 526
Cloud, Henry, 531
Coaching, 181, 342, 350, 360, 381–82, 386
*Code of Conduct for Canadian Forces Personnel*, 233
Cognition(s), 23, 171, 189, 289, 369, 374
Cognitive, 12–13, 23–24, 26, 95, 170–71, 174, 181–82, 188–89, 191, 217, 220, 279, 290, 322, 367–70, 392–93, 395, 438–43, 446–47, 453, 471, 473, 475–77, 481, 487–88, 497, 512; Appraisal, 481; Consistency, 23; Dissonance, 23–24; Processes, 95, 497
Cognitive Resources Theory, 512
Cohen, Eliot A., 367–68
Cohesion, 57–68, 73–74, 76–77, 86, 94, 98, 135, 137, 161, 163, 205, 209, 228, 230, 237, 248, 253, 293, 295, 297, 342, 403–04, 406–09, 414, 416, 424, 431, 433, 439, 447, 456, 497, 526, 535; Horizontal, 58–59, 62–64, 67; Vertical, 59, 62–64
Coles, John P., 191
Collective, 57, 61, 66, 94, 169, 174, 207, 229, 231, 235–36, 239, 244, 250, 342–43, 371, 384, 414, 425, 440, 447, 452, 494–97, 505
Combat, 60–64, 73–83, 86–87, 89, 91–96, 98–101, 103, 116, 132, 158–59, 161–62, 216, 229, 236, 240, 258, 276, 279–80, 285, 288, 290–95, 320–21, 330, 348, 397, 398, 402, 410, 423, 425, 429–30, 434, 439, 442, 454; Exhaustion, 92; Guilt, 91
Combat Motivation, 63, 73–77, 79–83, 86, 425; Definition of, 74
Combat Stress, 63–64, 91, 93–95, 98, 276, 402
Combat Stress Reactions (CSR), 64, 91–95, 98–101, 276
Command, 9, 11, 21, 25, 59–60, 63, 104–12, 125, 133, 135, 138, 150, 155, 173, 178, 193, 224, 234, 238–39, 281–82, 304, 332, 343–44, 351, 357, 386, 451, 458, 464, 507, 547–48
Commander's Intent, 105–07, 110–11, 241, 343
Communications, 36, 38–39, 54, 74, 79–80, 84, 95, 123, 278, 294–95, 332, 339, 404, 439, 442, 464, 502, 504, 533, 536–37
Competence, 49, 51–53, 67, 83–84, 207, 349, 366, 392, 396, 406, 439, 501, 508, 515, 531–32, 536
Competencies, 12, 32, 42, 179–81, 203, 438–39
Conference Board of Canada, 384
Confidence, 40, 53, 73–74, 76–81, 83, 85–86, 93–94, 107–08, 159–63, 189, 209, 290, 292–93, 297, 349, 356, 366, 373, 386, 403, 406–07, 409, 423–24, 431, 485, 502–03, 511, 517, 532
Conflict, 22, 31, 41, 60–62, 67–69, 81, 106, 129–38, 162–63, 171, 176, 204, 216, 225, 233–34, 240, 264, 273, 292, 320, 371, 395–96, 404, 410, 423, 440, 442, 444, 453–54, 464, 489, 504, 526, 537, 543, 547; Causes of, 129–30; Characteristics of, 130–31; Impact of, 131; Interpersonal, 129, 136, 504; Management, 134, 136
Conflict Resolution, 41, 130, 132–135, 137–38, 442, 444, 464, 537; Interest-Based, 134–38; Power-Based, 132–33, 136–38; Rights-Based, 132, 134, 136, 138; Selecting an Approach, 132

# INDEX

Confucius, 525
Contingency, 297, 515–16
Control, 21, 24–25, 43, 45, 64, 78, 95, 105, 107, 109, 111, 133, 138, 148, 161–63, 172, 175, 205, 209, 229–30, 238, 240, 288, 291–93, 295, 297, 318, 324, 327–28, 343, 348, 391, 396, 420, 446, 451, 457, 471, 481, 489, 515–16
Cooper, Joel, 24
Cooperation, 22, 76, 193, 196, 494, 517
Corporateness, 394
Cosgrove, Peter, 50
Counselling, 41, 140–45, 148–56, 361
Counsellor, 141, 143, 152–56, 380
Counterproductive Behaviours, 33
Courage, 17, 50, 55, 82, 158–164, 166, 231, 237, 288, 292, 295, 303–04, 398, 457; Moral, 50, 158–60, 164, 304, 457; Physical, 158–60, 164, 457
Cowdrey, Christian, 59
Creativity, 106, 129, 168–70, 172, 174–79, 181–82, 184, 186, 216, 254, 371, 527
Credibility, 36, 38, 40–41, 49, 51, 54, 77, 84, 116–17, 120–21, 125, 142, 144, 269, 439, 447, 501–02, 507
Criticism, 266, 304, 334, 357–59, 363, 405, 486, 506, 513, 534
Cultural, 118, 142–43, 145, 172, 176, 187–96, 200–10, 223, 245–46, 248, 252–53, 327, 392, 421, 447, 452, 472, 476; Awareness, 142–43, 191–92, 196; Change, 172, 208; Cross-Cultural, 188, 206–07, 248; Identity, 202, 209; Intelligence (CQ or CULTINT), 143, 187–197; Norms, 204
Culture, 26, 34, 39–42, 118, 122, 132, 137, 172, 175, 178, 180, 188–94, 200–10, 213, 248–53, 256, 263, 289, 302, 310, 313–16, 324, 342–43, 372, 380, 397, 416, 447, 455, 458, 464

Danger, 60, 66, 73, 78–79, 95, 101, 159–60, 230–31, 234, 236, 286–88, 291, 293, 296, 344, 348, 402, 423, 523
Davis, Karen D., 318
Death, 33, 63, 73, 77, 85, 93, 96, 196, 215, 235, 267, 276, 289, 320–23, 325–34, 372, 481, 485
Decision-Making, 105–07, 111, 130, 136, 149, 173, 176, 178, 200, 215–17, 219–21, 223–25, 233, 240, 272, 274, 281, 290, 349, 366, 368–71, 374, 421, 487, 500, 503–04, 532
Defence Ethics Program (DEP), 55, 375, 447

Defence Mechanism, 17
Defence Research and Development Canada (DRDC), 526, 536, 538, 545
Department of National Defence (DND), 168, 240
Deprivation, 79, 93–94, 161, 288, 293
Devine, Patricia, 474
Diet, 281–82, 406, 431–33, 465
Dilemma, 246–47, 263–65, 267–70, 272–74, 311, 487
Dinter, Elmar, 86, 285, 291
Directorate of Human Rights and Diversity (DRHD), 244
Directorate of Military Gender Integration and Employment Equity (DMGIEE), 254
Discipline, 13, 66, 73–74, 78, 84, 86, 130, 160, 162, 220, 228–41, 243, 281, 292–94, 300, 342, 407, 439, 446–47, 456, 468
Discretionary Knowledge, 392
Distress, 92, 321, 326, 357, 483–86, 488–89
Diversity, 200, 209, 244–54, 256–57, 259, 314, 421, 447, 505; and Canadian Legislation, 246, 254, 256; Definition of, 244; Demographic Impact of, 246–47, 250; and Leadership, 314, 505
Diversity Issues, 244, 314, 447; and Ethnicity, 245, 248–50; and Gender, 245, 248, 251–52, 254, 314; and Religion, 245, 248, 253–54; and Sexual Orientation, 248, 254; and Visible Minorities, 245, 247, 250–51
Dixon, Norman F., 367
Dolan, Shimon, 117
Dollard, John, 75, 83, 286, 290–91, 295–97, 300
Drucker, Peter, 41
Duckitt, John, 473
Duty, 17, 55, 73–74, 78, 80–81, 86, 92, 98–100, 149, 159, 229, 238, 254–56, 259, 261, 268, 270–71, 273, 303–06, 341, 371–72, 387, 397–98, 408, 422, 444, 456, 519, 545
Duty to Accommodate, 254, 256; and Bona Fide Justification (BFJ), 255; and Bona Fide Operational Requirement (BFOR), 255; and Health and Safety, 254–55; and Undue Cost, 255
*Duty with Honour: The Profession of Arms in Canada*, 17, 229, 397
Dyer, Gwynne, 405
Dynamics, 31, 178, 188, 312, 347, 351, 404, 407, 416, 437, 494, 496, 498–99, 504, 508, 531

Earley, P. Christopher, 188–89

Education, 23, 25, 118, 161–62, 166, 188, 191, 236, 245, 270, 292–93, 297, 299, 310, 315–16, 321, 332, 335, 342, 372, 383–84, 386, 391, 393–94, 397, 433, 441, 451, 453, 457, 461–64, 466

Elaboration Likelihood Model (ELM), 115, 127

Emergent Leadership, 340, 347–48

Emotion, 15, 18, 27, 33, 48, 77, 79, 82, 94–97, 117, 122, 124, 126, 130, 148, 155, 161, 166, 277, 280, 282, 285, 290–91, 293, 321–22, 324–25, 328–29, 331, 339, 346, 358, 403, 415, 417, 422, 424, 475, 482, 484, 487, 497, 515

Empathy, 142–43, 407

Employment Equity Act (EE Act), 244, 246, 251; Definition of, 251; and Designated Groups, 245; and Visible Minorities, 245, 247, 250–51

Entrepreneurialism, 441, 451

Environment, 9–10, 12, 24, 32–33, 49, 75, 78, 83, 107, 111, 115, 129, 131, 133, 135, 140, 150, 162, 169, 173, 177–78, 190, 192, 194, 201, 206, 208–10, 220–21, 224–25, 240–41, 245, 248–49, 264–65, 267–68, 270–72, 277, 282, 312, 320, 342, 369–70, 372–73, 378–79, 384–85, 394, 396, 402, 404, 407, 409–10, 414, 418–19, 421, 423, 425, 430–31, 433, 436, 439, 446–47, 459, 473, 477, 484–85, 488–89, 500, 508, 520, 529, 536, 543

Equality, 83, 207, 251–52

Esprit de Corps, 60, 248, 403–04, 406, 433, 501

Esteem, 419, 485

Ethical Conduct, 270–72, 372

Ethical Decision-Making, 272, 274

Ethical Dilemma, 262–65, 267–70, 272–73, 373

Ethical Foundations, 265

Ethics, 48, 50, 53, 202, 261–63, 265–67, 269–73, 371–73, 375, 464, 519

Ethnicity, 245, 248–50, 312, 315

Ethos, 17, 26, 49, 51, 53, 55, 73–74, 80–81, 86, 164, 209, 229–31, 236, 238, 241, 249, 261, 302–03, 307, 341, 371–72, 394–97, 431, 437, 439–40, 447, 453, 457–58, 464, 503, 505, 543

Euthyphro, 264

Experience, 9, 13, 15, 17, 19–20, 22–23, 26, 31, 34, 62, 77–78, 82, 92, 95–97, 104, 112, 143, 149, 161–63, 171, 173–74, 189, 196, 209, 220–21, 223, 231, 236, 238, 245, 287–88, 290, 305, 312, 314, 316, 321–25, 327–28, 331, 335, 369, 381, 384, 391, 397, 406, 417, 421, 424, 433, 438, 441, 453, 461–64, 466, 474, 483–85, 505, 512–13, 529

Expertise, 12–13, 36, 48–50, 52–53, 55, 80, 101, 116–17, 170, 176, 181, 228, 239, 241, 305, 340, 394–97, 399, 438–39, 441–43, 446, 451, 453, 457–59, 501, 507, 515, 532, 543

Eye Contact, 83, 122, 145–47, 296

Facial Expressions, 122, 145

Faith, 97, 255, 294, 322, 324, 327, 403

Fastabend, General David, 32

Fatigue, 33, 82, 92, 125, 162, 276–82, 288–89, 291–92, 322, 434, 487

Fazio, Russell, 19, 24, 474

Fear, 16, 22, 43, 60, 73, 75, 77–80, 82–83, 93–95, 103, 117, 122, 158–59, 161–64, 231, 285–97, 299–300, 322, 326, 328–29, 339, 415, 535

Feedback, 38, 40, 150, 152, 177, 235, 342, 355, 357–63, 381, 442, 506, 513, 532; Delivery of, 358, 361; and Follow-Up, 358; Negative, 177, 360, 362–63, 513; Preparation of, 358

Femininity (Feminine), 309–10, 314–15, 318

Feminist, 311, 313

Feminist Movement, 311

Festinger, Leon, 23

Fiedler, Fred, 515

Field, Richard, 119

Fishbein, Martin, 25

Fitness (see Physical Fitness)

Fleishman, E.A., 520

Focus Group, 35, 346, 408, 412

Follower, 173, 235, 302–07, 345–46, 351, 356, 417, 496, 499, 505, 511, 513–14, 516–21

Followership, 302–03, 307

Forward Treatment, 98

Fragging, 62–63

Franks, General Fred, 281

Freidson, Elliot, 392, 395, 451

Freud, Sigmund, 513

Frustration, 21–22, 33, 146, 535

Fukuyama, Francis, 528

Gal, Rueven, 58, 62–63

Gambetta, Diego, 525

Gardner, John, 525

Garvin, David, 175, 184

Gays, 309

Gender, 26, 117–18, 245, 248, 251–52, 254, 309–17, 328, 339, 378, 472, 474; Blind,

312; Free, 312; Inclusive, 312, 316; Neutral, 312, 316; Relationships, 312; Roles, 313, 316
General System of War and Conflict, 395, 440, 453–54, 464
Geneva Conventions, 233
Gilson, Lucy L., 176
Gooch, John, 367–68
Gosselin, Major-General J.P.Y.D., 10
Greene, Robert, 404
Greenleaf, Robert K., 519–20
Grief, 95–97, 320–35, 337; Definition of, 321; Ignoring of, 323, 329; and Leader Responsibilities, 334, 339; Process of, 324, 326–28, 330, 332–33; and Reactions, 95, 321–22, 332; and Survivor Guilt, 96–97, 102, 323, 331; Theories of, 324–25
Guilt, 95–96, 102, 331

Hackett, General Sir John, 390
Hague Conventions, 233
Hamner, Clay, 24
Harm, 96, 239, 264, 269, 289
Hart, B.H. Liddell, 104
Hart, Major-General T.S., 295
Health, 82, 116–17, 131, 177, 254–56, 321, 328, 330, 334, 337, 409, 430–31, 465, 468, 482, 484, 487
Hearts and Minds, 85, 193–94, 196, 249
Henderson, Darryl, 66
Heraclitus, 31
Hersey, Paul, 516
Herzberg, Frederick, 419
Hierarchy of Loyalties, 357
Hillier, General Rick, 171, 386
Homer, 378
Honour, 57, 73, 86, 229, 238, 457, 534
Horn, Colonel Bernd, 60, 174
Horth, David M., 179
Howell, J.P., 522
Hughes, Richard L., 170–71,
Human Relations, 416–17
Human Systems, 204
Humour, 84, 120, 278, 294
Huntington, Samuel, 390, 393–95

Identity, 22, 35, 43, 49, 55, 80, 180, 202, 204–05, 209–10, 250, 342–43, 371–72, 394–97, 399, 447, 451, 459, 534, 537
Ideology (*see also* Professional Ideology), 17, 80, 160, 371–72, 441, 447, 451–53, 457–59, 464

Ignatieff, Michael, 160
Immediacy, 98, 101
Inclusion, 55, 107, 205, 210, 309, 508
Incompetence, 60, 84, 356, 366–68
Inequality, 207
Inferences, 474
Influence, 15–18, 20–21, 23, 26, 35–36, 38, 63–65, 74–75, 78, 83, 93, 98, 101, 107, 110–11, 114–15, 117, 120, 126, 132, 143–44, 162, 164, 192, 201–04, 206, 208, 210, 223, 235, 267–68, 271, 285, 288, 293, 296, 313, 315, 327, 339–43, 345, 347–52, 371–72, 374, 384, 394, 407, 416, 418, 420–21, 424, 447, 453, 458, 472, 474–75, 477, 494–96, 499–500, 508, 512–18, 529–31; Direct, 341–42, 384; Indirect, 342–43, 351
In-Group, 22, 189, 514
Innovation, 12, 31, 40–41, 170, 175, 179, 245, 254, 374, 446
Inspirational, 346, 350
Instrumental, 16–17, 63, 77, 172, 262, 410
Integrity, 17, 40, 50–53, 55, 64, 67, 81, 117, 160, 238–39, 269, 271, 303–06, 381, 396, 398, 416, 457, 501, 511, 531
Intelligence, 48, 118, 143, 160, 187–91, 193, 511–12
Interdependence, 495
Inter-Governmental Organization (IGO), 193–94, 196
Intrinsic, 262, 417–18, 422–23, 482, 517

Janowitz, Morris, 61–62, 390
Jeffery, Lieutenant-General Michael, 448
Jenkins, Richard, 250
Job Satisfaction, 385, 403, 417, 419, 501, 527
John, Oliver, 477
Johns, Gary, 415, 421, 426
Johnson, James, 189
Judgment, 48, 99, 125, 148–49, 153, 172, 175, 184, 190, 217, 263, 278, 282, 366–75, 396, 439, 446–47, 474–75, 487, 502, 505, 515, 530, 532, 534
Jung, Carl, 513
Jurisdiction, 392
Jussim, Lee, 473

Kant, Immanuel, 266
Kanter, Rosabeth, 31
Katz, Daniel, 16, 18
Kegan, Robert, 372–73
Kellett, Anthony, 60, 74, 79, 290

Kelloway, Kevin, 482
Kirkland, Faris, 64
Klein, Gary, 221
Knackstedt, Janine, 367
Knott, S.W., 40
Knowledge, 9, 12–13, 16–17, 19–20, 26, 32, 42, 49, 52, 62, 67, 85, 94, 118–19, 121, 161–62, 170, 180, 184, 187, 189–91, 193, 195–96, 201, 207, 221, 236, 238–39, 265, 270, 292, 294, 321, 323–24, 332–33, 336, 339, 348, 357, 359, 368, 371–72, 374, 379, 381, 383–84, 391–93, 395, 397–98, 423, 438–441, 446, 451–54, 457–58, 461, 463–64, 466, 468–69, 475, 477, 488, 496, 504–05, 512, 516, 521, 529–30, 532, 543–44
Kotter, John, 32, 34, 37–38, 507
Kouzes, James M., 49
Kramer, Roderick, 527, 534
Krosnick, Jon, 20

Lagacé-Roy, Daniel, 372, 375
Laissez-Faire, 348, 513
Leader Framework, 437–39
Leader Member Exchange Theory (LMX), 514
Leader Metacompetencies, 369, 436, 438
*Leadership in the Canadian Forces: Conceptual Foundations*, 499, 508
*Leadership in the Canadian Forces: Doctrine*, 499
*Leadership in the Canadian Forces: Leading People*, 302, 499–500
*Leadership in the Canadian Forces: Leading the Institution*, 35, 499
*Leadership Network, The*, 53
*Leadership Quarterly, The*, 176, 498
Leadership Skills Model, 512, 521
Leadership Style, 10, 133, 172, 314–15, 348, 350–51, 516
Leadership Substitute Theory, 518
Learning Organization, 174, 184, 438–39, 442, 447, 462, 468–69
Learning Strategies, 13, 179, 181, 436, 442–43, 445
Least Preferred Co-Worker (LPC) Model, 515
Lencioni, Patrick, 525
*Les Aventures de Télémaque*, 379
Lesbians, 309
Levering, Robert, 536
Lippmann, Walter, 472
Loyalty, 17, 42, 53, 55, 58, 60, 63, 264, 268–69, 272, 303, 305–06, 398, 403–04, 406–07, 416, 456, 482, 501, 515

Maccoby, Michael, 415, 417–18
Machiavelli, Niccolò, 42
Makalachki, Alexander, 57
Management, 11, 25, 31–32, 36, 41, 53, 92, 98, 101, 106, 108–12, 118, 129, 131, 133–36, 188, 206, 223, 294, 326, 331, 333, 343–45, 351, 386, 394, 408, 416, 439, 458, 464–65, 481, 488–91, 513
Managerialism, 441, 451
Marshall, Colonel S.L.A., 82, 159, 163–64, 285, 289–91, 295
Masculinity (Masculine), 158, 309–10, 314–15, 318
Maslow, Abraham, 419
Mathews, Michael, 50
Mayer, Bernard, 137
McAllaster, Craig, 36
McCann, Carol, 538
McCauley, Clark, 473
McClelland, David, 420
McGregor, Douglas, 420
Meilinger, Colonel P.S., 303
Member Well-Being, 344, 372, 415, 422–23, 437, 447, 465, 543
Mentee, 380–83, 386
Mentor, 141, 378–83, 386
Mentoring, 181, 316, 350, 378–87
Metacompetencies, 13, 51, 53, 170, 369, 436, 438, 440, 442
Military Culture, 201, 203–04, 209, 397, 455, 458
Military Ethics, 50, 372, 548
Military Ethos, 17, 26, 55, 73–74, 80–81, 86, 164, 209, 229–31, 236, 238, 241, 249, 261, 302–03, 307, 341, 371–72, 394–97, 431, 437, 439–440, 453, 455, 457–58, 503, 505, 543
*Military HR Strategy 2020*, 306
Military Identity, 80, 209
Military Law, 232–34
Mill, John Stuart, 266
Miller, William, 159
Mills, Harry, 360
Mission Command, 105–07, 111, 178
Moghaddam, Fathali, 473
Montgomery, Field Marshal Sir Bernard Law, 78, 85, 162, 293, 402
Moral Temptation, 274
Morale, 58–61, 73–74, 77, 82, 86, 98, 107, 131, 134–35, 209, 237, 239, 276, 278, 281, 288, 290, 328–29, 342, 350, 373, 402–10,

# INDEX

412, 417, 424, 431, 433, 504, 526, 535; Assessment of, 60, 408, 412; Surveys of, 409
Moran, Lord Sir Charles Wilson, 160, 403
Mores, 229, 250, 262
Mosakowski, Elaine, 188
Moskos, Charles, 394-95
Motivation, 35-36, 38, 48, 63, 73-83, 85, 107, 111, 132, 138, 141-42, 173, 189-90, 207, 278, 280, 322, 341, 346, 348-50, 359, 387, 403, 414-25, 461, 482, 498-500, 519; Extrinsic, 418; Intrinsic, 417-18, 422, 482
Motowidlo, Stephan, 483
Multiple Linkage Model of Leadership, 508, 517
Mum Effect, 125
Murphy, Peter J., 412
Murphy, Robert, 176, 184
Mutiny, 62-63

Nasmyth, Guy, 183
Naturalistic Theories, 217, 220-21, 370, 374
Navy, 202, 204, 236, 397, 474
Needs Theory, 419-20
Nelson, Debra (D.L.), 484, 489
Neutralizer(s), 518
New Zealand Defence Force, 51
Non-Governmental Organization (NGO), 193, 196
Normative Decision Model, 518
Normative Ethics, 265, 267
Normative Theories, 216-17, 221, 369
North Atlantic Treaty Organization (NATO), 105, 193, 242, 403, 409, 429
Noy, Shabtai, 64

O'Toole, James, 31
Okros, Alan, 174
Open-Mindedness, 416, 421
Operation Desert Storm, 279, 281
Operational Art, 224, 440, 453-54
Operational Planning Process (OPP), 219-20, 225, 370
Organ, Dennis, 24
Organizational, 24-25, 31, 33-34, 36, 40-41, 43, 48-49, 52, 57-59, 61, 64, 66-67, 76, 109, 117, 119, 130, 132-34, 136-38, 149-50, 155, 172, 178, 183, 200-02, 204-10, 223, 246-48, 250, 256, 313-16, 342-43, 345-46, 367, 380, 382-85, 394, 402, 406, 408, 414-16, 418, 420-23, 425, 437, 439, 442, 447, 458, 483, 489, 496, 526-528-29, 536-37, 543; Citizenship, 527; Climate, 208; Culture, 40, 132, 137, 200-02, 204-05, 248, 314-15, 343, 380

Palus, Charles J., 179
Parsons, Talcott, 391, 452
Participative Leadership, 513
Path-Goal Theory of Leadership, 517
Performance, 33, 37, 57, 59-61, 64, 78, 94, 98, 107, 131, 140-41, 149-51, 153, 155, 169, 205, 208, 235, 238, 276-82, 290-91, 293, 306-07, 312, 314, 316, 341-42, 345-47, 358-360, 380, 382, 384, 387, 397, 402-03, 416, 420, 422, 424, 426, 445, 483, 486-87, 495-96, 498, 501, 506, 512, 517-19, 527, 532-33, 535; Standards of, 278
Persuasion, 115-16, 126, 190, 339
Peterson, Randall S., 188
Petty, Richard, 115
Philosophy, 105, 111, 137, 169, 174, 178, 202, 261, 266, 398, 439, 446-47
Physical Fitness, 288, 429-31, 433-35, 465
Pinder, Craig, 58
Plato, 264, 273
Position-Based Leadership, 340, 347-48
Posner, Barry Z., 49
Post-Traumatic Stress Disorder (PTSD), 92, 100, 102, 330, 490
Power, 24, 33, 36, 38, 66, 77, 104, 109, 125-26, 132-33, 136-38, 159-60, 178, 230-32, 238, 288, 290, 294, 318, 339-40, 420, 430, 453, 458, 483, 507-08, 516, 544
Pratt, Laurie, 483
Predictability, 53, 485, 531
Predisposition, 371, 530
Prejudice, 18, 21-23, 472
Pride, 33, 58-60, 78, 163, 205, 232, 235-36, 246, 281, 403, 405-06
Primary Intervention, 488-89
Problem-Solving, 109, 133, 135, 149, 170, 172, 177, 182, 215, 223, 225, 370, 439, 446, 497, 512
Professional Body of Knowledge, 461, 463-64, 468-69
Professional Development, 12-13, 39, 41, 51, 53-54, 151, 154, 170, 172-76, 179-80, 186, 369, 386, 397, 408, 414, 422, 436-38, 440-47, 461-64, 467, 533, 536; Methodologies, 443; Strategies, 13, 179, 436, 442-43, 445
Professional Development Framework (PDF), 12-13, 51, 53, 170, 369, 436, 438, 440-46, 464

Professional Ideology, 12, 49, 53, 170, 181, 371–72, 393, 395–97, 399, 437–43, 446–47, 451–53, 455, 459, 463, 469, 544
Professional Military Education (PME), 188
Professionalism, 13, 48, 181, 215, 367, 390–95, 397, 399, 436, 444, 451–53, 457, 459, 464
Protege, 387
Public Service of Canada, 168
Puertas, Lorenzo, 190

Quillian-Wolever, Ruth, 482
Quinn, Robert E., 183

Race, 26, 117, 208, 245, 250, 312
Rachman, S.J., 288
Rando, Theresa, 325
Reciprocity, 398–99, 421
Relativism, 263
Religion, 117, 191, 245, 248, 253–55, 263, 272, 294, 310, 312
Resistance, 19–20, 25, 36, 39, 42–44, 60, 236, 240, 276, 339, 481, 487, 506
Respect, 23, 49–50, 52–53, 59, 62, 64, 66, 84, 101, 114, 116–17, 125, 133, 142–44, 148, 172, 178, 196, 220, 228, 233–34, 238–39, 244, 246, 248–49, 251, 254, 291, 315, 351, 367–68, 372, 379, 415–16, 419, 421–22, 424, 432, 441, 473–74, 502, 516, 531, 533
Revenge, 535
Risk, 37, 63, 80, 83, 100, 106–08, 112, 133, 160, 162, 164, 169, 176, 178, 240, 278, 280, 296, 327, 330–31, 356, 367, 406, 430, 434, 439, 457, 519, 528–30
Risk-Taking, 34, 40, 528
Role Compliance, 529
Role Model, 65, 380–81, 383
Rothbart, Myron, 476
Rules of Conduct, 262

Saks, Alan M., 415, 421
Salas, Eduardo, 495
Satisficing, 218, 221, 370
Scales, Jr., Major-General Robert H., 288
Schmidtchen, David, 178–79
Schmitt, John, 221
Scientific Management, 416
Second World War (*see also* World War II), 60, 62, 75–77, 82, 92, 159, 162, 223, 248, 288, 290–91, 294–96, 339, 366, 379, 402, 454
Secondary Intervention, 488
Self-Actualization, 419
Self-Confidence, 49, 73–74, 79, 86, 146, 161, 293, 356, 383, 406, 431, 511, 516
Self-Development, 13, 51, 180, 439, 441, 447, 461–68
Selye, Hans, 483–84, 487–88
Servant Leadership, 519–20
Sex, 309–11
Sexual Orientation, 248, 253–54, 312
Shalit, Ben, 159, 289
Shalley, Christina E., 176
Shell Shock, 92
Sherblom, John, 123
Shils, Edward, 61
Simmons, B.L., 484, 489
Simon, Herbert, 217, 223, 375
Singer, Peter, 263
Skill(s), 12, 41, 51, 77–79, 81, 106, 110–11, 126, 133, 144, 147, 149, 154, 170–73, 177–78, 184, 189–92, 196, 203, 215, 217, 236, 238–39, 248, 270, 279, 281, 321, 327, 342, 344, 348, 351, 357, 359, 370, 379–82, 387, 392, 394, 396–97, 406–09, 423, 433, 439, 441, 446–47, 454, 456–57, 461, 463–66, 481, 483, 485, 488, 500, 505–06, 512, 516, 521, 528, 537, 543
Sleep, 82, 94–95, 277–283, 288–89, 322, 328, 402, 404, 406, 431, 434
Slim, Field-Marshal William, 78, 158, 164, 292, 410
Smith, Robert, 63
Snider, Donald, 50
Sociability, 511
Social, 12–13, 16, 20–26, 31, 53, 57–59, 63, 76, 92, 95, 115, 129, 131, 134, 136, 159, 169–70, 181, 189, 191–92, 200–01, 203, 205, 209, 216–17, 232, 245, 250, 266, 286, 309–15, 317–18, 322, 326–27, 334–35, 339, 349, 351, 361, 367, 385, 408, 417, 424, 438–39, 441–43, 447, 452, 472–74, 481, 483, 497, 512, 514–15, 520–21, 525–26, 535, 545–46, 549–50; Construction, 209, 315; Exchange Theory, 514; Norms, 20, 24, 26, 134; Sanction, 25
Socrates, 158, 160, 264, 273
Spectrum of Leader Influence Behaviours, 340–41, 348–49, 351
Stability, 20, 48, 203–05, 207, 258, 332, 341, 419, 447
Starbucks, 421
Stereotype(s), 15, 27, 130, 148, 168, 314, 471–78
Stevenson, Eric, 173
Stewardship, 49, 52, 344, 439, 443, 447, 458

# INDEX

Stewart, Nora, 61
Stouffer, Samuel, 75–76, 83, 159, 163, 288, 290, 296
Strategy, 34, 120, 150, 152, 205, 217, 247, 251–52, 256, 316, 339, 369, 379, 386, 388, 415, 440, 453–54, 537
Stress, 32, 63–64, 75–77, 79, 91–95, 97–99, 101, 103, 156, 162, 209, 231, 236, 276, 289–90, 292–94, 330–33, 402, 404, 424, 431, 433–34, 465, 481–91, 493, 501, 512
Stress Management, 481, 488–90
Stressor(s), 91, 93–94, 288, 409, 481–82, 484–88, 490
Subcultures, 194, 202, 204, 209
Subjectivism, 263
Subordinates, 26, 35, 38, 40, 42–43, 59, 64, 75, 78, 82–84, 105–10, 112, 116, 140–45, 147, 149–52, 155–56, 164, 173, 178, 230, 232–39, 241, 281, 292, 297, 303, 323, 332, 334, 343, 349, 363, 383–87, 407–08, 417, 424, 429, 433, 442, 447, 474, 481–83, 487–90, 494, 496, 501, 536
Substitute(s), 110, 232, 237, 309, 329, 518
Survivor Guilt, 96–7, 323, 331
Systemic Operational Design (SOD), 224–25, 370
Systems Theory, 223
Systems Thinking, 170, 177, 217, 223–25, 370–71, 374

Tactics, 77, 116, 161, 292, 402, 440, 446, 453–54, 526
Taliban, 194
Taylor, Don, 473
Taylor, Frederick, 416
Taylorism, 416–17
Team Building, 51, 143, 407, 416, 500
Team Cohesiveness and Teamwork, 503
Team Commitment and Support, 505
Team Effectiveness, 205, 494, 497
Team Leader, 380, 495, 497, 501, 506, 508, 549; Attributes, 501, 508; Functions, 497, 508; Processes, 497, 508
Team Leadership, 423, 494–96, 498–99, 508; CF Definition of, 499; Definition of, 494–95
Team Performance, 205, 495–96, 498, 527
Team Trust, 439
Terrorism, 270
Thomas, David C., 189
Thompson, Megan M., 538
Threat, 18, 24, 42, 93–94, 117, 162, 231, 289, 453, 482

Total Quality Management, 32
Towne, Neil, 359
Traditions, 43, 202, 209, 232, 529
Training, 17, 23, 63, 73–74, 78–79, 81, 86, 98, 133, 154, 160–62, 166, 170, 188, 190, 192, 196–97, 221, 228–32, 235–36, 238, 241, 255, 290, 292–93, 297, 299, 327, 333–35, 342, 348, 369, 383, 385–86, 391–93, 397, 405, 407, 422, 432–33, 441, 453, 461–64, 466, 481, 529, 546
Trait(s), 49, 159, 175, 315, 339, 472, 474, 503, 511–12, 515
Trait Theory, 511–12
Transactional, 18, 341, 345, 347–48, 351, 488, 513, 519
Transactional Leadership, 345, 513, 519
Transformational, 18, 173–74, 341, 345–47, 350–51, 373, 447, 481, 498, 519–20, 522, 532
Transformational Leadership, 18, 174, 341, 345–48, 350–51, 373, 519–20, 532
Trauma, 100, 102, 334, 547
Trust, 36, 39–40, 43, 48, 50–51, 53, 58, 64, 83, 85, 93, 105, 107–08, 111, 144, 178, 207, 248–49, 269, 296, 303–04, 351, 379, 393, 398, 404, 439, 455–56, 500–05, 525–37, 545, 549–50
Trustee, 530–31
Trustor, 530–31

Ulmer, Lieutenant-General W.F., 532
Unit Climate, 58, 208, 407–08
Unit Cohesion, 60, 65, 94, 98, 135, 137, 248, 297, 414, 416
United States Marine Corps, 59, 190
Unity of Command, 106, 193
Unity of Effort, 105–06, 111, 193, 199
Ury, William, 134
Utilitarianism, 265

Values, 16–19, 24, 26, 41, 48–51, 55, 59, 63, 66–67, 80, 118, 130, 142–43, 158, 169, 174–75, 179, 187, 189–91, 193–96, 200–04, 206, 208–10, 213, 229–30, 232–33, 236, 239, 241, 244–46, 249–50, 253, 255, 2–65, 268–73, 303, 307, 312, 314, 327, 341, 343, 346, 351, 355, 371–73, 381, 387, 396–98, 415–16, 439–40, 447, 455–57, 471, 503, 529, 531, 533–34, 536
Van Baarda, Th. A., 274
Vander Zanden, James, 23
Verweij, D.E.M., 274

Virtue, 50, 105, 109, 158, 261, 266, 372, 458, 527
Vision, 34–38, 41, 95, 106, 111–12, 173, 277, 346, 407, 452, 505, 519
Vroom and Yetton's Normative Decision Model, 518
Vroom, Victor H., 518
Vulnerability, 528, 530

Walker, Dr. Robert W., 193
Weisinger, Hendrie, 357, 362
Welch, Jack, 41
Well-Being, 64, 82, 141, 255, 295, 320–21, 344, 350, 372–73, 404, 407, 415–16, 421–23, 425, 427, 430–34, 437, 447, 465, 482, 502, 533, 543
West Point, 50
Wild, Bill, 208
Willpower, 160
Wolever, Mark, 482
Wong, Leonard, 50, 60, 181, 191
Wood, John T., 287
World War II (*see also* Second World War), 475

XY Theory, 420

Yerkes-Dodson Law, 486
Yetton, Phillip, 518
Yukl, Gary, 494–95, 508, 522

Zaccaro, Stephen J., 169, 496–97
Zanna, Mark, 19
Zeman, P.M., 190